HYBRIDIZING SURFACE PROBE MICROSCOPIES

Toward a Full Description
of the Meso- and Nanoworlds

HYBRIDIZING SURFACE PROBE MICROSCOPIES

Toward a Full Description of the Meso- and Nanoworlds

SUSANA MORENO-FLORES
JOSE L. TOCA-HERRERA

CRC Press
Taylor & Francis Group
Boca Raton London New York

CRC Press is an imprint of the
Taylor & Francis Group, an **informa** business

CRC Press
Taylor & Francis Group
6000 Broken Sound Parkway NW, Suite 300
Boca Raton, FL 33487-2742

First issued in paperback 2018

© 2013 by Taylor and Francis Group, LLC
CRC Press is an imprint of Taylor & Francis Group, an Informa business

No claim to original U.S. Government works

ISBN-13: 978-1-4398-7100-3 (hbk)
ISBN-13: 978-1-138-37458-4 (pbk)

Visit the Taylor & Francis Web site at
http://www.taylorandfrancis.com

and the CRC Press Web site at
http://www.crcpress.com

Contents

Preface..xi

Authors...xiii

Chapter 1 Introduction 1

Observing Nature: Sequentiality and Simultaneity.............................1
    Combination Macroscopic/Microscopic: Extending
    Measurable Range to Better Understand the Connection
    between Ultrastructure and Function 4
    Combination Model-Based/Direct Observation............................4
        Making the Right Choice 4
        How to Judge the Applicability of a Model................................5
    Combination Preparative/Measuring Technique...........................5
    Combination of Local Techniques to Track the Dynamics of
    Processes 6
    Combination of Local Techniques of Different Contrast...............6
Systems That Can Profit from Combined Techniques 6
    Living Cells 6
    Films..8
Final Remarks 8
References 9

Chapter 2 Scanning Probe Microscopy as an Imaging Tool: The Blind
Microscope 11

Probe Imaging 11
    The SPM Family...24
    The Scanning Probe Microscope 25
        AFM 26
        STM 26
    Probes and Tips 26
        STM Probes 28
        AFM Probes 30
    Scanners..33
    Detection of the Tip's Position..36
        Tunneling Detection 36
        Optical Beam Detection 37
        Piezoresistive Detection...38
    SPM Imaging Modes...38

Feedback Mechanism: Controlling the Property X 39
Feedback in Contact Mode 41
Feedback in Noncontact Mode 41
Feedback in Intermittent-Contact Mode................................. 43
Topography and More: Feedback on X Mapping Y 44
Artifacts in SPM Images 46
Time Resolution in SPM Imaging 49
Summary 51
References 53

Chapter 3   What Brings Optical Microscopy: The Eyes at the Microscale 55

Fundamentals of Optical Microscopy 55
Essential Parts of an Optical Microscope: The Image and
Diffraction Planes.. 64
Diffraction Sets the Limit of Detection, Spatial Resolution,
and Depth of Focus 66
Interference Sets Image Formation 69
Optical Contrasts 69
Phase Contrast 70
Differential Interference Contrast 71
Fluorescence Microscopy: Bestowing Specificity 72
Optical Microscopy of Fluorescent Objects 77
Light Source.. 78
Collecting the Emitted Fluorescence: The Dichroic
Mirror and the Emission Filter 78
Drawbacks of Fluorescence Microscopy: The Ever-Present
Photobleaching 80
High-Performance Modes of Fluorescence Microscopy.................. 81
Confocal Laser Scanning Microscopy (CLSM)........................... 81
Fluorescence Lifetime Imaging Microscopy (FLIM) 83
Total Internal Reflection Fluorescence (TIRF): A Near-Field
Microscopy 86
TIRF Is Based on the Optical Principles of Total Internal
Reflection 86
An Evanescent Wave Acts as the Excitation Source in TIRF ... 86
Fluorescence Scanning Near-Field Optical Microscopy
(Fluorescence SNOM).. 89
More than Near-Field Fluorescence Imaging 93
Photobleaching or Background—Despite Weak and
Highly Localized Illumination 93
Optical Microscopies: Summary.. 95
Combined OM-SPM Techniques: Eyesight to the Blind.................. 95
SPM and Optical Fluorescence Microscopy 98

Fluorescence SNOM and Single-Molecule Detection................ 106
    Single Fluorescent Molecules as Test Samples...................... 107
    Single-Molecule Imaging: Obtaining Molecular
    Orientations 108
    Conformations of Single-Polymer Chains in Films 109
    Single-Molecule Diffusion 111
    Distribution of Molecules and Molecular Complexes of
    Biological Interest in Synthetic and Native Membranes 112
References 115

Chapter 4   What Brings Scanning Near-Field Optical Microscopy: The
               Eyes at the Nanoscale.. 119

               Fundamentals of Scanning Near-Field Optical Microscopy............ 119
                   Background Suppression 124
                   The Nature of the SNOM Probe: Aperture and Apertureless
                   SNOM.. 126
                       Aperture SNOM 126
                       Apertureless SNOM... 128
                   The History of SNOM Is the History of Its Probes.................... 131
                       Aperture Probes.. 132
                       Apertureless Probes.. 135
                       Tip-on-Aperture Probes.. 135
                   Feedback Modes in SNOM 135
                   Contrast Mechanisms in SNOM.. 137
                       Polarization Contrast 140
                       Refractive Index Contrast 142
                       Infrared-Vibrational Contrast (Infrared Apertureless
                       SNOM).. 142
                   Summary 142
               Applications of SNOM.. 143
               References 149

Chapter 5   Adding Label-Free Chemical Spectroscopy: Who Is Who?............ 153

               Chemical Spectroscopy.. 153
               Raman (and IR) Microscopy 166
                   Raman Microscopy beyond the Diffraction Limit: Near-Field
                   Raman Spectroscopy 168
                       Aperture Near-Field Raman Spectroscopy........................... 169
                       Tip-Enhanced Near-Field Raman Spectroscopy (TERS)....... 169
                       Tip-Enhanced Coherent Anti-Stokes Raman Scattering
                       (CARS) 173
                       Sources of Enhancement of the Raman Signal in Near-
                       Field Raman Spectroscopy 173

IR Microscopy beyond the Diffraction Limit: IR Near-Field
Spectroscopy... 178
Aperture SNOM + IR Spectroscopy 179
Apertureless SNOM + IR Spectroscopy............................ 180
Another Source of Enhancement in Near-Field IR
Microscopy: Phonon Excitation.. 182
Summary 183
Applications of SPM + Raman Spectroscopy............................... 185
(Sub-) Monolayers of Dyes 185
Single-Wall Carbon Nanotubes 185
Inorganic Materials: Silicon and Semiconductors 187
Biomolecules 187
Virus, Bacteria, and Human Cells.. 188
Applications of SPM + IR Spectroscopy 194
Inorganic Materials 194
Polymers 194
Viruses and Cells... 195
References 199

Chapter 6   Combining the Nanoscopic with the Macroscopic: SPM and
Surface-Sensitive Techniques..203

Model-Based Surface Techniques 203
Fundamentals of Surface Plasmon Resonance............................... 214
Surface Plasmons on Metal Surfaces 214
Photon Creation/Excitation of SPPs 217
Experimental SPR 219
Fundamentals of Ellipsometry 221
Basic Equation of Ellipsometry.. 221
Single-Layer Model 222
Experimental Determination of Ellipsometric Angles 225
Rotating-Element Techniques...226
Rotating Polarizer/Analyzer 226
Rotating Compensator 227
Phase Modulation 227
Null Ellipsometry 228
Fundamentals of Quartz Crystal Microbalance............................229
Electromechanical Analogy of a QCM Sensor: Equivalent
Circuit of a Resonator.. 231
Small-Load Approximation: Connecting Frequency Shifts to
Load Impedance 233
Measuring Frequency Shifts and More: Modes of Operation
in QCM.. 235
Impedance Analysis 235
Ring-Down Technique.. 236

Frequency Shifts and Viscoelastic Parameters: Some Case
Examples 238
The Special Case of Rigid Films in Air: Sauerbrey
Equation and Film Thickness 238
Viscoelastic Films and the Importance of Measuring at
Different Overtones 239
Main Drawback: The Importance of Qualitative Interpretation ... 241
Summary 241
The Combination SPM and Model-Based Surface Techniques 241
SPM + SPR 241
SPM + Ellipsometry 244
SPM + QCM 248
References 254

Chapter 7 Scanning Probe Microscopy to Measure Surface
Interactions: The Nano Push–Puller 257

Force Curves: Surface Forces and More 257
Measuring the Probe–Sample Interaction as a Function of the
Relative Displacement 268
Noise and Artifacts 272
Quantitative Determination of Forces: Instrument and
Cantilever Calibration 273
Voltage to Deflection: Determination of the Sensitivity 273
Deflection to Force: Determination of Spring Constant 273
The Issue of Getting Absolute Distances 277
Qualitative Interpretation of Force Curves 280
Chemical Force Microscopy 281
The Science of Pulling Single Molecules or Ligand–Receptor
Pairs 282
Statistical Description of Bond Rupture: Two-State Model 283
Particularities of Molecular Recognition Spectroscopy 286
Negative Controls 286
Inferring Single Ligand–Receptor Interactions from
Adhesion Events 287
Particularities of Molecular Unfolding 288
The Science of Pushing: Contact Nanomechanics 290
A Short Note on Probes for Nanomechanics 290
Contact Region of the Force Curve: Beyond the Point of
Contact 291
Mapping Interactions 294
Force Curve Maps 294
Pulsed Force Mode 296
Molecular Recognition Imaging 297
Summary 298
References 300

Chapter 8   Tidying Loose Ends for the Nano Push–Puller:
                Microinterferometry and the Film Balance 303

                Microinterferometry.. 303
                Fundamentals of Reflection Interference Contrast Microscopy 307
                    Lateral and Vertical Resolution in RICM 311
                    RICM Setup.. 312
                    Dual-Wavelength RICM (DW-RICM) 314
                    Applications of RICM 315
                    RICM: Summary.. 316
                The Combined SPM/RICM Technique.. 318
                The Film Balance and Air–Fluid Interfaces 321
                Fundamentals of the Film Balance 325
                    The Transfer of Monolayers onto Solid Supports....................... 328
                        The Langmuir-Blodgett Technique.. 328
                        The Langmuir-Schaeffer Technique 328
                    The Film Balance: Summary 329
                The Combined AFM + Film Balance: The Monolayer Particle
                Interaction Apparatus (MPIA) 329
                References 333

Index... 335

Preface

Many are the books and reviews about scanning probe microscopies that cover the basics of their performance, novel developments, and state-of-the-art applications. This book may appear to be another of this kind. But it is not.

Indeed, this is not another book about scanning probe microscopy (SPM). As authors, we do not aim to focus on what SPM *can* do, but rather on what SPM *cannot* do and, most specifically, on presenting the experimental approaches that circumvent these limitations.

The approaches are based on the combination of SPM with two or more techniques that are complementary, in the sense that they can do something that the former cannot. This serves a double purpose; on the one hand, the resulting hybrid instrument outperforms the constituent techniques, since it combines their individual capabilities and cancels out their individual limitations. On the other hand, such an instrument allows performing experiments of dissimilar nature in a simultaneous manner.

But to understand the limitations of any technique also means to understand how this technique works. We do not skip this essential point; on the contrary, we have devoted a considerable amount of book space in explaining the basics of each technique as they are introduced. In the case of SPM, we have endeavored to present its fundamentals from a different, rather intuitive, perspective that, in our opinion, makes it distinctive from previous literature on the topic, and it ultimately serves a pedagogical purpose. At the same time, we have tried to avoid explaining the particularities of each SPM-based technique and opted for a rather generalized approach that may *suit everyone*. In this context, each member of the SPM family is presented as a particular case.

Though the contents of the book are organized in eight chapters, they follow a rather subtle scheme that revolves around the two main capabilities of SPM: that of imaging and that of measuring interactions. The fundamentals of SPM imaging are explained in Chapter 2 and followed by suitable combinations of SPM imaging with far-field and near-field optical microscopy, spectroscopy, and surface-sensitive techniques. These combinations are described in Chapters 3, 4, 5, and 6, respectively. Likewise, the fundamentals of SPM to measure interactions are explained in Chapter 7, followed by corresponding combinations with microinterferometry and film balance, which are both described in Chapter 8.

Each chapter that deals with one or more combinations of the type *technique A–technique B* starts with a historical account of the technique that is being introduced (*B*), followed by some background information that may be required to fully understand the contents of the chapter. The chapter then continues with the fundamentals of the *technique B* and the description of the combined *technique A–technique B* setup. The chapters end with some case examples where the combination has been applied in research. Research topics range from materials science to biosciences.

Because *this is not another book about scanning probe microscopy*, the content is addressed to a wide spectrum of readers, from researchers to teachers, without forgetting the student community. We hope that the text will fulfill their expectations. This would make us happy.

Last but not least, we would like to greatly thank Dr. Rafael Benitez, Uwe Rietzler, and Dr. Enrico Gnecco as well as the respective libraries of the University of Extremadura, the Max Planck Society, and the Spanish CSIC for providing useful literature references. Likewise, we are greatly thankful to Jacqueline Friedmann, Dr. Roberto Raiteri, and Professors Helmut Möhwald, Andreas Fery, and Georg Papastavrou for their valuable comments and proofreading.

Authors

Susana Moreno Flores graduated in chemistry at the Complutense University (Madrid, Spain), where she also did her doctorate studies at the Department of Physical Chemistry. Thereafter she did postdoctoral stays at the Department of Statistical Physics of the Ecole Normale Supérieure (Paris, France), at the Max Planck Institute of Polymer Research (Mainz, Germany), and at the Chemistry Department of the University of Basel (Basel, Switzerland) before working as a researcher in the Biosurfaces Unit at CIC BiomaGUNE (San Sebastián, Spain). She is currently assistant professor at the Department of Nanobiotechnology of the University of Natural Resources and Life Sciences (Vienna, Austria).

José L. Toca-Herrera is professor of biophysics at the University of Natural Resources and Life Sciences (BOKU-Vienna, Austria). After receiving his degree in physics from the University of Valencia, he entered a 12-month research training program at the Max Planck Institute for Polymer Research (Mainz, Germany) with Prof. Wolfgang Knoll. He completed his Ph.D. at the Max Planck Institute of Colloids and Interfaces (Golm) under the supervision of Prof. Helmuth Möhwald. After several postdoctoral stays with Prof. Regine von Klitzing (TU-Berlin), Prof. Jane Clarke (University of Cambridge), and Prof. Uwe B. Sleytr (BOKU-Vienna), he joined in 2004 the Department of Chemical Engineering of the Rovira i Virgili University (Tarragona) as RyC research professor. In 2007, he moved to CIC BiomaGUNE (San Sebastián), led by Prof. Martin-Lomas as group leader. In September 2010, he joined the Department of Nanobiotechnology at BOKU (Vienna), where he leads the Laboratory of Biophysics.

1 Introduction

Experimental science is the framework we use to observe and understand nature. It comprises a series of instrumental techniques through which we constantly put nature to the test; mathematical tools to quantitatively describe the instrumental output; and our capacity of reasoning, background knowledge, and common sense to find a rational explanation out of it. The first two steps, namely measurement and analysis, are repeated a number of times to ensure that the observed result is reproducible before thinking of a possible explanation.

Nature is rather complex, so we humans have compartmentalized it in a series of objects, processes, and diverse phenomena that we can tackle separately. The division of nature has led to the division of knowledge in a series of scientific disciplines, each with its own battery of technical and theoretical tools and even with its own terminology and concepts. This parallel and, in some aspects, unequal run in understanding nature has ruled the way of doing science since its origins.

This remained the case until disciplines stopped and stared at what others were doing, and they realized that our understanding of nature could be largely benefited from the exchange of information between the different disciplines. Nowadays, interdisciplinarity dictates the tempo of science. The various fields of knowledge borrow concepts, theories, and even techniques from one another with the goal of fusing the diverse knowledge about nature.

The next step toward this goal may involve exactly this word, *fusing*—fusing concepts, fusing theories, fusing techniques. This book is about the latter. The integration of techniques, either from one discipline (intradisciplinary techniques) or from different ones (interdisciplinary techniques) may evolve our way of doing science and significantly improve our knowledge of nature.

But there is no need to wait that long. Indeed, combining techniques may have an immediate impact in the way we do experiments, and that is through what may be called simultaneous measuring.

OBSERVING NATURE: SEQUENTIALITY AND SIMULTANEITY

Obvious as it may appear, the study of processes or phenomena occurring in nature should be as complete as possible, and this mainly requires investigating them from different points of view. Each so-called point of view may focus on one characteristic property, usually obtained through experiment with an instrumental technique. Though convenient and informative in most cases, a single property may not suffice a comprehensive description of phenomena. Rather, it is preferred to gather as many characteristic properties as possible—therefore the need to employ more than one technique.

Observing nature with several techniques can be done either sequentially or simultaneously. The sequential approach may as well follow the "one for one" maxim: one sample for one technique at a time. The strategy is depicted in the upper-left part of Figure 1.1. The object of study (the sample) is measured with a first technique; after completion, the very same one, or another, which should be a duplicate of the first one, is measured with a second technique, and so on until the sample, or another identical, is measured with the ith-technique. We have already commented on the importance of obtaining reproducible results. In sequential measurements, the reproducibility follows the "all for one" maxim: all samples for one technique at a time. Each technique demands to repeat the experiment with several identical samples (upper-right part of Figure 1.1). If N is the number of samples per technique, a total of $N{\cdot}i$ measurements is required. The procedure can thus be demanding, lengthy, and prone to experimental mistakes, for example if errors occurred in the duplication of samples (as many as $N{\cdot}i$ in total). On the other hand, the method proves robust, since it thoroughly tests the reproducibility of the sample and the instruments.

However focusing on one single property at a time may make us not see the wood for the trees. A naturally occurring event of scientific interest may be efficiently described if it is simultaneously *covered* by a set of techniques—just as a program or event is broadcast on television, where a few cameras and microphones simultaneously track the images and sounds of the whole event from different angles and positions. Regardless of the preference of the stage manager or the likely bias of the broadcast channel, the wider the employment of technical means, the more accurate and faithful the coverage will be. The high-quality information that results from that is received by either the journalist or the audience, who will, in principle, be capable of making a good analysis.

The analogy serves us to introduce the simultaneous approach in observing data. Parallel measurements may thus obey the "one for all" motto: one sample for all techniques. The set of techniques is employed simultaneously to investigate the different aspects of the sample, as shown in the lower-left part of Figure 1.1. Reproducibility of results is accounted for by repeating the parallel measurements with several samples (see lower-right part of Figure 1.1). The total number of results will be still $N{\cdot}i$, but they will be attained in much shorter time with fewer samples (N).

If the *sample* is rather a dynamic process where various properties may change in parallel at differing time scales, the concept of reproducibility gains a new dimension. Processes are tested as a whole; results are not considered individually but rather in relation to others; and therefore reproducibility may as well be referred to whole processes rather than to single results.

The aim of this book is to describe a series of combined techniques that make parallel measurements possible. All combinations referred to in this book share one common partner: a scanning probe microscope (SPM). This technique is young, although it has quickly become essential in the study of the morphology of most materials with a resolution that allows the detection of molecules and even atoms. The technique is not restricted to that, though. It can also be used to track processes occurring at the boundary of two materials—interfaces—as well as to measure surface interactions. The technique is thus versatile, but it is not exempted from limitations. The combination with other techniques can greatly reduce these limitations as well as increase the

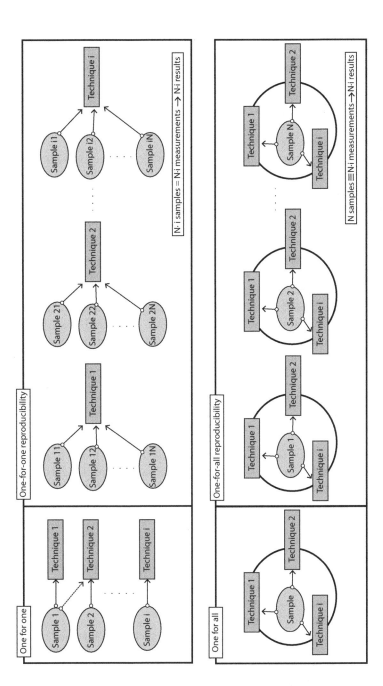

FIGURE 1.1 Measuring strategies. The upper half of the figure illustrates the "all for one" approach for sequential measurements. The same sample or several samples are used in turn with a particular technique to obtain a definite result one after the other. Repeating the same experiment with several samples (*N*) for each technique will prove the reproducibility of the results. The lower half of the figure illustrates the "one for all" approach for parallel measurements. In this case, a single sample is used by several techniques that simultaneously provide information about it. Reproducibility is checked by repeating the simultaneous measurement with several samples (*N*).

range of applicability of all techniques involved. In the following discussion, we will briefly comment on the possible synergic consequences of such combinations.

COMBINATION MACROSCOPIC/MICROSCOPIC: EXTENDING MEASURABLE RANGE TO BETTER UNDERSTAND THE CONNECTION BETWEEN ULTRASTRUCTURE AND FUNCTION

"The whole is more than the sum of its parts"...or not? Can we say more than "more"? We might dare answer the question by combining global and local techniques in order to investigate processes or characterize materials at the macroscopic, microscopic, or nanoscopic scales simultaneously.

Macroscopic techniques provide magnitudes that are actually average values over large sample regions. They include "classical" methodologies such as optical reflectometry, ellipsometry, or surface plasmon resonance. They all produce optical constants, such as thickness and/or refractive index, that is, single values to characterize samples as a whole. The combination of macro and micro techniques can shed light on how those magnitudes are affected by the microscopic and even nanoscopic nature of matter, a question of fundamental character in materials science.

An archetypical microscopic technique is optical microscopy, which is able to display images of specimens with a resolution of a few hundred nanometers. Although the technique is clearly outperformed by SPM or scanning near-field optical microscopy (SNOM) in terms of lateral resolution, its relevance is not at all undermined. Far from being mutually exclusive, optical microscopy and SPM are rather complementary. The latter significantly extends the measurable range of spatial dimensions down to very small objects, but the former ensures that we can still see the wood for the trees.

COMBINATION MODEL-BASED/DIRECT OBSERVATION

There are techniques that provide results as long as a mathematical model is applied. The model normally assumes that the sample fulfills a series of morphological characteristics, among other properties, that on the whole can be formulated mathematically. The assumptions oversimplify in the sense that they overlook to a certain extent the complexity of the real-life scenario. Consequently, models are intrinsically inaccurate.

Making the Right Choice

The key issue here is the right choice of a model. If the latter is based on assumptions of questionable validity or if it lacks in rigor, the results may be either false or inaccurate. The results are then affected by a certain uncertainty of analytical source, in addition to the usual instrumental uncertainty.

However, this inaccuracy may not stop us from using such techniques and obtaining reasonable results. Usually a model's complexity scales with accuracy, but sometimes this is not the case, and we can still use relatively simple models as long as the assumptions are "reasonable."

How to Judge the Applicability of a Model

No doubt, a previous knowledge of our sample is a must to tackle such a question. A direct, assumption-free observation of our sample's morphology may dissipate most of the reasonable doubts, if not all. Then an imaging technique such as SPM steps in. Now we can observe the sample's morphology, and this helps us to decide whether the chosen model is wrong or not, or if it rather oversimplifies the problem and needs some refinement. The refinement may come in the shape of an SPM result: the numerical determination of sample's roughness, topological profile, spatial heterogeneity, or stiffness, which can be introduced as additional parameters in the model. The latter in turn produces a result that may be the sample's thickness, its refractive index, or its viscoelastic parameters. By using particular results and applying the refined model, we may lose generality in the short term. Yet, we immediately gain in accuracy and knowledge, which in the long run may pave the way for the generation of models of wider applicability.

COMBINATION PREPARATIVE/MEASURING TECHNIQUE

Usually sample preparation precedes the experiment, and this procedure is considered to be critically important. The procedure is essential, no doubt; without a sample there is no experiment to perform. But it is also critical in the sense that it often determines whether an experiment is a complete success or a complete failure. The protocol of sample preparation as well as the sample's transfer and mounting onto the measuring device are key steps that can simply go wrong. Sometimes we are lucky enough to detect the source(s) of error and find a solution; at other times, however, we cannot. Or, rather, we cannot find a solution.

One typical example may be the preparation of immobilized proteins onto solid supports. A simple method consists of casting one or more drops of protein-containing solution onto a clean surface, waiting a specific amount of time to allow the molecules to adsorb on the support, and rinsing the excess with water or buffer solution afterwards. The sample must be kept always in a liquid state, since the proteins may just degrade (denature) otherwise. Even if the proteins and solvents used are of the highest purity, there are many factors that may influence the sample's quality. The environmental conditions of the laboratory may bring contaminants to the exposed surface; even the sample handling by a not-too-careful experimentalist may be a source of contamination. Keeping the sample clean and wet at the same time as it is being transferred and mounted onto the instrument can be most challenging, and the process may be quite unsystematic. Given these difficulties, sample preparation can have a serious impact on the reproducibility of the results.

Combining a preparative measuring technique with SPM makes it possible to prepare the sample and perform the experiment in one pot. Using a film balance* in combination with the SPM allows to simultaneously produce monomolecular layers and study them in situ, skipping the most disturbing transfer step that may bring about

---

* For a detailed description of a film balance, the reader is referred to Chapter 8.

utter sample modification or even sample damage. Sensoring techniques such as quartz crystal microbalance (QCM) or surface plasmon resonance (SPR) used in combination with perfusion methods can also be integrated to track surface changes as the sample is being prepared. The number of unsystematic errors derived from the preparation of more than one sample or from mishandling will be significantly reduced.

COMBINATION OF LOCAL TECHNIQUES TO TRACK THE DYNAMICS OF PROCESSES

If one combines various local techniques that are sensitive to distinctive interfacial properties and have relatively high time resolution, it is possible to track the dynamics of processes at interfaces in a comprehensive way. For example, measuring surface interactions or the morphology of monomolecular layers as they are being formed, expanded, or contracted is possible with a combination of an SPM and a film balance; on the other hand, tracking in real time the kinetics of molecular physisorption or electrochemical deposition of matter is at hand with a combination of an SPM and an electrochemical technique and/or QCM or SPR.

COMBINATION OF LOCAL TECHNIQUES OF DIFFERENT CONTRAST

Being able to thoroughly characterize matter by measuring different properties at the same time is a goal easily achieved by combining SPM with chemical spectroscopic techniques. Mechanical contrast is the source of the first technique to detect sample features, but it undoubtedly lacks the specificity needed to be used as an identifying tool of sample constituents. Chemical spectroscopy provides the chemical contrast required in those cases, but it lacks lateral resolution. The combination of SPM and chemical spectroscopy as one type of scanning optical near-field microscopy (SNOM) makes it possible to perform local spectroscopy at a resolution of a few nanometers.

This addresses the techniques and their interrelations. But what about the objects of study? We should not forget that the ultimate goal is to attain a comprehensive view of nature. We give now two examples of systems worth studying with integrated techniques.

SYSTEMS THAT CAN PROFIT FROM COMBINED TECHNIQUES

LIVING CELLS

Cells are the building blocks of living organisms. Just as the atom is the smallest constituent of matter with chemical identity, a single cell may be considered the smallest living constituent of *living* matter. And yet it is one of the most complex single entities encountered in nature.

In a living cell, myriad processes occur simultaneously at diverse scales. Molecules and ions are constantly moving in and out of the cell through pores in the cell membrane; inside, chemicals are transported from one side of the cell to another wherever they are needed; vesicles are formed either at the cell membrane as a result of intake of nutrients (endocytosis) or inside the cell to serve as vehicles for biomolecules; the cell can move, spread, or contract as a whole in its interaction with the environment; or it may grow, duplicate, and eventually die according to its living cycle. Each process is in itself a collection of events that involve molecular signaling, trafficking of self-organized structures, and orchestrated movements or deformations of large parts of the cell.

Single cells are certainly systems from which much can be learned if studied with combined techniques. Figure 1.2 illustrates a single cell in action while being simultaneously measured by different techniques. These techniques give complementary information, such as that provided by different cameras broadcasting a huge event. As an example, the process of cell adhesion on a solid support can be tracked in all its stages—from the anchoring step where the cell, initially suspended in solution, lands onto the support, until the spreading and growing steps where the cell significantly changes its morphology and, as well, its interaction with the solid support. Local changes at the basal membrane in close proximity with the support can be detected by techniques such as total internal reflection fluorescence (TIRF), reflection interference contrast microscopy (RICM), or QCM. If a microelectrode array is used as support, it is possible to measure the cell's extracellular potential or to locally electrostimulate the cell (Shenai et al. 2004; Saenz Cogollo et al. 2011). On the other hand, the apical membrane exposed to the surrounding fluid can be scanned by the probe of an SPM. Likewise, the probe can exert a local compression at a certain region of the cell membrane, and the transmission of this local pressure on other parts of the cell down to the basal membrane can be studied with confocal optical microscopy as well as with the other techniques mentioned previously.

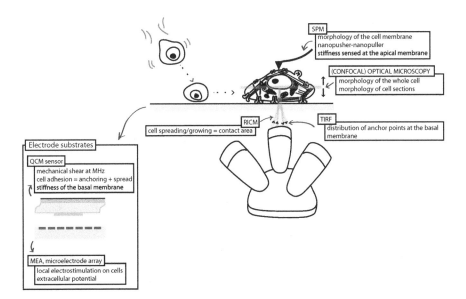

FIGURE 1.2 Samples for simultaneous measurements: living cells. Living cells are not only morphological entities, but the scenario of a plethora of interrelating processes that can simultaneously be studied by an appropriate set of combined techniques.

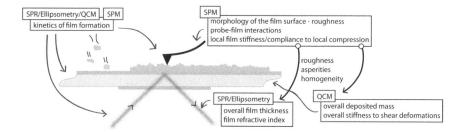

FIGURE 1.3 Samples for simultaneous measurements: films. Films can be fully character-ized in terms of surface topology, surface roughness, thickness, reflectivity, and stiffness in a simultaneous manner by employing an appropriate set of combined techniques. The instru-ments can also be used as time-resolved techniques to track the process of film formation.

FILMS

Films made of either organic or inorganic materials can be fully characterized in a single combined experiment, as hinted in Figure 1.3. This may include the very pro-cess of film formation on a solid support, which can be tracked by techniques such as QCM, ellipsometry, or SPR. They all can record changes that occur at the boundary between the solid support and the surrounding medium. On the other hand, time-resolved SPM can be employed to observe the morphological changes of the inter-face as the film forms, which can greatly help to interpret the changes observed by the other techniques. Once the film is prepared, the morphology of the film surface can be scanned by SPM, whereas SPR or ellipsometry can provide optical constants of the film, such as thickness and refractive index. As we will see in Chapter 6, these magnitudes are calculated from models that assume a definite film morphology. However, if the techniques are combined with SPM and applied to single-layer films, film morphology is a direct result that can be used to validate the chosen model. An example of that is the ever-challenging issue of understanding the effect of roughness of surfaces. In this regard, topography images of the surface obtained with an SPM can be used as reference in the simulation of contact and in the determination of real contact areas between two rough surfaces that exhibit multiple asperities.

FINAL REMARKS

The book is divided into two main sections that refer to the two main capabilities of SPM: imaging and measuring interactions. Each section opens with a chapter where the fundamentals of SPM in each of its performance modes are described. The sec-tions are completed with chapters devoted to a particular technique with which SPM has been combined. Each of those chapters starts with an account of the fundamen-tals of the technique, followed by a description of the combined setup(s), and ends with a few examples of scientific works where such combination has been employed.

Though we explain the basic working principles, the book is not intended to be a reference for each of the techniques mentioned here. For a deeper insight, the interested reader is referred to more comprehensive and specialized books and reviews listed in the reference section at the end of each chapter. The figures serve a didactic purpose and are not meant to be faithful representations of real life, inasmuch as they clarify ideas or concepts referenced in the text.

REFERENCES

Saenz Cogollo, J. F., M. Tedesco, S. Martinoia, and R. Raiteri. 2011. A new integrated system combining atomic force microscopy and micro-electrode array for measuring the mechanical properties of living cardiac myocytes. *Biomed. Microdevices* 13:613–621.

Shenai, M. B., K. G. Putchakayala, J. A. Hessler, B. G. Orr, M. M. Banaszak Holl, and J. R. Baker Jr. 2004. A novel MEA/AFM platform for measurement of real-time, nanometric morphological alterations of electrically stimulated neuroblastoma cells. *IEEE Trans. Nanobioscience* 3:111–117.

2 Scanning Probe Microscopy as an Imaging Tool

The Blind Microscope

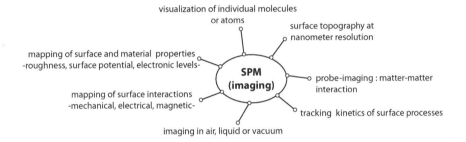

PROBE IMAGING

Just as an optical microscope produces a two-dimensional image through the interaction of light and matter, the scanning probe microscopies (SPMs) produce three-dimensional images through the interaction of matter and matter, the first "matter" being the sample and the second "matter" being a small, usually sharp probe.

The interaction between sample and probe can be of various types: mechanical, electrostatic, electronic, or magnetic, to mention a few. Those interactions are in most cases the source of image contrast in SPM. The information that one can extract from the sample is consequently as different as the interactions: morphology, topography, stiffness, compliance, electrical conductivity, magnetization. The results come in the shape of three-dimensional images, where each property is plotted as a function of the sample position and the lateral resolution is dependent on the probe size. To put it in numbers, ultimate lateral and vertical resolutions (in the case of topography images) amount to 5–10 nm and 1 nm, respectively. The SPM-generated imagery is routinely used in the most varied research fields, and their applicability is enormous. From micro-objects to nano-objects down to atoms—in air, in liquid, or in ultra-high vacuum—practically everything small can be *superficially yet utterly* imaged.

The history of SPM starts in 1982 with the invention of the first member of the family, the scanning tunneling microscope (STM) by Gerd Binnig and Heinrich Rohrer in the research laboratories of IBM. The revolutionary microscope was capable of displaying the topography of surfaces down to atomic resolution. The invention constituted a major breakthrough in the field of physics, for which both scientists received the Nobel Prize just a few years later, in 1986. In that very same year, the same group published the invention of the atomic force microscope. Since then and mainly until the end of the past century, there has been a continuous flow of instrumental inventions that, in one way or another, derive from either an STM or an AFM (atomic force microscope) setup. (A detailed account of these inventions is shown in the timeline of Figure 2.1.) On the whole, the techniques constitute what has become known as the ever-expanding family of scanning probe microscopies. The twenty-first century starts off with the development of high-speed atomic force microscopy by the group of Toshio Ando. The work of these scientists has considerably improved the time resolution of the technique that now can operate at speeds comparable to video rates.

The discussion in the following background section will help the reader to understand the basics of scanning probe microscopy.

Background Information

PIEZOELECTRICITY AND PIEZO ACTUATORS

Piezoelectricity is an effect encountered in a few materials that develop an electric potential when mechanical pressure is applied to them.[*] Conversely, such materials can experience a mechanical deformation if an electric potential is applied instead. The deformation can be very precise and minute, on the order of nanometers, which makes them most suitable to be employed as actuators in nanopositioning and motion-control devices. This capability is of primordial importance in the field of scanning probe microscopies.

Naturally occurring piezoelectric materials such as quartz, tourmaline, or Rochelle salt exhibit very weak piezo effects, which motivated the development of synthetic polycrystalline ferroelectric ceramic materials such as barium titanate or lead (plumbum) zirconate titanate (PZT) with enhanced piezoelectric properties.

We will concentrate on PZT, since it is the material mostly employed in the SPM instrumentation. The crystalline structure of PZT is characterized by a cubic unit cell shown in Figure 2.2a that becomes deformed and asymmetric below a certain temperature, called the *Curie temperature*. Under these conditions, each individual crystallite has an associated electric dipole and is piezoelectric. Macroscopically, however, the material does not exhibit piezoelectricity as long as its microscopic electric dipoles are randomly orientated. To make it piezoelectric, it is required to induce a permanent polarization in the material by a process called *poling*, depicted in Figure 2.2b. During poling, the material is

---

[*] Unless otherwise stated, this information has been extracted from the Web page of the company Physik Instrumente (PI GmbH & Co., Karlsruhe, Germany): http://www.physikinstrumente.com/en/products/piezo_tutorial.php#DetTOC.

1982 -- G. Binnig and H. Rohrer invent the scanning tunneling microscope (STM)

Nobel Prize(1986)

 Binnig, G., Rohrer, H. 1982. Surface studies by scanning tunneling microscopy. *Phys. Rev. Lett.* 49:57-61

1986 -- G. Binnig and coworkers invent the atomic force microscope (AFM)

 Binnig, G., Quate, C.F., Gerber, C. 1986. Atomic force microscope. *Phys. Rev. Lett.* 56:930-933

1987 -- Y. Martin and coworkers and J.J. Sanz and coworkers present in parallel the magnetic force microscopy (MFM)

 Martin, Y., Wickramasinge H.K. 1987. Magnetic imaging by "force microscopy". *Appl. Phys. Lett.* 50:1455-1457

Sanz, J.J. , García, N., Grütter, P., Meyer, E., Heinzelmann, H., Wiesendanger, R., Rosenthaler, L., Hidber, H.R., Güntherodt, H.J. 1987. Observation of magnetic forces by the atomic force microscope. *J. Appl. Phys.* 62:4298-4295

1988 -- The method of optical beam detection is developed and implemented in AFM setups.

Y. Martin and coworkers present the electric force microscope and applied the Kelvin effect on a scannng probe microscope

 Martin, Y., Abraham, D.W., Wickramasinge, H.K. 1988. High-resolution capacitance measurement and potentiometry by force microscopy. *Appl. Phys. Lett.* 52:1103-1105

1989 -- P. K. Hansma and coworkers invent the scanning ion conductance microscope (SICM)

A.J. Bard and coworkers present the scanning electrochemical microscopy (SECM)

 Hansma, H.K., Drake, B., Marti, O., Gould S.A.C. Prater, C.B. 1989. The scanning-ion conductance microscope. *Science* 243: 641-643

 A.J. Bard , Fan, F.-R.F., Kwak, J. Lev, O. 1989. Scanning electrochemical microscopy. Introduction and principles. *Anal. Chem.* 61:132-138

1993 -- L.P. Ghislain and W.W. Webb present the concept of combining optical tweezers with a scanning probe microscope in a new technique called photonic force microscopy (PFM)

 Ghislain, L.P., Web, W.W. 1993. Scanning force microscope based on an optical trap. *Optics Letters* 18:1678-1680

2001 -- T. Tadao and coworkers implemented a new atomic force microscope that can operate at very high speeds, yielding images at a rate comparable to those of video cameras

 Tadao, T., Kodera, N., Takai, E., Maruyama, D., Saito, K., Toda, A. 2001. A high-speed atomic force microscope for studying biological molecules. *PNAS* 98:12468-12472

FIGURE 2.1 History of SPM. The SPMs are a collection of very young techniques with a great impact in the field of nanoscience.

subjected to a strong electric field that induces the alignment of the electric dipoles along a specific direction—that of the electric field—and hence produces an intense polarization. After poling, the PZT has a diminished though remnant polarization, which confers the material piezoelectric properties.

A piezo actuator is a piece of piezoelectric material with defined geometry. Most common geometries include the cylindrical in the case of tube piezo actuators and the rectangular in the case of laminar piezo actuators, among others. The piezo actuator has electrodes attached at specific places to generate deformations

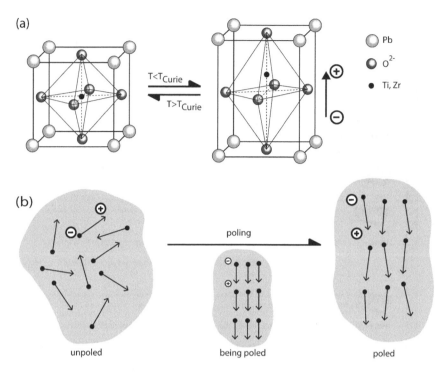

FIGURE 2.2 (a) Unit cell of a piezoelectric ceramic (PZT). Above the Curie temperature, T_{Curie}, the unit cell is cubic and has no associated dipole moment. Below T_{Curie} the unit cell deforms and so does the spatial distribution of the atoms. This results in the generation of a dipole moment, which is oriented as depicted in the figure. (b) Process of poling a piezoelectric ceramic. The dipole moments in the unpoled material are randomly oriented. During poling, the dipoles align along the applied electric field. After poling, the dipoles relax to a certain extent but remain aligned, creating a macroscopic polarization that makes the material piezoelectric.

along either one, two, or three main axes, x, y, z, denoted here as 1, 2, and 3, respectively, as shown in Figure 2.3a.

Tube Piezo Scanners

Tube piezo scanners consist of a thin-walled tube of ceramic material, silvered in and out to hold the electrodes. Two electrode pairs as long as the tube are attached at the outer wall of the cylinder, each pair oppositely placed and 90° from one another, as Figure 2.3b shows. An additional electrode is at the inner wall of the tube. An electric potential applied at each electrode pair of the outer wall results in a lateral displacement, either along the x-axis or the y-axis. To generate a tensile deformation along the z-axis (either elongation or contraction), an electric potential is applied between the outer and the inner walls of the tube. This is the *transverse piezo effect*.

The deformation along the z-axis, or in other words, the change of length of the tube, can be estimated by the following expression:

$$\Delta L \approx d_{31} L \frac{U}{d} \tag{2.1}$$

where d_{31} is the strain coefficient along the 3-axis when the potential is applied along the 1-axis (perpendicular), also called *piezo gain*. L is the original length of the tube; U is the applied potential; and d is the wall thickness. Values for d_{31} are negative and typically lie in the range of -180 to -262 pm/V (E. Meyer, Hug, and Bennewitz 2007).

Lateral displacements are achieved through tube bending. This is accomplished by applying electrical potential on the opposing electrodes at the outer walls, as Figure 2.3b shows. When the voltage is applied along one pair of electrodes, the tube flexes along one direction (e.g., along the x-axis); if the voltage is instead applied along the pair of opposed electrodes at 90° relative to the former pair, the tube will bend along the perpendicular direction, e.g., along the y-axis. The bending range ΔB, either along the x- or y-axis, can be approximated by

$$\Delta B \approx \frac{2\sqrt{2} \cdot d_{31} \cdot L^2 \cdot U}{\pi \cdot D_{inner} \cdot d} \tag{2.2}$$

where D_{inner} is the inner diameter of the tube.

Most SPMs rely on the lateral and vertical displacements of tube piezo actuators to move along the three spatial directions. Due to the tube geometry, lateral displacements are always accompanied by a displacement along the z-axis (see Figure 2.3b). This results in a vertical offset that is usually negligible for small lateral displacements compared to the tube length, i.e., $\Delta B/L \approx 10^{-7}$ (E. Meyer, Hug, and Bennewitz 2007).

STACK PIEZO SCANNER

A stack piezo scanner consists of a pile of ceramic disks separated by thin metallic electrodes (Figure 2.3b). The disks form a cylinder-like structure that elongates along its axis. This elongation of a stack piezo scanner can be estimated by the following expression:

$$\Delta L \approx d_{33} \cdot n \cdot U \tag{2.3}$$

where d_{33} is the strain coefficient (the displacement occurs alongside the field direction), n is the number of ceramic disks, and U is the applied voltage. These are the stiffest of all piezo actuator designs.

LAMINAR PIEZO ACTUATOR

Laminar piezo actuators are thin, laminated pieces of ceramics where the electrodes are placed at both sides of the slab in a sandwich-like configuration. An electric potential applied between the electrodes generates an elongation of the piezoelectric material parallel to the electrode plane (see Figure 2.3b).

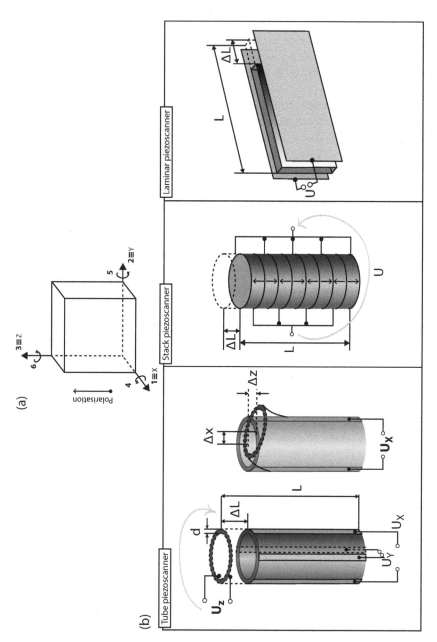

FIGURE 2.3   (*See facing page.*)

The displacement along the major axis of the laminar piezo actuator can be estimated by Equation (2.1); in this case, L is the major side of the slab and d its thickness.

LIMITATIONS

Piezo actuators are affected by several problems that limit their performance. These are mainly nonlinearity, hysteresis, and creep or drift.

Nonlinearity

Piezo scanners do not linearly deform with the applied voltage but exhibit certain deviation. The upper left graph of Figure 2.4 shows the elongation of a piezoelectric material when it is subjected to a gradually increasing voltage; the curve is not a straight line but, rather, it has an S shape. The ratio of the maximum deviation Δy to the ideal linear extension y, $\Delta y/y$, is called *intrinsic nonlinearity*, and it is expressed as a percentage. It typically ranges from 2% to 25% (Howland and Benatar 2000).

Hysteresis

Hysteresis is a feature that piezoelectric materials share with magnetic materials when they are subjected to a voltage cycle. As the voltage increases linearly from zero to a certain value and returns to zero in the same fashion, the actuator first elongates and then contracts. However, the position of the actuator during elongation does not coincide with that experienced during contraction.

The upper right graph in Figure 2.4 shows the change of length of the piezo actuator as a function of increasing and decreasing voltage: It follows two different paths. The hysteresis is defined as the ratio of the maximum deviation between the two curves to the maximum extension, $\Delta y/y_{\text{max}}$. It increases with voltage, and it can be as high as 15%–20% (Howland and Benatar 2000; E. Meyer, Hug, and Bennewitz 2007), which greatly impairs the position repeatability.

Creep or Drift

Creep or drift is the time change in the piezo actuator deformation with no accompanying change in the voltage. This is especially evident after a sudden change of voltage, where the piezo actuator continues to move or deform well

---

FIGURE 2.3 *(See facing page.)* (a) Notation of axis and rotations. Axes x, y, and z are denoted 1, 2, and 3, respectively. Rotations around x, y, and z are referred as 4, 5, and 6, respectively. (b) Types of piezoscanners. (*Left*): The tube piezo scanner elongates its length when a potential (U_z) is applied between an outer and inner electrode by the *traverse piezo effect*. Lateral displacements are achieved by applying a potential between the oppositely oriented lateral electrodes attached at the outer side of the tube, either along the x-axis (U_x) or the y-axis (U_y). In the second case the pair of electrodes affected is that positioned at 90° with respect to the x-pair. (*Middle*): The stack piezo scanner is a pile of disk-like piezoceramics where the polarization is alternated, between upwards and downwards. The potential is established between the set of down-oriented and the set of up-oriented disks to produce an elongation along the axis of the stack. (*Right*): The laminar piezoscanner is sandwiched between two electrodes and it elongates along the direction parallel to the electrodes.

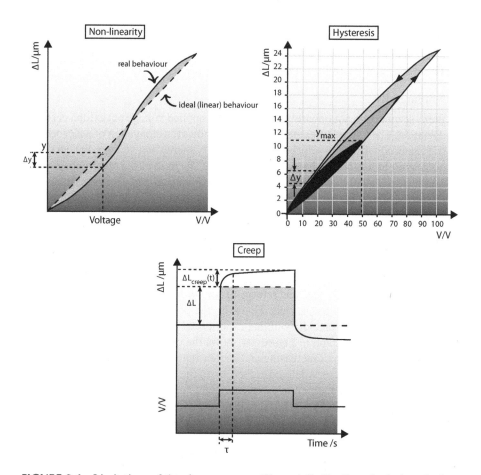

FIGURE 2.4 Limitations of the piezo scanners. (*Upper left*): Nonlinearity is described as a deviation from the ideal behavior, represented as a dashed line in the figure. The actual behavior of the piezoelectric follows an S shape. Quantitatively, it is expressed as the ratio $\Delta y/y$, Δy and y being the magnitudes depicted in the figure. (*Upper right*): Hysteresis loops where elongation is registered as the voltage increases and decreases back again to its original value. The piezoceramic does not exactly follow the same path. Hysteresis is quantitatively defined as a ratio between the greatest position difference between the two paths, Δy, divided by the maximum voltage applied, y_{max}. (*Lower graph*): Creep is the change of length experienced by the piezoceramic when no changes of voltage occur. Creep is mostly evident short after the application of a step voltage. The length of the piezoceramic continues to change even though the voltage is constant. The change of length, ΔL_{creep}, depends on time in a logarithmic manner, with a characteristic time constant, t.

after the voltage has been applied (i.e., time zero), as shown in the lower graph of Figure 2.4. Creep is enhanced at high temperatures and decreases logarithmically with time according to the expression,

$$\Delta L(t) \approx \Delta L_{t=0.1}\left[1 + \gamma \cdot \lg\left(\frac{t}{0.1}\right)\right] \tag{2.4}$$

where ΔL is the change of position as a function of time, and $\Delta L_{t=0.1}$ is the change of position 0.1 s after the voltage has been applied; γ is the creep factor that depends on the actuator properties, and it is usually 0.01–0.02.

AGING

Far from being constant, the strain coefficient d_{31} of piezoelectric materials changes exponentially with time and use. The aging effect is linked to the polycrystalline nature of the material and the fact that each crystallite has an associated dipole moment. With use and repeated application of a voltage in the same direction, more and more dipoles are aligned. The response of the piezoelectric material increases with the number of aligned dipoles, and therefore the former will deform more with time. Contrarily, if the scanner is not often used, more and more dipoles will lose their orientation, and the piezoelectric response will diminish with time.

CLOSED-LOOP ACTUATORS

Nonlinearity, hysteresis, and creep hamper accuracy of piezo actuators, but they can be virtually eliminated by employing a *servo-control* on the performance of the piezo actuator. Figure 2.5 shows the principle of closed-loop actuators. A sensor attached to the moving stage registers the position of the stage, y, after the application of a certain voltage V to the piezo actuator. A controller receives the sensor signal and compares it with the set value (called reference value), y_{ref}, which results in a position difference, $\Delta y = y - y_{ref}$. If Δy is not zero, the controller converts the magnitude into a voltage difference $\Delta V(\Delta y)$. The amplifier applies the new voltage signal $V + \Delta V(\Delta y)$, thus compensating for any deviations to the expected value. Closed-loop actuators rely on either strain gauge sensors or capacitor sensors to monitor the position of the piezo actuator. Capacitive sensors provide the best accuracy.

TUNNELING EFFECT

It is evident that a solid body cannot traverse walls. This is what we observe in everyday life. The physics of macroscopic objects such as those we can observe or feel is ruled by classical mechanics. However the world of very small objects, such as molecules or atoms, is ruled by quantum mechanics. And in this world, traversing walls is a likely event.

Tunneling is a quantum mechanical phenomenon in which a particle may penetrate a potential barrier even if it has not enough energy to surmount it. If the potential barrier has a height V and a thickness L and the energy of the

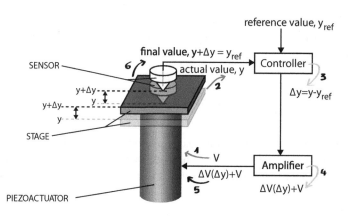

FIGURE 2.5 Closed-loop actuators. Any hysteretic effect, creep or nonlinearity can be compensated if the actuator is equipped with a servo-control. To an applied voltage, V the piezo actuator responds by changing its length (step 1). A sensor attached to a stage on top of the piezo actuator registers the position, y (step 2). The controller receives the sensor signal and compares it with the reference value y_{ref}. If the difference $\Delta y = y - y_{ref}$ is not zero, the controller sends a signal proportional to the difference to the amplifier (step 3), which in turn applies to the piezoactuator a compensating voltage $V + \Delta V(\Delta y)$ (step 5). The new position, $y + \Delta y$ now coincides with the expected value, y_{ref} (step 6).

particle is E, such as $E < V$, the probability of the particle traversing the barrier can be expressed as (Atkins and de Paula 2006)

$$T = \left[1 + \frac{\left(e^{\kappa L} - e^{-\kappa L}\right)^2}{16 \cdot \frac{E}{V} \cdot \left(1 - \frac{E}{V}\right)}\right]^{-1} \quad (2.5)$$

where $\kappa \hbar = [2m(V - E)]^{1/2}$ and m is the mass of the particle. If the barrier is high and/or thick so that $\kappa L \gg 1$, the expression in Equation (2.5) simplifies to

$$T = 16 \cdot \frac{E}{V} \cdot \left(1 - \frac{E}{V}\right) \cdot e^{-2\kappa L} \quad (2.6)$$

As Equation (2.6) shows, the transmission probability exponentially decreases with the thickness of the barrier as well as with $m^{1/2}$. This means that tunneling is likelier to occur through thin barriers than through thick ones and that light particles are more able to tunnel than heavy ones. Tunneling is thus more important for electrons than for protons or neutrons, which are considerably heavier.

Electron tunneling is behind one modality of scanning probe microscopy called *scanning tunneling microscopy* (STM). In STM, an electric potential is applied between a sharply pointed metal wire and a conducting solid, i.e., the sample. The tip of the wire is positioned very close to the sample's surface so that electrons may tunnel across the barrier, which is the small space gap between the wire tip and the sample's surface.

As we will see later in this chapter, the amount of electrons that may flow from the wire tip to the sample or vice versa is a complex function of the applied potential, the tip–sample separation, and the density of quantum energy states at the *Fermi level* of the sample.

ENERGY STATES IN METALS

The structure of metals can be viewed as a network of N atoms of the same kind surrounded by a "cloud" of electrons. The electrons are distributed in certain space regions called *molecular orbitals*, each of them characterized by an energy level. The number of molecular orbitals is N, as many as atoms. Each molecular orbital can be in turn occupied by one or two electrons at most or remain unoccupied. In metals, the energies of these orbitals are so close together that they form bands. Different bands result from the overlapping of different types of orbitals, and they may be separated by energy gaps of diverse magnitude.

In metal conductors, electrons partly occupy a band of closely spaced energy levels, so that the band contains unoccupied levels as well as occupied levels, as Figure 2.6a shows. The electrons can *easily* move to the unoccupied levels if these are close enough to one another—high density of states—and if the electrons have enough thermal energy; this movement confers the material its electrical conductivity. Consequently, the highest occupied molecular orbital (HOMO) depends on temperature. At $T = 0$, all electrons occupy the lowest energy levels, and the HOMO is the Fermi level.

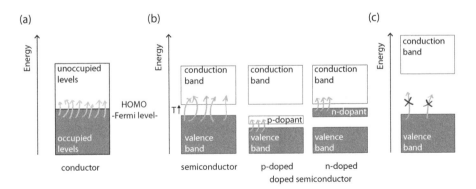

FIGURE 2.6 Band diagrams and electron distributions in metals. (a) In a conductor. The band is partially occupied with electrons that can easily promote to the unoccupied levels. The Fermi level is the highest occupied molecular orbital (HOMO) at $T = 0$. In semiconductors (b) and insulators (c), the Fermi level lies at the top of the band. The valence bands are fully occupied. To enhance material's conductivity, electrons should be promoted to the conduction band, i.e., by increasing the temperature. Alternatively, the energy gap between the bands can be reduced by introducing impurities or dopants, either electron donors (n-doping) or electron acceptors (p-doping). In insulators, only a few electrons, if any, can surmount the big energy gap.

In semiconductors and insulators, all the molecular orbitals that conform the lowest energy band are already occupied at $T = 0$. In these cases, the Fermi level lies on the top of this band, which is called the *valence band*. The electrons could only leave the valence band and reach the next unoccupied band, the *conduction band*, if they can overcome the energy gap between them. In semiconductors, the electrons can gain sufficient energy to reach the conduction band if the temperature is moderately increased (see Figure 2.6b). If the band gap is large so that at moderate temperatures there are very few electrons that promote to the conduction band, the material is an insulator (see Figure 2.6c) (Tipler and Mosca 2003).

HARMONIC OSCILLATOR

Let us assume we have a mass connected to the end of a spring, whose other end is attached to a solid wall. If we displace the mass a bit, the spring will oscillate around the equilibrium position. If the mass-spring system were undamped, it would move endlessly, and its motion would be described as a sinusoidal oscillation, with constant amplitude and a constant, *natural frequency* ω_0 or f_0.[*] That is the motion of a *harmonic oscillator*.

Real-life oscillators, such as the mass-spring system or a pendulum, do not move endlessly, but stop after a certain time. The motion is said to be damped, usually by frictional forces proportional to the velocity, and therefore they are called *damped oscillators*. To maintain the harmonic motion of an oscillator, it is necessary to apply an external force that varies sinusoidally with time. The purpose of such a force is to supply the oscillator with energy that otherwise would be lost during motion due to friction. In this case, the oscillator is said to be a *driven oscillator*.

Driven oscillators may be used as probes in SPM, as we will see later. In this case, the driving force comes from a piezo actuator that is attached to the oscillator, a bending spring. The piezo actuator is fed with a sinusoidal voltage and thus deforms in a sinusoidal manner. If the frequency at which the piezo actuator operates, ω, coincides with that of the oscillator, ω_0, the transfer of energy will be maximal and the amplitude of the oscillator will reach a highest value. This phenomenon is called *resonance*.

If the driving force depends on time according to the following expression,

$$F_{\text{driving}} = F_0 \cos \omega t \tag{2.7}$$

the motion of a driven oscillator at the stationary state can be described as

$$x(t) = A \cdot \cos(\omega t - \varphi) \tag{2.8}$$

---

[*] The variable f is the linear frequency and ω is the angular frequency, related through the expression $\omega = 2\pi f$. Both linear and angular frequencies are indistinctively used throughout this chapter.

where A is the amplitude and φ is the phase shift of the oscillation relative to the piezo actuator's deformation. A and φ are, in turn, functions of the driving force and of the motional parameters of the oscillator according to

$$A = \frac{F_0}{\sqrt{m^2\left(\omega_0^2 - \omega^2\right)^2 + b^2\omega^2}}$$

(2.9)

and

$$\tan\varphi = \frac{b\omega}{m\left(\omega_0^2 - \omega^2\right)}$$

(2.10)

respectively. The variable b is the damping coefficient or the proportionality constant of the frictional force, $F_{fr} = bv$, that tends to stop the motion. Figure 2.7 shows a typical curve of the amplitude and phase of a driven oscillator as a function of the driving frequency. When $\omega = \omega_0$, the amplitude reaches its maximum value ($A = F_0/b\omega$), and the phase shift φ is equal to $\pi/2$ ($\tan\varphi = \infty$). The width of the resonance curve scales with the damping of the oscillator, and it is characterized by a parameter called *quality factor* or simply Q *factor*. The Q factor is proportional to the ratio of the energy stored to energy lost per oscillation cycle;

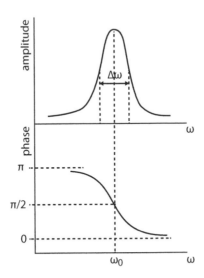

FIGURE 2.7 Resonance curves of a driven oscillator. The graphs show the dependence of the vibration amplitude and phase of a vibrating oscillator as a function of the driving frequency. The resonance occurs when $\omega = \omega_0$. At this frequency, the amplitude reaches a maximum and the phase is equal to $\pi/2$. The width of the resonance, $\Delta\omega$, is another characteristic parameter of the resonator, which is inversely proportional to the Q factor of the resonator.

for small damping, the Q factor is the ratio of the natural resonance frequency ω_0 to the resonance width $\Delta\omega$,

$$Q = 2\pi\left(\frac{E}{|\Delta E|}\right)_{cycle} = \frac{\omega_0}{\Delta\omega} \qquad (2.11)$$

Oscillators that exhibit sharp resonances have very high Q factors. In SPM, probes with higher Q factors exhibit better sensitivity than probes with low Q values. Image quality and resolution usually improve with high-Q probes (i.e., $Q > 100$) (Tipler and Mosca 2003).

THE SPM FAMILY

Since its invention in the early 1980s, more and more techniques have been incorporated to the SPM family, each one under the name of a scanning-X-microscopy, where X is a new property that could be imaged. With time, SPM has turned into a comprehensive concept that encompasses a considerable collection of scanning-X-microscopies, as Figure 2.8 shows. Cumbersome as it may be to describe each one of them, they all share a few common basics that greatly simplify the task. Those basics are described in the next section.

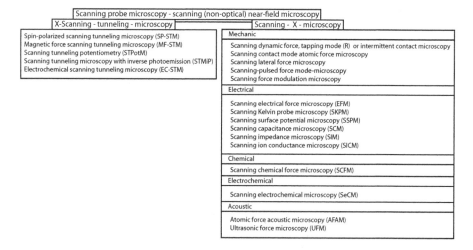

FIGURE 2.8 The SPM family. SPM encompasses a wide collection of techniques that can be arranged in two main groups: those based on the setup and performance of scanning tunneling microscopy (STM) and atomic force microscopy (AFM).

The Scanning Probe Microscope

The scanning probe microscope can be considered as the technical analogy of a human deprived of eyesight. Under this condition, we humans sense nearby objects with the help of our hands or sticks. As we move forward we intuitively probe our immediate environment by moving our hands or sticks in a systematic manner, for example, from left to right or up and down. Whenever we encounter an obstacle, we follow its contour by sliding or tapping our sticks or fingers over its surface. If the object were a hot plate, we would in this case not touch it and burn ourselves after sensing the heat, preferably at a safe distance.

A scanning probe microscope shares the human intuition in sensing objects and so behaves as a blind device that can detect micro- and nanosized objects, as illustrated in Figure 2.9. The instrument's stick is a *probe* ended in a nanometer-sized *tip* that is brought in contact or in close proximity to the sample surface. The probe is moved over the

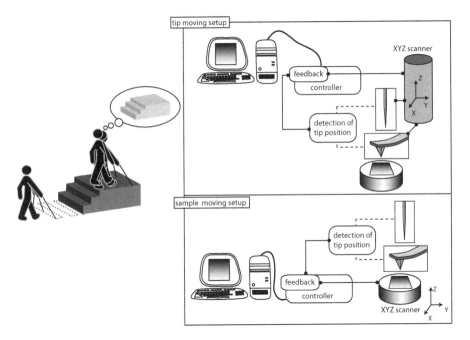

FIGURE 2.9 The blind analogy. A scanning probe microscope performs quite similar to a human deprived of his eyesight to detect the shape and size of objects. The drawing at the left-hand side shows a person who with the help of a stick goes up a staircase. By moving the stick left and right, back and forth, the person can figure out how the object looks. The right-hand side shows the general configuration of an SPM, either in a tip-moving setup (upwards) or in a sample-moving setup (downwards). The instrument is equipped with a sharp pointed probe (the stick) that can be moved in any direction by a piezo scanner. The probe can be a metal wire (in the STM-based setups) or a sharp object attached to a cantilever (in the AFM-based setups). In addition, the instrument includes a detector to register the tip's position (x,y,z) at all times and a feedback mechanism that controls the interaction of the tip with the sample.

sample surface in a *probe-moving setup* (or vice versa, in a *sample-moving setup*) with great precision by means of a piezo-driven device. Finally, the instrument is equipped with a *means to get the tip's (relative) position* at any time and a *feedback mechanism* to control either how "hard" the tip should slide or tap or how "close" the tip should be kept from the sample surface. In both cases, the tip will follow the contour of the object by scanning it from left to right, from right to left, and up and down in a raster manner.

All the SPM variants are based on one of the two most typical instrumental configurations: that of the scanning force microscope, also called atomic force microscope (AFM), and that of the scanning tunneling microscope (STM).

AFM

The probe consists of a sharp tip, with a typical radius of curvature (ROC) of 10–30 nm that is attached to the free end of a micromachined cantilever. The instrument is also equipped with an optical mechanism to detect the tip's position with a four-quadrant position-sensitive detector (PSD) that registers the reflection of a laser focused on the cantilever's free end.

STM

The probe is a sharp-ended, conductive wire, generally made of Pt or a Pt/Ir alloy, and there is no PSD detector. Instead, the probe's location is indirectly obtained by measuring the tunneling current between the probe's tip and the conductive sample as long as an electric potential is applied. Under these conditions, the value of the tunneling current controls how far the tip should approach the sample's surface, since it exponentially increases with decreasing tip–sample separation. When the tunneling current is sufficiently high or, more precisely, when it reaches a preset value, the tip does not further approach the sample and stays in that position as long as the tunneling current does not change.

PROBES AND TIPS

In an ideal world where the probes were infinitely sharp and the surfaces infinitely flat, the SPM images would show the true profile of the samples, unaffected by any probe's influence whatsoever. However, in the real world, the probes have a finite geometry, and though they are sharp, they always have a certain width and bluntness at the very end, which is defined by their radius of curvature (Figure 2.10a). Likewise, the real-world samples are far from being atomically flat; rather, they are more or less corrugated. Consequently, the tip may interact with different parts of the sample, and the sample may in turn interact with different parts of the tip. All in all, this often results in a nonnegligible—and probably complex—convolution of the sample features with the tip shape that affects every SPM image.

The probe's tip influence, though concomitant to all SPMs, can (and should) be minimized. Moreover, the shape of the tip must be accurately controlled and its geometry known in order to determine the true profile of the sample. To do that, the probe should fulfill the following requirements:

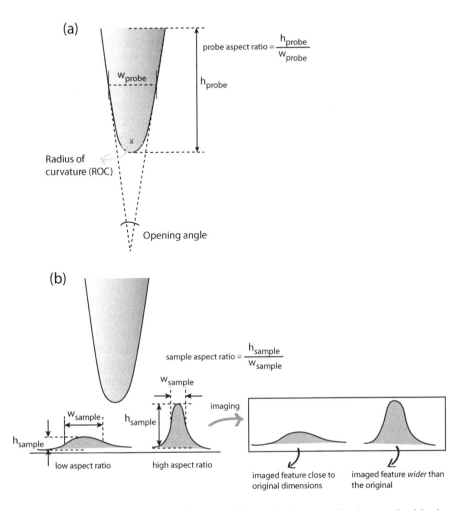

FIGURE 2.10 SPM probes. (a) Probe shape. The probe is generally characterized by its height, h_{probe}, its width, w_{probe}, its opening angle, and its radius of curvature (ROC). A good probe should have a high aspect ratio, a small opening angle and a small ROC to image the sample's true topography at the highest resolution. (b) Probe imaging. If the sample features are not sharp compared to the tip (low aspect ratio), the imaged profiles resemble the true sample topography; conversely, if the features are sharp compared to the tip (high aspect ratio), the imaged profiles look distorted due to the influence of the tip, or *tip convolution*.

1. Smallness: The smaller the probe, and the probe's tip, the better. Or, at least, it should be negligibly small compared to the sample features. Lateral resolution will consequently increase by decreasing the probe size.
2. Sharpness: The edge around the very end of the probe should be as normal to the sample surface as possible. This means that the opening angle of the probe shown in Figure 2.10a has to be very small, or at least small compared to the lateral dimensions of the sample features. Figure 2.10b shows

the original and imaged profiles of a sample that has wide and narrow features compared to the tip dimensions. The images of samples with lower aspect-ratio features are more representative of the true sample's topography than those obtained from samples showing higher aspect-ratio features.

3. Length: The probe should be long enough so that the sample features do not interact with parts of the probe far from the tip, which will negatively affect the vertical resolution. The probe should thus be sharp alongside its length. Thus requirements 2 and 3 can be combined to state that the aspect ratio of the tip should be as large as possible and in any case large compared to the aspect ratio of the sample features. The aspect ratio is defined as the quotient between the probe's length and its width.

4. Softness (especially with mechanical probes): The probe should be soft enough not to produce sample deformations due to excessive contact pressure.

5. Durability: This especially concerns tips required to be in contact with the sample during scanning. A tip should resist possible wear and damage after continuous scanning.

STM Probes

These probes are shown in the left-hand side of Figure 2.11 and are basically metal wires, made of either tungsten (W) or a platinum–iridium alloy (Pt/Ir). Tip sharpening is not an easy task, though, and several methods have been developed that differ in sophistication. The simplest of all consists of shear-cutting the wire with household scissors to produce a sharp end along the cut. The procedure is far from being

---

FIGURE 2.11 *(See facing page.)* STM-based probes *(left)*. Metal wires whose end can be either electrochemically etched or mechanically cut to produce a sharp tip. Electron micrographs show that electrochemical etching produce rather smooth, highly symmetric probes, while cutting may produce nonsymmetric tips with irregular faces and edges. (Micrograph of the etched STM tip is reprinted with kind permission of Prof. Milunchick, Milunchick's group at the University of Michigan, http://www.mse.engin.umich.edu/research/highlights/electrochemical-etching-of-ultrasharp-tungsten-stm-tips. Micrograph of the mechanically cut STM tip is reprinted with kind permission from Scanning Probe Solutions, Ltd., Cambridge, U.K., http://www.scanningprobesolutions.com/Products.php.) AFM-based probes *(right)*. These probes consist of pyramidal tips attached at the free end of rectangular or triangular cantilevers. The pyramids may be triangular, square, or rhomboidal based. In the latter case, the tips exhibit a front and a back half-opening angle, θ_{front} and θ_{back}, respectively, that differ in a few degrees. When mounted in the AFM-setup at a tilt angle α, the front and back edges may have the same inclination to the surface normal or one of the edges may be very close to it (*quasi-normal* edge). The first case occurs when θ_{front} is larger than θ_{back}, and the opposite holds for the second case. The first case will produce images with symmetric edges when the tip scans alongside the cantilever long axis back and forth. The second case will produce asymmetric edged-sample features, but closely resembling the real ones when scanned by the quasi-normal edge. The electron micrographs show a few examples of AFM-type tips, pyramidal-type, carbon-nanotubes and an apex showing a high-aspect ratio spike. The two last micrographs are examples of high-aspect-ratio probes. The probes are in this case positioned at the apex of the pyramids. (Micrograph of the carbon nanotube attached to an AFM tip is reprinted with permission of Nanosensors™, Neuchâtel, Switzerland, http://www.nanosensors.com/CNT-FM.htm.)

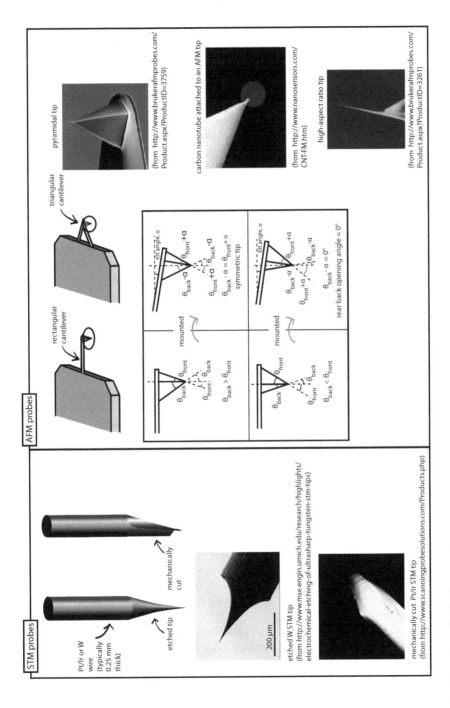

FIGURE 2.11   (*See facing page.*)

reproducible, but its simplicity makes it one of the most commonly used. The cuts are normally not smooth enough, but they may produce sharp protuberances that can as well act as tips. An alternative method is electrochemical etching of tungsten wires that usually produces smooth tips with curvature radii on the order of 10–100 nm (E. Meyer, Hug, and Bennewitz 2007). The wire, which is also an electrode, is immersed in an aqueous solution of sodium hydroxide for a few seconds. A piece of stainless steel or platinum acts as a counterelectrode. The wire thins at the air–water interface and eventually breaks in two, cutting off the current between the electrodes. Both tips, that of the part that remains fixed over the interface and that of the part that breaks off, can be used for scanning. But before that, it is required to wash away the NaOH residuals from the tips and to remove the metal oxide from the surfaces that have been produced during etching.

AFM Probes

Considered to be the most versatile probe type for present and future SPM implementations, the AFM probe is a *cantilever* with or without an integrated sharp *tip* "hanging" from its free end. The fixed end is attached to a relatively thick, millimeter-sized slab, the *chip*, to ease handling. The right-hand side of Figure 2.11 shows the appearance of such probes.

In the SPM setup, these probes rest on their chips so that the cantilever is not perfectly horizontal to the sample plane but slightly tilted. The tilt angle α, usually 10°, ensures that only the tip and no other part of the probe stays in immediate proximity to the sample surface.

The chip and the cantilever are generally made of the same material, usually doped or undoped silicon or silicon nitride, but other materials such as gallium arsenide or diamond are also employed. Though the tips are most frequently made from the same material as the cantilever, they can be produced from a different material, for example, nanotube-like carbon or diamond-like carbon on silicon cantilevers or silicon tips on silicon nitride cantilevers. Diamond tips are particularly durable and resistant to wear. Silicon and silicon nitride probes are batch-produced with high reproducibility and high throughput. Bulk micromachining and anisotropic etching of whole silicon wafers are employed for the silicon variants, whereas molding from pyramidal-shaped pits is the method of choice to produce silicon nitride probes (Oesterschulze et al. 2006). The tips can be further sharpened by repeated oxidation at reduced temperatures, with subsequent removal of the oxide layer.

The cantilevers are micrometer-sized beam springs that are able to bend vertically or laterally. The shape of these cantilevers may be either rectangular or triangular. The apex of the triangular cantilever is the free end where the tip is positioned. These cantilevers are usually 100–200 μm long from chip to free end, 10–50 μm in width, and 500–700 μm thick.

Tips may adopt different aspects and geometries, the most typical being the triangular or the square pyramids. The inclined sides of these tips define identical front, back, and side angle, and hence they produce sample profiles that are affected by the same type of artifact, regardless of the scanning direction. Rhomboidal-based or asymmetric pyramids with different side inclinations are also very frequently encountered, though, where the front and back angles differ by a few degrees, as the right-hand side of Figure 2.11 shows. They are convenient in cases where accurate measurements of sample sharp edges are required. Since the cantilever probe is mounted with a slight inclination, at least one side of the pyramid will be normal to the sample plane, minimizing the tip artifact along that particular scanning direction. High-aspect ratio tips are at hand by attaching carbon nanotubes at the apex of the pyramidal tips, or by creating a bunch of spikes, either of the same material as the tip or of carbon out of contaminated probes in an electron microscope.

The cantilever probe is basically a force transducer. It senses the interaction between the tip and the sample and transforms it into a measurable bending of the cantilever. For small deformations, the force is obtained by applying Hooke's law:

$$F = k \cdot \Delta x \qquad (2.12)$$

where k is the spring constant of the cantilever and Δx is the magnitude of the cantilever's bending. Calculation of the cantilever spring constant is the core of cantilever calibration and key to the quantitative determination of tip–sample interactions. Its description lies beyond the scope of this chapter, but it will be revisited in Chapter 7, dedicated to forces.

However, cantilever probes can be used to sense forces other than mechanical. Indeed, the most typical electric and magnetic probes are cantilever-based sensors coated with a conductive or a magnetic material, respectively.

Electrical Probes

The cantilevers, made of doped silicon, are coated in this case with a layer of 20–100 nm of a conductive material like platinum, platinum-iridium (Pt/Ir) alloy, cobalt, or doped diamond. Cobalt and doped diamond coatings serve a double purpose. On the one hand, cobalt confers the tip with magnetic properties, since it is ferromagnetic; on the other hand, doped diamond is used to produce tips with enhanced wear resistance, since diamond is the hardest natural material known.

Magnetic Probes

The most common coating material is a cobalt-based alloy with some additives, such as chromium or platinum (Oesterschulze et al. 2006). The ferromagnetic coating is arranged in separated grains over the tip's surface, and this results in different lateral resolution of some tips compared to others. The resulting tips are said to be hard, since they have high coercitivity; in other words, they exhibit a high resistance to demagnetization and are best aimed for strong magnetic signals. Soft magnetic coatings with reduced coercitivity are also at hand by choosing other coating materials such as cobalt, nickel–iron alloy, iron, and granular iron (silicon oxide) (Oesterschulze et al. 2006).

High-Aspect Ratio Probes

These are aimed to further push the limit of lateral resolution. In this case, the length:width ratio is increased to the extreme by using carbon nanotubes or ultrasharp spikes at the apex of the tip pyramids. The right-hand side of Figure 2.11 shows electron micrographs of both types of probes. Carbon nanotubes are known for their great stiffness and well-defined geometry, which consists of hollow cylinders with single or multiple walls, each one made of an atomic network of sp^2-bonded carbons. The diameters are in the nanometer range, whereas the length is on the order of micrometers up to centimeters; this makes them one of the structures with the highest aspect ratio known (at least 1000) and hence ideal to work as SPM probes (Marsaudon et al. 2008). The probes are highly resistant to wear, but they are very stiff and therefore not optimal for imaging soft samples.

Spiked tips made of carbon or silicon with aspect ratios ranging from 10–100 are an alternative to carbon nanotubes when softer tips are required. The radii of curvature are within 1 and 15 nm for the carbon-made and the silicon-made ones, respectively. The tips are suitable for soft and flat samples, but they are relatively easily altered and damaged, and thus prone to cause artifacts.

A rather imaginative alternative to those high-aspect ratio probes consists of using the edge of a thin plane as tip (van den Bos et al. 2002). This is of special advantage for electrical and magnetic probes, where coatings enlarge the tip dimensions and hence worsen the tip's performance. The tip plane has a triangular shape and it hangs from the free end of edge-on-resting cantilevers, perpendicular to the cantilever plane, as seen in Figure 2.12. The magnetic coating is then deposited on the thin side of the tip plane, thus producing a magnetic sensor with extremely small lateral dimensions and hence suitable to perform high-resolution magnetic force microscopy.

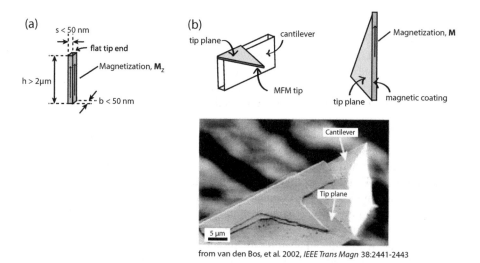

from van den Bos, et al. 2002, *IEEE Trans Magn* 38:2441-2443

FIGURE 2.12   The cantilever. An exotic design of a probe for magnetic force microscopy (MFM). (a) Requirements for an MFM probe to achieve high lateral resolution: a high-aspect ratio probe ($h/(a \times b) > 200$) with a flat, narrow base and made of a magnetic material whose polarization is along the long axis of the probe. (b) The probe is produced at the edge of the cantilever as a triangular platform of small thickness. The free-hanging apex of the triangle is the tip of the probe. An electron micrograph shows the appearance of such a probe. (Figures a and b were redrawn and reprinted, respectively, from van den Bos et al. 2002. With permission.)

SCANNERS

An SPM scanner should move with high precision and accuracy along the three spatial directions (x, y, and z). Along the x- and y-axes, the scanner systematically rasters a rectangular area by moving back and forth along one line or *row* (the *fast axis*) before moving along the perpendicular direction (the *slow axis*) and scanning the next row, as seen in Figure 2.13. The scanner moves along the z direction following the topography of the surface, its movement controlled by the feedback mechanism. Most of the SPM scanners are made of lead zirconium titanate (PZT), a piezoelectric material where the transverse piezo effect applies. The transverse piezo coefficient, d_{31}, for such material is typically −262 pm/V.

Some Numbers: Scanners Need High-Voltage Amplifiers

In this section we estimate the voltage required to generate typical displacements in SPM. We will consider elongations as well as lateral displacements of a tube PZT piezo scanner of $d_{31} = -262$ pm/V, with a ceramic layer thickness d of 1 mm and length L of 1 and 5 cm. Applying Equations (2.1 and 2.2), it is possible to calculate the required voltage to produce a given deformation. Figure 2.14 shows two plots where voltage amplitudes are plotted as a function of the elongation (Figure 2.14a) and as a function of the lateral displacement (Figure 2.14b) of the piezo scanner. Elongations on the order of several hundred nanometers to 1 μm require voltages around 80 V and 400 V; however, lateral displacements of a few micrometers require voltages of one to several kilovolts. These values are high enough to require high-voltage amplifiers. Since they can be operated at such high voltages, scanners can thus be dangerous, and handling must be done with care.

The scanners for SPM should be rapidly responsive, highly linear, creep- and hysteresis-free, and thermally stable. Rapid response refers to the time required to respond to a sudden voltage change, if needed. This allows scanning at higher speeds with greater accuracy in determining sample topography. Typically, a piezo actuator can reach a certain displacement in approximately one-third of the period of its resonant frequency

$$T_{min} = \frac{1}{3f_0} \tag{2.13}$$

Piezo scanners should be operated well below their resonance frequencies in order to avoid ringing and unwanted resonances.

Nonlinearity can be compensated by proper calibration or virtually eliminated by using closed-loop scanners. The latter applies for hysteresis and creep; hysteresis becomes an important issue in raster scanning, since the piezo scanners are subjected to repetitive lateral motion, which is achieved by the application of

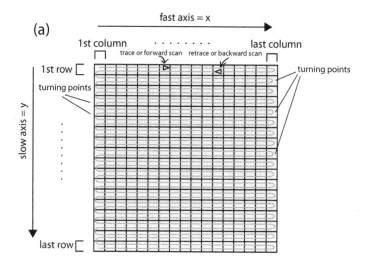

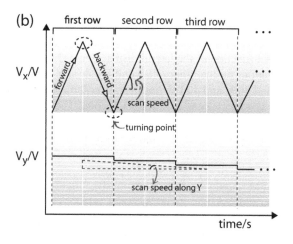

FIGURE 2.13 Raster scanning. (a) The figure shows a square-like area divided in rows and columns, defining an array of *pixels*. The trace line indicates the path followed by the probe during the scanning: At each row, the probe moves from left to right (*forward*) and from right to left (*backward*), scanning twice along a line. The probe then moves one row down along the y-axis before scanning the following row. The process is repeated until the whole area is scanned. The lateral sizes of the scaned area are turning points where the scanner must change direction. This implies that the sign of the electric potential that makes the scanner move has to suddenly change sign. The discontinuity in the applied voltage can thus originate scanner-induced artifacts, such as creep or hysteresis. (b) Voltage signal applied to the piezoscanner assuming that no hysteresis or creep occurs. The voltage applied along the x-axis is periodic, sawtooth shaped. The period and the amplitude of the signal define the scan speed along the x-axis. The turning points are the discontinuities in the voltage signal between row scans. The voltage applied along y changes in a steplike manner when the probe moves from row to row; otherwise it is constant. The overall variation of y with time defines the average scan speed along the y-axis.

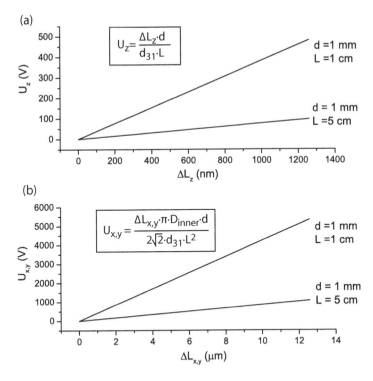

FIGURE 2.14 Voltages required to operate a PZT tube piezo scanner. (a) Elongations. The figure shows a graph of voltages applied to generate the typical displacements encountered in SPM (from nanometers to micrometers). High voltages are required to produce elongations of the order of 1 mm in the case of 1- and 5-cm-long scanners. (b) Lateral displacements. They are on the order of a few micrometers and voltages are 10 times larger than those required to produce elongations.

periodic, sawtooth-like potential signals. Most of the SPM scanners are tube piezo actuators, which are affected by vertical–lateral crosstalk when large lateral displacements, e.g., on the order of $10–10^2$ μm, are required (see Figure 2.3). In these cases, the scans are affected by a significant curvature. An alternative is to use flexures as guides for the piezo actuator movement. A flexure is a frictionless, stictionless device based on the elastic deformation (flexing) of a solid material, for example, steel. These devices are capable of producing highly straight motions in the nanometer range or below.

Thermal stability of the PZT ceramics is relatively good, compared to other materials; the strain caused by temperature variations can be as high as 0.11% if the temperature changes from 100°C down to −70°C. These conditions are clearly not met in everyday life, and therefore one can say that the scanners are fairly stable. Nonetheless, the strain coefficients are temperature dependent; as a matter of fact, at liquid helium temperature, the piezo gain drops to approximately 10%–20% of its room-temperature value. The operating temperatures must not exceed the Curie temperature though, since above such temperature the scanners may depole and thus lose their piezoelectric properties. Curie temperatures for PZT actuators are on the order of 150°C–350°C, and piezo scanners can typically work up to 80°C–150°C.

Last but not least, the resolution of piezo scanners is in principle unlimited; there are no threshold values above which strain occurs. However, this makes them extremely sensitive to electronic noise, since it can bring about noticeable variations in their position. To ensure continuous, smooth motion at atomic resolution, which is a must in scanning probe microscopy, the piezo scanners should be driven with low-noise amplifiers.

Some Numbers: How Fast Scanners Can Be

Typical resonance frequencies of standard scanners are in the range of 0.1–10 kHz.[*] As a first approximation, the resonant frequency of a piezo scanner, f_0, can be described as an unloaded spring rigidly attached at one of its ends,

$$\omega_0 = 2\pi f_0 = \sqrt{\frac{k}{m}} \tag{2.14}$$

where k is the stiffness of the piezo actuator and m is the effective mass. If the scanner is set to operate at scan frequencies close to its resonance frequency, the rapid response may bring about overshoots that seriously impair the performance of the scanner. To increase the operational scan speeds, it is thus necessary to work with scanners of high resonance frequency and to employ light, stiff scanners (such as the stack piezo actuators). Maximal operational scan speeds should then be well below 0.1–10 kHz.

DETECTION OF THE TIP'S POSITION

We have already mentioned that SPM technologies rely on two main basic configurations: that of the STM and that of the AFM. Each setup has its own methods of detecting the tip position: tunneling detection in the case of STM and a few more in the case of AFM, such as optical beam detection (G. Meyer and Amer 1988; Alexander et al. 1989), capacitive (Neubauer et al. 1990) and piezoresistive detection (Tortonese et al. 1991; Minne, Manalis, and Quate 1995), or optical interferometry (Rugar et al. 1988; Rugar, Mamin, and Guethner 1989; Schönenberger and Alvarado 1989). In the following discussion, we will briefly comment on the tunneling detection as well as the optical beam and piezoresistive detection.

Tunneling Detection

This method relies on the dependence of the tunneling current on the tip–sample distance. The tunneling effect is a quantum-mechanical phenomenon of small but

---

[*] Catalog of PI instruments: Planar AFM Piezo Scanner Stages & Controllers. 2011. Physik Instrumente (PI) GmbH & Co. KG. Downloadable from Internet at: http://www.piezo.ws/pdf/Catalog_AFM_Planar_Piezo_Stage_Scanner_C.pdf.

nonnegligible probability. Contrary to what it may be expected in everyday life where classical physics apply, particles with energy E can penetrate barriers of higher energy, ϕ, otherwise insurmountable. In STM, the particles are electrons, and the barrier is the gap (i.e., vacuum or fluid filled) between the sample and the tip. The tunneling current, which is the flow of electrons between the tip and the sample separated by a small distance, is thus a consequence of the tunneling effect. The tunneling current can be calculated from the density of states of the sample at the Fermi edge, $\rho_s(E_F)$

$$I_t \propto V\rho_s(E_F)\exp\left[-2\frac{\sqrt{2 \cdot m \cdot (\Phi - E)} \cdot z}{\hbar}\right] \propto V\rho_s(E_F) \cdot e^{-1.025\sqrt{\Phi}z} \qquad (2.15)$$

where V is the applied potential between the tip and the sample; m is the electron mass; $\hbar$ is the reduced Planck's constant, $h/2\pi$; Φ is the barrier height in eV; and the tip–sample separation, z, is in angstrom units. The tunneling current depends exponentially on the tip–sample separation, and hence it is extremely sensitive to this magnitude. Indeed, for $\Phi = 5$ eV, which is the case for gold, the tunneling current decays one order of magnitude when the distance is changed by 1 Å.

As mentioned previously, the tunneling effect is an unlikely event, and therefore the tunneling currents are very low, normally on the order of 1–10 pA. Voltages on the order of a few hundred millivolts are typically applied between tip and sample, which should both be conductive, and the tunneling current is measured with pre-amplifiers, which are I-V converters with resistors on the order of 100 MΩ to 1 GΩ (E. Meyer, Hug, and Bennewitz 2007).

Optical Beam Detection

This method is the most widely used in commercial AFM-based setups, due to its simplicity and ease-of-use (Jarvis, Sader, and Fukuma 2008). The method can be employed in air and in liquid, since it is physically detached from the sample-probe environment. Figure 2.15 shows a scheme of the optical beam detection system. A focused laser beam impinges on the back side of the cantilever, close to its free end. The reflected beam is detected by a position-sensitive photodetector (PSPD). Upon cantilever deflection, the reflected beam changes its trajectory, reaching the PSPD at a different position. The detector then produces a differential current signal that is proportional to the displacement of the reflected beam relative to the original position. The current signal is transformed into a voltage signal with an I-V converter and fed into a differential amplifier. This generates a voltage signal that is ultimately proportional to the cantilever deflection.

The optical arrangement upon which this method is based can convert extremely small cantilever deflections into detectable displacements of the laser spot in PSPD. The resulting magnification attained with such a system, M, can be expressed as follows:

$$M = \frac{d}{\Delta z} = \frac{2L}{l} \qquad (2.16)$$

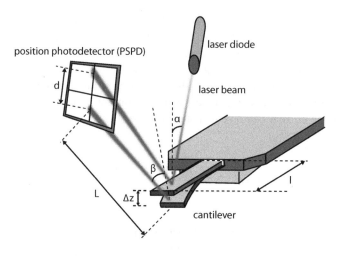

FIGURE 2.15 The optical beam detection system (OBD). A laser beam is reflected by the backside of the cantilever. The incident angle is α. The laser reflection is detected by a position-sensitive photodetector (PSPD). When the cantilever deflects, the reflected beam changes its trajectory, since the incident angle β is different. The reflected beam hits the PSPD at a different position. The difference d between the original and the deflected beam is a magnification of the actual cantilever's deflection Δz.

where Δz is the position change of the cantilever's free end (i.e., due to deflection or vibration), d is the corresponding displacement of the laser spot at the PSPD, l is the length of the cantilever, and L is the distance between the cantilever and the PSPD (see Figure 2.15). The magnifications are typically from ×100 to ×1000, which enables the detection of very small cantilever deflections or vibrations, in the order of 0.1 nm (Sone and Hosaka 2008).

Piezoresistive Detection

In this case, a piezoresistor is embedded in the cantilever, acting as a sensor. When the cantilever bends, the piezoresistor experiences a change of electrical resistance ΔR that can be measured and used to calculate the cantilever deflection according to

$$d = \frac{t \cdot K \cdot R}{\Delta R} \sin^2\left(\frac{l \cdot \Delta R}{t \cdot K \cdot R} \right)$$

(2.17)

where t is the cantilever thickness, K is the gauge factor of the piezoresistor, and R is the piezoresistance. ΔR is measured with Wheatstone bridge circuits engraved in the cantilever chip. The sensitivity of the piezoresistive detection is approximately 10 times lower than that of the laser beam deflection (Sone and Hosaka 2008).

SPM IMAGING MODES

Following the blind-human analogy, we can sense the objects around us if we can touch them in one way or another. SPM can likewise sense objects by sliding or tapping the probe over the sample surface. The first one is denoted scanning in *contact mode*, whereas the second case is referred as scanning in *tapping* or *intermittent contact* mode. But are we able to detect the presence of objects without touching them? As mentioned previously, that would be possible if the objects were bestowed with a property that we could detect from a distance, such as the heat released from a warm object. We approach the hot body with our hands ahead as long as we do not detect any heat (or if the heat is bearable), and step back if the heat is too high for us to withstand. By following this strategy, we get an idea of the object's position and dimensions. Bats, on the other hand, emit ultrasonic waves that are reflected by objects, so that they detect their presence before crashing onto them. In SPM it is also possible to sense objects without probe-touching them as long as there is an object-related property that can be detected from a distance. That is the *noncontact mode* of SPM imaging.

FEEDBACK MECHANISM: CONTROLLING THE PROPERTY X

Imaging with a sharp tip is not enough to obtain high-quality images, though. Feedback in SPM is crucial to achieve optimal performance, i.e., maximal sensitivity and resolution. A sharp tip may produce images with deficient contrast and low resolution if the feedback has not been properly set. Contrarily, a blunt tip can surprisingly produce good-quality pictures if the feedback is optimal. Setting the feedback to its optimal performance is not routine or easy, as it depends on a number of factors, the sample being one of them as well as the scanning conditions. It is the user's task to optimize the performance of the feedback, and it is usually a matter of experience and "feeling" to achieve it.

A feedback mechanism in SPM needs a property or magnitude, X, to keep at constant value, the so-called *set point*, while the tip is scanning the sample. The property X has to fulfill certain requirements:

- It has to be generated from the tip–sample interaction.
- It has to depend on the separation between the tip and the sample. The stronger the dependence is, the better.

Figure 2.16 depicts the elements of the feedback loop and its effects when the X property decreases or increases at low tip–sample separations. The detection system registers the changes of the property X with time, $X(t)$, and sends this information to the feedback. It is a passive element, since it does not alter the property X. The feedback receives this information and compares it with the set point, the value of X to keep constant, X_{sp}. If $X(t)$ is equal (within the instrumental noise) to X_{sp}, the feedback will not actuate, and neither the tip nor the sample will change their respective positions. However, if $X(t)$ differs from X_{sp} by a larger amount than the instrumental noise, then the feedback will change the relative position of the tip or the sample to restore the value of X back to X_{sp}. If X increases with the tip–sample separation (right-hand side of Figure 2.16), an increase in X ($X - X_{sp} > 0$) will result in the tip and sample approaching one another to restore X to X_{sp}; conversely, if X decreases with

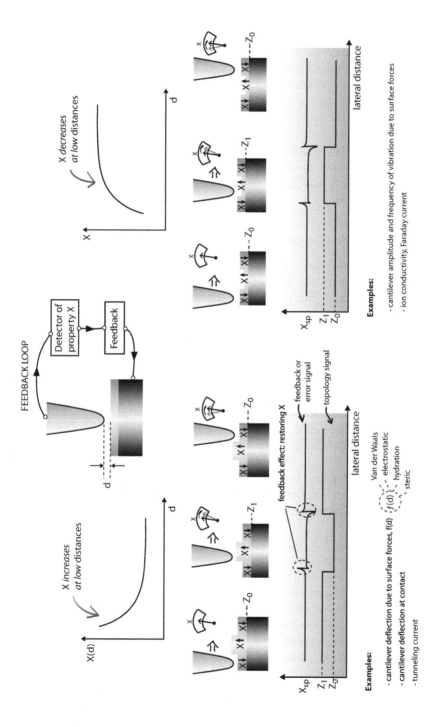

FIGURE 2.16 (*See facing page.*)

tip–sample separation (left-hand side of Figure 2.16), an increase of X will result in moving the tip and sample apart to retrieve X_{sp}.

For the feedback to be effective, it must respond to changes in X *as fast as possible*, or at least faster than the time required for the tip to change from one position of the sample to the adjacent one. In other words, the feedback should respond faster than the inverse of the scan speed at each sample position. The feedback's efficiency also profits from the dependence of X on the tip-to-sample separation: For a strong dependence, a considerable change in X may correspond to a small change in the tip-to-sample distance, enabling the detection of small objects and thus increasing sensitivity.

Feedback in Contact Mode

As mentioned previously, the probe in contact mode is touching the sample, and therefore the tip–sample distance is zero. The property X is, in this case, the *force* that the probe exerts on the sample, which increases or decreases by moving the piezo scanner either toward or away from the sample. Under these conditions, the vertical movement of the piezo scanner is not accompanied by a vertical displacement of the probe, but it may lead to a deformation of the probe or of the sample, if the latter is soft enough. Since the probe presses against the sample surface, this imaging mode can be harsh, especially with soft, deformable samples, where plowing and wear can occur. It is thus essential to image at low forces to prevent or at least minimize sample damage or unwanted sample deformation. It is equally important to keep the force constant during scanning.

The most suitable probe is that attached to a cantilever; in this case, the force applied is directly related to the cantilever deflection through Equation (2.12). Therefore, keeping the imaging force constant is equivalent to keeping the cantilever deflection constant. Contact-mode images thus show topography data at *constant deflection*.

Feedback in Noncontact Mode

The probe and sample are separated by a defined distance z that is kept constant throughout the scanning. There are several methods to achieve this, and all of them rely on the detection and control of a property that monotonically varies with the distance and can be detected by the probe without the need to contact the sample. Two of the most common methods are the use of frequency modulation in AFM-based setups and tunneling current in STM-based setups. The first method is generally

---

FIGURE 2.16 (*See facing page.*) The feedback loop on the property X. The property relates to a specific tip-sample interaction that decreases or increases with tip–sample separation. The feedback loop is composed of the probe connected to the X detection system, which in turn is connected to the feedback electronics and this to the piezo scanner. Any variation on $X(t)$ will be detected and compensated to restore the original value, X_{sp}. The feedback is a PI controller, characterized by proportional (P) and integral (I) gains, which are set by the user. The proportional gain is sensitive to the $X - X_{sp}$, while the integral gain is sensitive to $\int X(t)dt$. The response of the feedback to the changes of X determines the quality of SPM imaging. (*Left*): X increases at low distances. Any increase or decrease in X with respect to the set value X_{sp} will be compensated by lowering or lifting the piezo scanner position, respectively. (*Right*): X decreases at low distances. The increase or decrease in X with respect to the set value will be compensated by lifting or lowering the position of the piezo scanner, respectively.

used in ultrahigh vacuum (Sugawara 2007) to detect extremely small forces, usually between single atoms. The cantilever is set to vibrate along the surface normal at its resonant frequency, f_0, or overtones, which causes a slight modulation of the tip–sample distance. The amplitude of vibration should be very small (below 1 nm) compared to the tip–sample separation so that the tip never reaches the sample surface and the likely changes of the tip–sample interaction arising from the distance modulation are minimal. In other words, the force gradient should be constant throughout the vibration. Upon approaching the surface, interaction forces cause a shift in the resonance frequency, Δf, that is proportional to the force gradient, which is in turn a function of the tip–sample distance

$$\frac{\Delta f}{f_0} \cong -\frac{1}{2k} \cdot \frac{\partial F}{\partial z} \tag{2.18}$$

where k is the spring constant of the cantilever and F is the interaction force between tip and sample. Equation (2.18) holds for frequency shifts that are small compared to the resonance frequency ($\Delta f \ll f_0$) (Schirmeisen, Anczkowski, and Fuchs 2004). Scanning thus at *constant frequency shift* is equivalent to preserving the tip-distance separation. Under these circumstances, it is possible to obtain topography images with subatomic resolution.

The second method is based on the fact that the tunneling current increases with decreasing tip–sample separation in an exponential manner (Equation 2.15), which means that keeping the tunneling current constant throughout scanning is equivalent to keeping the tip–sample distance constant. In this case, topography effects can be inferred by imaging at *constant current*. It is not a direct measure of the true topography of the sample, though. In fact, STM images are generally difficult to interpret, since not only the topography but also the local density of energy states have an influence on the tunneling current (E. Meyer, Hug, and Bennewitz 2007).

Constant-current feedback is also employed in other SPMs; however, the nature of the current differs. In scanning ion conductance microscopy, it is rather the ionic current between the probe (a hollow micropipette with an attached electrode) and a reference electrode immersed in the sample medium, usually an aqueous ionic solution (Schäffer, Anczykowski, and Fuchs 2006). The ionic current depends strongly on the tip–sample distance by the so-called *current squeezing effect*: As the tip–sample distance decreases and the current flow is squeezed against the sample, the conductance path gets narrower. Consequently, the ionic resistivity increases, resulting in reduced ionic current. A similar scenario occurs in scanning electrochemical microscopy (Oesterschulze et al. 2006; Wittstock, Burchardt, and Pust 2007); in this case, a *Faraday current* resulting from the electrolysis of a compound dissolved in an electrolyte solution is measured between an amperometric microelectrode surrounded by an insulating glass coating, which comprises the tip, and a reference electrode immersed in the solution. As long as the tip is far away from the sample, the current measured at the tip, $i_{T,\infty}$ is diffusion-limited

$$i_{T,\infty} = 4nFDr_T c^* \tag{2.19}$$

where n is the number of electrons transferred in the electrochemical reaction R $\leftrightarrow$ O + ne⁻, F is the Faraday constant, D is the diffusion coefficient, r_T is the radius of the tip, and c^* is the concentration of the electrochemical compound in the bulk solution. The presence of the sample modulates the Faraday current in different ways, depending on the electrochemical nature of the sample. If the latter is an inert and insulating surface, the sample blocks the diffusion of the electroactive compound between the electrodes. Consequently, the Faraday current decreases with tip–sample distance with respect to the bulk value. Conversely, when the tip approaches a conductive or even an electrochemically active sample, the electroactive species can be reoxidized or rereduced at the sample surface, recycling the compound and hence enhancing the Faraday current with respect to the bulk value. In this case, the current increases with decreasing tip–sample distance.

Electrostatic forces and electrical capacitances are among other types of non-contact interactions upon which feedback can be applied. In electrostatic force microscopy (Tikhomirov, Labardi, and Allegrini 2006), the *coulombic interaction* between an electroactive tip and sample is measured through cantilever-type probes upon the application of a dc bias voltage. The electrostatic force is a monotonic function of the tip–sample separation, according to Coulomb's law ($F_{electr} \propto 1/z^2$). Analogously, in scanning capacitive microscopy (E. Meyer, Hug, and Bennewitz 2007), the electrical capacitance, C, between the tip and sample is measured upon the application of a dc bias voltage. The magnitude monotonically increases with decreasing tip–sample distance ($C \propto 1/z$). Thus keeping either the electrostatic force or the electrical capacitance constant is equivalent to keeping the tip–sample distance constant.

Feedback in Intermittent-Contact Mode

The mode operates in AFM-based setups in a similar way as the frequency-modulated noncontact mode. The cantilever is set to vibrate to and from the sample at its resonance frequency or at one of its overtones. The tip–sample distance is thus set to vary from $z - A$ to $z + A$, where A is the vibration amplitude. However, it differs from the noncontact counterpart in that the tip contacts the surface for a short period of time. Contact takes place at sufficiently small distances (z) and in the course of the cantilever oscillation, as the tip travels toward the sample and reaches the turning point (at $z - A$) before moving away from it. Since this occurs periodically during each vibration cycle, the tip intermittently contacts or taps the sample. In this mode, it is possible to feedback different parameters of the vibration such as the amplitude and the phase. When a vibrating cantilever approaches the sample, the interaction forces may produce a shift in the vibration frequency, which has implications in the amplitude as well as in the phase, as seen in Figure 2.17. If the cantilever is set to vibrate at a constant frequency, f_{set}, equal to the resonance frequency of the free cantilever, f_0, then the presence of the sample and hence of interaction forces produces a reduction in the vibration amplitude; the phase, though, may decrease or increase if the forces are attractive (left of Figure 2.17a) or repulsive (right of Figure 2.17b), respectively.

Keeping the *vibration amplitude constant* at a particular value implies keeping the interaction and hence the average tip–sample separation constant during the imaging process, which enables one to obtain topography data. The so-called set

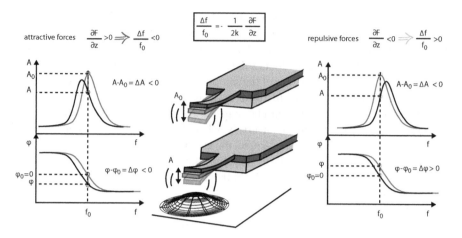

FIGURE 2.17 Changes in resonance due to interaction forces. The change in the resonance frequency is proportional to the gradient of the interaction force, according to the highlighted equation in the figure. The changes observed are relative to the vibration properties of the free cantilever, that oscillates far away from the sample at a frequency f_0 ($\omega_0 = 2\pi f_0$). The resonance frequency decreases when the forces are attractive and increases when the forces are repulsive (*Left*): Effect of attractive forces on the resonance curves: The amplitude at a fixed frequency (the natural frequency of the free-vibrating cantilever) decreases and so does the phase. (*Right*): Effect of repulsive forces on the resonance curves: The amplitude decreases whereas the phase increases.

point amplitude, A, should be smaller than the amplitude of the free vibration[*] of the cantilever, A_0, to detect any interaction; the smaller A is, the stronger the interaction, since the tip–sample separation gets smaller. Feedback can also be applied to the phase in addition to the amplitude. Setting the phase at zero allows one to map the frequency shifts as a function of the sample position. Frequency shifts are proportional to the gradient of the interaction force and therefore are more easily interpreted, even on a quantitative basis.

TOPOGRAPHY AND MORE: FEEDBACK ON *X* MAPPING *Y*

The great potential of scanning probe microscopies is their versatility in sensing most surface interactions and properties. This is possible by choosing one property, X, to control the tip–sample separation with parallel tracking of the variations of another property, Y, as a function of the sample position. In this way, it is possible to obtain simultaneously topography and Y-maps in a single scan. A typical example in contact-mode imaging is monitoring the lateral deflection of the cantilever while feedback is actuating upon the cantilever's vertical deflection. Scanning at constant deflection can thus provide maps of the lateral deflection of the cantilever as it slides over the sample. This lateral deflection is proportional to the lateral or *dragging*

---

[*] We denote *free vibration* to the oscillation experienced by the cantilever in the bulk medium, far away from the sample surface and hence unaffected by any surface-induced interactions.

force between two surfaces in relative motion: tip and sample in this case. In other words, it senses the friction force, and this microscopy is thus denominated *lateral force* or *friction force microscopy*. An example in noncontact mode or intermittent-contact mode imaging is the so-called *phase-contrast imaging*. In this case, feedback actuates on the amplitude of the oscillating cantilever, which is kept constant during scanning while a lock-in amplifier measures the phase shift of the cantilever's vibration as a function of the sample's position. Phase shifts of the cantilever's vibration are mainly a consequence of sensing areas of differing viscoelasticity. The resulting images are particularly difficult to interpret, though, since phase contrast not only depends on the surface properties of the sample, but also on experimental parameters such as the set frequency of the cantilever's vibration or the amplitude ratio A/A_0. However, in a few cases, especially on flat samples composed of more than one component with distinctive mechanical properties (i.e., hardness or elasticity), phase images are especially informative.

Another convenient approach consists of using a feedback mechanism that modulates either X or Y to separate the different contributions of the tip–sample interaction or to decouple the Y-property from stray effects. In these circumstances, either the property X or Y is set to oscillate at, close to, or well below the resonance frequency of the probe (i.e., the cantilever), while Y is registered using lock-in techniques.[*] As mentioned previously, in contact mode imaging, the tip and the sample are brought into contact, and the scanning force is kept constant by maintaining a certain cantilever deflection. At the same time, either the cantilever or the sample may be set to oscillate along the surface normal at frequencies in the range of 5–20 kHz, generally below the resonance frequency of the cantilever. This means that the deflection and hence the force at contact are modulated. The feedback keeps the average (dc) deflection constant; however, the modulation (ac) deflection varies too rapidly for the feedback electronics to buffer it, and therefore both the amplitude and phase of this oscillation are not feedbacked.

Maps of amplitude and phase changes are acquired together with the topography images in a technique called *force modulation microscopy* (FMM). Amplitude and phase contrast in FMM can be related to the viscoelastic properties of the sample (Radmacher, Tillmann, and Gaub 1993). In the so-called *Kelvin force microscopy*, the contact potential difference, V_{cpd}, between tip and sample can be mapped by applying a modulation of a certain frequency ω in the bias voltage between tip and sample. At the same time, the cantilever is being oscillated at its resonance frequency f_0, and a typical noncontact mode or intermittent contact mode feedback actuates on the phase so that the frequency shift can be registered. If the bias voltage depends on time according to the following expression

$$V_{bias} = V_{DC} + V_{AC} \sin \omega t \qquad (2.20)$$

with a steady dc component (V_{DC}) and an oscillating ac component, the frequency shift exhibits two harmonic components, at ω and 2ω, namely Δf_ω and $\Delta f_{2\omega}$, respectively,

---

[*] Lock-in techniques are very important in the treatment of the near-field signal in scanning near-field optical microscopy. They are thus explained in more detail in Chapter 4.

$$\Delta f_{\omega} \propto \left(V_{DC} - V_{cpd}\right) \cdot V_{AC} \sin \omega t$$

$$\Delta f_{2\omega} \propto V_{AC}^2 \cos 2\omega t$$

(2.21)

A second feedback mechanism actuates on Δf_{ω} by changing V_{CD} so that Δf_{ω} is zero. Under this condition, V_{CD} is equal to V_{cpd}, the magnitude of interest.

Alternatively, lift-off or double-scanning techniques can produce simultaneous images of different properties. The techniques consist of twice scanning the sample line by line. In the first line scan, topography data is obtained by employing one of the standard feedback mechanisms in contact or intermittent contact; in the second line scan, the probe is retracted from the sample surface (*lifted off*) a certain distance, d, defined by the user, and the property Y is simply monitored as a function of the sample position *at constant height* without any applied feedback. A prominent example is the detection of magnetic forces; the distance d at which the tip scans the sample is large enough to avoid the influence of nonmagnetic tip–sample forces (E. Meyer, Hug, and Bennewitz 2007). Magnetic force microscopy mostly relies on lift-off techniques, but the tip–sample distance can also be controlled through tunneling current or tip–sample capacity (E. Meyer, Hug, and Bennewitz 2007).

Table 2.1 lists some of the SPM techniques that derive from the "feedback on X mapping Y" strategy.

ARTIFACTS IN SPM IMAGES

In spite of their great advantages, the imaging modes of SPM have their limitations. Image quality and accuracy strongly depend on the tip's shape, size, geometry, and texture. In all cases, the object's image is heavily influenced by the tip; in other words, sample topographical features appearing in SPM images are convoluted with the tip shape.

In addition, using either nonoptimal or damaged tips lead to artifacts that should not be wrongly interpreted. It is thus essential to carefully consider the likely effect of tip artifacts in SPM images before assigning any displayed features to a sample's attributes.

Especially acute are the tip artifacts that may appear when imaging sample features whose aspect ratio is large compared to that of the tip; Figure 2.18a shows a sample composed of an ordered array of needlelike structures and a tip characterized by a curvature radius, r. The tip broadens the topography features, which appear as wide spots of larger dimensions than the actual ones. Likewise, if the separation between sample features is small compared with the curvature radius of the tip, the former cannot be properly imaged, since the tip cannot reach the deepest regions. This results in the underestimation of corrugation amplitudes or hole dimensions. The spacing between needles (w) in Figure 2.18b is small compared to r, and therefore the line profile does not resemble the actual profile. In addition, the needle height is clearly underestimated. The effect is analogous to that produced by a blunt tip.

Probes having multiple tips as well as unsmoothed or crashed tips produce a repetitive pattern on sample protrusions that are small compared to the tip dimensions. In these cases, it is the topography of the tip rather than the sample that is

TABLE 2.1

Relation of SPM Techniques Where the Feedback Control Applies on the Property X While Imaging Property Y

Feedback	SPM Technique	Feedback property, X	Mapped property, topography + Y
contact feedback	Lateral (friction) force microscopy	vertical cantilever deflection	cantilever lateral deflection
	Scanning Capacitance Microscopy		differential conductance, impedance
	Current sensing –AFM (conductive AFM)		capacitance
	Scanning spreading resistance microscopy		spreading resistance
int. contact feedback	Phase contrast – AFM	amplitude | of cantilever vibration	phase of cantilever vibration
	Electrostatic force microscopy	amplitude or phase |	second lift-off scan: electrostatic force at an applied bias potential
	Magnetic force microscopy		second lift-off scan: magnetic force
mixed feedback	Scanning ion conductance microscopy	amplitude of vertical/lateral cantilever vibration current (non-contact feedback)	ionic current / topography
tunneling feedback	Scanning tunneling potentiometry	dc tunneling current	surface potential
	Spin polarized-STM		spin polarization
	STM with inverse photoemission		plasmon excitation / photoemission
	Electrochemical STM		material deposition/depletion (relative redox potentials)
double feedback	Scanning Noise Microscopy	feedback 1 : noise of tunneling current $I_t(t)^2 - \langle I_t(t)\rangle^2$ feedback 2: dc tunneling current $\langle I_t\rangle$	thermovoltage
	Kelvin Force Microscopy	feedback 1: amplitude/phase of cantilever vibration or cantilever vertical deflection feedback 2: resonance frequency shift	surface potential
other feedback	Scanning electrochemical microscopy	Faraday current (non-contact feedback)	material deposition/depletion (relative redox potentials)
	Scanning thermal microscopy	thermovoltage	thermal conductivity / local temperature

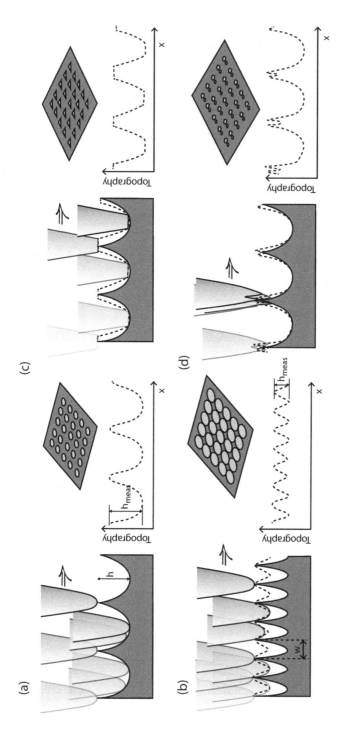

FIGURE 2.18 Some common tip artifacts. (a) Imaging tip shape on sharp sample features. The line profile differs from the sample profile since the curvature radius of the tip is larger than the size of the sample feature. The ordered pattern of spikes appear as round spots in the resulting SPM image. However the measured height (h_{meas}) is close to the actual height. (b) If the pattern pitch w is smaller than the curvature radius of the tip, the measured height underestimates the actual height. The spots in the SPM image appear larger and tight to one another. (c) A tapered tip may produce either square or triangular features identically oriented throughout the SPM image. (d) A double-tip probe generates line profiles and images exhibiting doubled features.

actually imaged. Indeed, sample features imaged by crashed tips may appear triangularly or squarely shaped, or they may appear doubled if imaged by double-tipped probes (Figures 2.18c and 2.18d, respectively). In the case of dirty or structured tips of irregular shape, the image shapes may no longer be symmetrical but may be repetitive enough to reveal a tip-induced artifact.

Other image artifacts are caused by the piezo scanner. Especially in tubelike piezo actuators, xy-scanning is concomitantly accompanied by a certain displacement in the z-direction, as mentioned earlier in this chapter. Consequently, planar surfaces appear curved in the SPM images; the so-called *image bow* is more pronounced at large scan sizes. Piezo hysteresis and creep are two additional sources of image artifacts. The first one causes disparity of line profiles obtained along the forward and the reverse directions. Likewise, sample features that exhibit axial symmetry relative to the z-axis, such as holes or protrusions, may appear otherwise asymmetric or tilted. If the sample features "move" in repetitive scans of the same sample's region, it is a symptom of instrumental drift. The same effect causes the sample features to appear either elongated or shortened if consecutive images are compared where the slow scan direction is reversed. Hysteresis and creep can be eliminated by using closed-loop scanners equipped with position sensors that measure the actual position of the piezo scanner, which can be in turn corrected. However, this type of correction reduces the signal-to-noise ratio, which makes high-resolution imaging difficult. In practice, correction for hysteresis and creep is enabled for low-resolution imaging at relatively large scan sizes and disabled otherwise.

Noise can appear either as high-frequency or low-frequency (i.e., bands) oscillations in SPM images. It can have different sources: High-frequency noise can be due to either electronic noise or to excessive feedback gains, among other causes. Low-frequency noise can be caused either by mode-hopping of the laser in the AFM-based setups (E. Meyer, Hug, and Bennewitz 2007) or by mechanical vibrations. They may appear as wide bands in the images, oriented either horizontally or vertically, respectively.

TIME RESOLUTION IN SPM IMAGING

The time required to acquire an SPM image defines the time frame of the technique. If SPM is to be used to follow dynamic processes on surfaces—such as corrosion, molecule transfer, reorientation, or reorganization—it is necessary to consider the time scale of such processes beforehand. Processes that are too fast compared to the time window of the technique simply cannot be monitored, since they are completed before one could obtain a single image. If the processes are slow compared to the time window of the technique, the acquisition of sequential images at different times will register the changes occurring on the sample surface from the beginning of the process to its completion.

Recording an SPM image generally takes time: The scanner as well as the mechanical parts of the instrument cannot operate above their respective resonance frequencies. Likewise, the electronics are not infinitely fast, but they are characterized by a bandwidth of approximately 30 kHz. Typical feedback-loop electronics cannot track changes of frequencies above a few kilohertz. Lock-in techniques in noncontact and intermittent contact mode need to register both the amplitude and phase of the cantilever oscillation, which requires acquisition times of at least one period, T. In other words, they need to "wait" at least one cycle to characterize the oscillation. All these factors contribute to limit the speed at which the SPMs can perform.

Some Numbers: Time Frame of Commercial SPMs

Most commercial instruments operate at scan frequencies of 1–30 Hz and scan sizes from 1 to 1000 nm. The scan frequency refers to the frequency with which a line is scanned, $f_{\text{line scan}}$; thus in order to calculate the time required to scan a defined area of the sample, we should apply the following equation:

$$t_{\text{image}} = A \cdot N \cdot \frac{1}{f_{\text{line scan}}} \tag{2.22}$$

where A is an integer that can be either 1 or 2, depending on whether the scan line is done in one direction (i.e., forward) or in both directions (i.e., forward and backward); N is the number of lines per image.

A typical value for N is 512, although higher or lower values are also frequently used. N defines the number of lines and hence the image resolution along the slow axis. Times required for completing a scan thus range from 20 s, typical for small scan sizes and molecular or atomic resolution, to 9 min, typical of intermittent contact mode imaging in liquids.

Judging from the typical time frames of SPM images, imaging at video rates of several frames per second is thus a challenge, and it is just out of reach for conventional setups. However, in the last decade there has been a considerable scientific effort to achieve this objective. Especially impressive is the pioneering work of the group of the Japanese scientist T. Ando, who presented a homemade setup in which all the key elements were optimized to reduce the time response of the technique without compromising the quality of instrumental performance. The technique was originally applied to image the motions of molecules of biological interest in a liquid environment (Ando et al. 2001). The chosen imaging mode was intermittent contact, the most appropriate for imaging such kinds of samples. Small cantilevers were employed that had unusually high resonance frequencies (1.3–1.8 MHz in air, 450–650 kHz in water) and small spring constants (0.15–0.28 N/m). In addition, a special type of scanner was used, which consisted of a set of stack-type piezoelectric actuators along the three axes with resonance frequencies at least 10 times higher than the conventional ones and driven by megahertz-fast electronics. This setup widened the bandwidth of the feedback loop to 60 kHz (limited by the bandwidth of the z-scanner), which resulted in scan rates of 1.25 kHz, tip speeds of 0.6 mm/s, and frame rates of 12.5 Hz. The setup was applied to track the motion of single myosin V molecules on mica surfaces.

Since that time, there has been a continuous effort to optimize the high-speed performance of SPMs. On the one hand, feedback controls of higher sensitivity (Ando et al. 2005) as well as new z-scanners have been developed that could push the speed limit to higher values and be applied to higher scan sizes (Ando et al. 2005; Schitter, Thurner, and Hansma 2008; Disseldorp et al. 2010).

SUMMARY

Table 2.2 comprises a brief account of the capabilities and limitations in scanning probe microscopy. The crucial elements that determine the vertical and lateral resolution of the technique are the piezo scanner and the probe. Choosing the right probe for each application and employing a highly linear (or linearized), properly calibrated piezo scanner are the main requirements to obtain high-quality, reliable SPM images. However, optimizing the operational performance of an SPM relies on the human factor, since the user should choose the proper feedback settings that determine the ultimate image quality.

TABLE 2.2
Summary of Imaging SPM

CHARACTERISTICS	SCANNING PROBE MICROSCOPY			
	SCANNING FORCE MICROSCOPY or ATOMIC FORCE MICROSCOPY			SCANNING TUNNELING MICROSCOPY
	CONTACT MODE	INTERMITTENT CONTACT MODE	NON-CONTACT MODE	
Lateral resolution	>1-5 nm (molecular resolution)	<1 nm (atomic resolution)	<1 nm (atomic resolution)	<1 nm (atomic resolution)
Vertical resolution	1 nm			
Sample requirements	solid-supported, atomically flat background			
	low roughness (μm, limited by vertical range of piezoscanner)			low roughness (< 1μm), conducting
Local/global	local, surface techniques			
Contrast	mechanical, electrical		morphological (topological) electrostatic, magnetic electrochemical	electronic : density of energy states at the Fermi level
Other features	feasible in a variety of sample environments: air, in-liquid, in vacuum; imaging – 3D representation of topography and/or other surface properties			
Limitations				
Sample damage	+ → -			
Artifacts	tip-induced and piezoscanner-related (creep, hysteresis, non-linearity); feedback-induced oscillations or noise; electronic noise; laser mode-hopping; thermal drift			
Time resolution	limited by bandwidth electronics (usu 30 kHz); limited by resonance of piezoscanner (approx. 10 kHz)			

REFERENCES

Alexander, S., L. Hellemans, O. Marti, et al. 1989. An atomic-resolution atomic-force microscope implemented using an optical lever. *J. Appl. Phys.* 65:164–67.

Ando, T., N. Kodera, E. Takai, D. Maruyama, K. Saito, and A. Toda. 2001. A high-speed atomic force microscope for studying biological macromolecules. *Proc. Natl. Acad. Sci. USA* 98:12468–72.

Ando, T., N. Kodera, T. Uchihashi, et al. 2005. High-speed atomic force microscopy for capturing dynamic behavior of protein molecules at work. *e-Journal of Surface Science and Nanotechnology* 3:384–92.

Atkins, P., and J. de Paula. 2006. *Physical chemistry.* New York: Oxford University Press.

Disseldorp, E. C. M., F. C. Tabak, A. J. Katan, et al. 2010. MEMS-based high speed scanning probe microscopy. *Rev. Sci. Instrum.* 81:043702-1/043702-7.

Howland, R., and L. Benatar. 2000. A practical guide to scanning probe microscopy. ThermoMicroscopes. http://web.mit.edu/cortiz/www/AFMGallery/PracticalGuide.pdf.

Jarvis, S. P., J. E. Sader, and T. Fukuma. 2008. Frequency modulation atomic force microscopy in liquids. In *Applied scanning probe methods VIII—Scanning probe microscopy techniques*, ed. B. Bhushan, H. Fuchs, and M. Tomitori, 315–50. Berlin/Heidelberg/New York: Springer-Verlag.

Marsaudon, S., C. Bernard, D. Dietzel, et al. 2008. Carbon nanotubes as SPM tips: Mechanical properties of nanotube tips and imaging. In *Applied scanning probe methods VIII—Scanning probe microscopy techniques*, ed. B. Bhushan, H. Fuchs, and M. Tomitori, 137–81. Berlin/Heidelberg/New York: Springer-Verlag.

Meyer, E., H.-J. Hug, and R. Bennewitz. 2007. *Scanning probe microscopy: The lab on a tip.* Berlin/Heidelberg/New York: Springer-Verlag.

Meyer, G., and N. M. Amer. 1988. Novel optical approach to atomic force microscopy. *Appl. Phys. Lett.* 53:1045–47.

Minne, S. C., S. R. Manalis, and C. F. Quate. 1995. Parallel atomic force microscopy using cantilevers with integrated piezoresistive sensors and integrated piezoelectric actuators. *Appl. Phys. Lett.* 67:3918–20.

Neubauer, G., S. R. Cohen, G. M. McClelland, D Horne, and C. M. Mate. 1990. Force microscopy with a bidirectional capacitance sensor. *Rev. Sci. Instrum.* 61:2296–308.

Oesterschulze, E., L. Abelmann, A. van den Bos, et al. 2006. Sensor technology for scanning probe microscopy and new applications. In *Applied scanning probe methods II—Scanning probe microscopy techniques*, ed. B. Bhushan and H. Fuchs, 165–203. Berlin/Heidelberg/New York: Springer Verlag.

Radmacher, M., R. W. Tillmann, and H. E. Gaub. 1993. Imaging viscoelasticity by force modulation with the atomic force microscope. *Biophys. J.* 64:735–42.

Rugar, D., H. J. Mamin, R. Erlandsson, J. E. Stern, and B. D. Terris. 1988. Force microscope using a fiber-optic displacement sensor. *Rev. Sci. Instrum.* 59:2337–40.

Rugar, D., H. J. Mamin, and P. Guethner. 1989. Improved fiber-optic interferometer for atomic force microscopy. *Appl. Phys. Lett.* 55:2588–90.

Schäffer, T. E., B. Anczykowski, and H. Fuchs. 2006. Scanning ion conductance microscopy. In *Applied scanning probe methods II—Scanning probe techniques*, ed. B. Bhushan and H. Fuchs, 91–119. Berlin/Heidelberg/New York: Springer-Verlag.

Schirmeisen, A., B. Anczkowski, and H. Fuchs. Dynamic force microscopy. In *Applied scanning probe methods I—Characterization*, ed. B. Bhushan, H. Fuchs, and S. Hosaka. 3–39. Berlin/Heidelberg/New York: Springer-Verlag.

Schitter, G., P. J. Thurner, and P. K. Hansma. 2008. Design and input-shaping control of a novel scanner for high-speed atomic force microscopy. *Mechatronics* 18:282–88.

Schönenberger, C., and S. F. Alvarado. 1989. A differential interferometer for force microscopy. *Rev. Sci. Instrum.* 60:3131–34.

Sone, H., and S. Hosaka. 2008. Self-sensing cantilever sensor for bioscience. In *Applied scanning probe methods VIII—Scanning probe microscopy techniques*, ed. B. Bhushan, H. Fuchs, and M. Tomitori, 219–45. Berlin/Heidelberg/New York: Springer-Verlag.

Sugawara, Y. 2007. Noncontact atomic force microscopy. In *Applied scanning probe methods VI—Characterization*, ed. B. Bhushan and S. Kawata, 247–55. Berlin/Heidelberg/New York: Springer-Verlag.

Tikhomirov, O., M. Labardi, and M. Allegrini. 2006. Scanning probe microscopy applied to ferroelectric materials. In *Applied scanning probe methods III—Characterization*, ed. B. Bhushan and H. Fuchs, 217–59. Berlin/Heidelberg/New York: Springer-Verlag.

Tipler, P. A., and G. Mosca. 2003. *Physics for scientists and engineers*, 5th ed. New York: W. H. Freeman.

Tortonese, M., H. Yamada, R. C. Barret, and C. F. Quate. 1991. Atomic force microscopy using a piezoresistive cantilever. *Proceedings of Transducers, IEEE* 448–51.

Van den Bos, A., I. Heskamp, M. Siekman, L. Abelmann, and C. Lodder. 2002. The Cantilever: A dedicated probe for magnetic force microscopy. *IEEE Trans. Magn.* 38: 2441–43.

Wittstock, G., M. Burchardt, and S. E. Pust. 2007. Applications of scanning electrochemical microscopy (SECM). In *Applied scanning probe methods VII—Biomimetics and industrial applications*, ed. B. Bhushan and H. Fuchs, 259–99. Berlin/Heidelberg/New York: Springer-Verlag.

3 What Brings Optical Microscopy

The Eyes at the Microscale

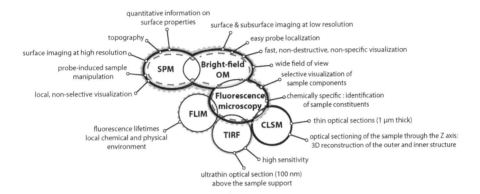

FUNDAMENTALS OF OPTICAL MICROSCOPY

Optical (also called light) microscopy makes use of a defined light path and linear optics to enlarge the size of tiny objects and produce an image. In this case, *tiny* means submillimeter size. However, the object should not be smaller than the wavelength of light, as will be shown later. This limitation thus imposes the minimum detectable size attained by this technique, i.e., a few hundreds of nanometers.[*]

Image formation requires the following events to occur: the object is illuminated and *diffracts* part of the light; both the diffracted and nondiffracted light are collected by an appropriate lens; and through *interference* they produce the image at a certain position—the image plane.

If illumination and the collection of light occur on the same side of the object plane, the microscope operates in *reflection mode*; conversely, if the collection of light occurs on the other side of the object plane, the microscope operates in *transmission mode*.

---

[*] We distinguish here between conventional optical microscopy and super-resolution microscopy. The latter includes a few optical microscopies, mostly based on fluorescence microscopy, capable of breaking the optical limit of detection (i.e., the diffraction limit, as we will see in this chapter) and of detecting even smaller objects: 4Pi microscopy, stimulated emission depletion microscopy (STED), stochastic optical reconstruction microscopy (STORM), or scanning near-field microscopy (SNOM), to name a few. This chapter is rather dedicated to standard optical microscopy, with a reference to two near-field optical microscopies that have been combined with SPM.

The interaction of light with matter has been a focus of interest of humankind for centuries, and the history of optical microscopy is just as old. Figure 3.1 illustrates it as a timeline starting at the end of the thirteenth century, when the Italian Salvino D'Armate invented eyeglasses, until the first decade of the twenty-first century, with the ultimate achievements on high-resolution optical microscopy.

The discussion in the following background section will help the reader to understand the basics of optical microscopy.

Background Information

Visible light has a double nature. On the one hand, it is a transversal electromagnetic wave whose wavelength is in between 390 nm and 720 nm; on the other hand, it is a bunch of minute particles or *photons*, each one carrying a minute amount of energy or *energy quantum* of magnitude $E = hc/\lambda$ that ranges between 2.5×10^{-19} and 5.2×10^{-19} J. Light that only has one wavelength is said to be *monochromatic*; white light, though, comprises waves of wavelengths covering the 390–720-nm range, and hence it is said to be *polychromatic*. The wave nature of light lies behind most of the optical phenomena we will be reviewing here. However, other phenomena such as fluorescence can only be understood if the particle nature of light is taken into account.

As electromagnetic waves, the electrical (**E**) and magnetic (**B**) fields are mutually perpendicular and oscillate in a sinusoidal fashion. The oscillations are normal to the direction of light displacement or *propagation*, as seen in Figure 3.2. The points that have a defined value of **E** and **B** at a particular position and time lie on planes of constant phase called *wave fronts*. These wave fronts are thus perpendicular to the direction of propagation. In particular, a point light source emits light in all directions, and the wave fronts are thus spherical. However, if these wave fronts propagate a relatively large distance, the radius of the wave front will be so large that eventually it will behave as though it were a planar wave front.

For simplicity light will be graphically depicted as lines with arrows pointing in the direction of propagation. To explain phenomena such as reflection, refraction, or diffraction, the light source is assumed to be far enough so that light behaves as a planar wave front. In this case, the incoming rays will be depicted as parallel lines.

WHEN LIGHT APPROACHES THE FRONTIER BETWEEN TWO TRANSPARENT MEDIA: REFLECTION AND REFRACTION

Light rays approaching a boundary or *interface* between two homogeneous, transparent media with different densities may undergo two transformations. Part of the light may traverse the interface and thus propagate through the material, and part of the light may be bounced back. In the former case, the light undergoes *refraction*; in the latter, *reflection*. Both phenomena may change the trajectory of light under certain conditions, which will be explained later.

1280 approx -- the italian Salvino D'Armate invents the first eye glasses.

1600 approx -- Zacharias Jansen and his father Hans make the first compound microscope.

1665 -- Robert Hooke observes the pores of a sliver of cork with a a leather-and-gold compound microscope constructed by Christopher White.

1674-1682 -- Anton van Leeuwenhoek builds a one-lens microscope to examine muscle fibers, bacteria, spermatozoa and blood flow in capillaries. He obtained the best available lens at that time through polishing, with a magnification of 250X.

1729-1733 -- Chester Moore Hall invents the achromatic lens.

1760-1790 -- Jan and his son Harmanus van Deyl make the first high power achromatic lenses.

1807 -- Ernst Abbe conceives a mathematical formula that calculates the maximum resolution of a microscope. He points out the importance of light diffraction in resolving object´s details. He introduces the concept of numerical aperture of an objective and developed oil-immersion lenses.

1830 -- Joseph Jackson Lister reduces spherical aberration of lenses by combining two different achromatic lenses. This finding led to the development of high-powered microscope lens.

1885 -- Paul Nipkow patents the Nipkow disc or spinning disc, a disc with holes arranged in a spiral pattern.

1893 -- August Köhler develops a key technique for sample illumination, overcoming many of the limitations of older techniques.

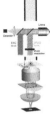

1903 -- Richard Zsigsmondy develops the ultramicroscope that resolves object´s details below the wavelength of light. (Awarded the Nobel Prize in Chemistry in 1925.)

1932 -- Frits Zernike invents the phase contrast microscope to study colourless, low-contrast biological materials. (Awarded the Nobel Prize in Physics in 1953.)

1953 -- Georges Nomarski develops differential phase contrast microscopy to visualize living or stained specimens with reduced or no optical contrast when viewed using bright-field illumination.

1957 -- Marvin Minsky patents the confocal microscope.

1965 -- Mojmir Petran and Milan Hdravsky present the scanning reflected-light microscope using a Nipkow disc as image scanning device.

1978 -- Thomas and Christoph Cremer develop the first practical confocal laser scanning microscope.

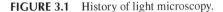

1994-1999 -- Stefan Hell conceives and constructs the stimulated emission depletion (STED) microscope.

2006 -- Three research groups parallel develop a new fluorescence microscope. The group of Eric Betzig and the group of Harald Hess coin it Photoactivated Localization Microscopy (PALM). The group of Xiaowei Zhuang coin it Stochastic Optical Reconstruction Microscopy (STORM).

FIGURE 3.1   History of light microscopy.

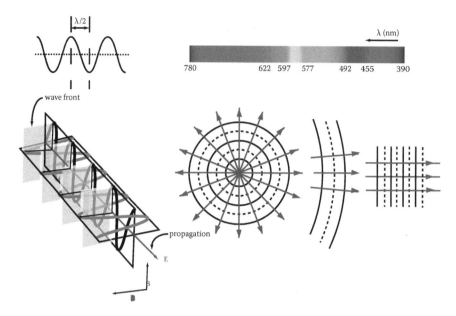

FIGURE 3.2 *(See color insert.)* Light as an electromagnetic wave (above). The wavelengths of visible light range from 390 nm (blue light) to 780 nm (red light). Wavefronts and light rays (below). Both electric and magnetic fields oscillate perpendicularly to the direction of propagation (transversal wave), defining the wave fronts as surfaces of constant phase. Light emerging from a punctual source propagates in all directions and the wave fronts are spherical. When the source is sufficiently far away, the wave front turns into planar and the light rays are parallel from one another.

REFLECTION

Perpendicular light rays impinging on the interface between air and glass will be bounced back along the normal to the interface. When the rays come from air to a higher density medium such as glass, they undergo *external reflection*; when the incident rays come from glass to air, they are bounced back to the glass through *internal reflection*. External and internal reflection are illustrated in Figure 3.3

If light reaches the interface at a certain angle θ_i, called the incidence angle, light will be reflected at an angle θ_r, called the reflection angle, in the fashion shown in Figure 3.3. The first law of reflection states that the angle of incidence is equal to the angle of reflection,

$$\theta_i = \theta_r \tag{3.1}$$

The second law of reflection states that the incident light, the reflected light and the interface all lie on the same plane, the *plane of incidence* (Figure 3.3).

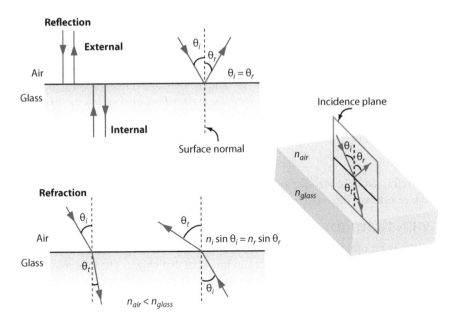

FIGURE 3.3 Reflection of light between air and glass (top left). Internal and external reflection occur when light comes from either air or glass, respectively. The angle of incidence θ_i is always equal to the angle of refraction θ_r. Refraction of light between air and glass (bottom left). The refracted light path always deviates from the incident light path according to the Snell-Descartes law. The interface as well as the incident, reflected, and diffracted light all lie on the sample plane (right).

When the interface is completely flat and flawless, well-defined light rays are produced upon *specular reflection*; however, if the interface is rough, light is reflected in different directions, which hinders the formation of well-defined rays. In this case, the reflection is *diffuse*. Most surfaces are not purely specular and lie in between these two extremes.

REFRACTION

Light propagates in vacuum at a speed c; however, when it traverses transparent media, which are in all cases denser than vacuum, light travels at a smaller speed, v. The ratio between the speed in vacuum and the speed in the medium is always higher than 1, and it is called the *refraction index, n*

$$n = \frac{c}{v} \tag{3.2}$$

When light rays come from a medium of low refractive index, such as air, to a medium with higher refractive index, such as glass, the light rays change direction in such a way that the transmission angle θ_t, that is, the angle formed between the refracted light and the surface normal, is smaller than the incident angle, and the light trajectory approaches the surface normal. Conversely, if the incident light comes from glass to air, the transmission angle will be larger than the incident angle, and the light beam diverges from the surface normal. Both cases are illustrated in Figure 3.3. The incident and transmission angles are related to each other through the *Snell-Descartes law*, also called the first law of refraction

$$n_i \sin \theta_i = n_t \sin \theta_t \qquad (3.3)$$

The refracted light also lies on the same plane of incidence as the interface, the surface normal, the incident light, and the reflected light.

WHEN LIGHT GOES THROUGH A LENS

A lens is an optical device that deviates light rays coming from an object in a defined way. As a result, it produces an image of that object. The lens consists of a disk of glass with at least one curved surface. The lens surfaces have a common symmetry axis, which is called the *principal* or *optical axis* of the lens, and they can be flat, convex, or concave. As Figure 3.4 shows, lenses with two convex surfaces are called convergent, since they bring light rays together; lenses with two concave lenses are called divergent, since they bring light rays apart. Since the performance of optical microscopes relies on convergent lenses, we will focus on this particular type of lens from now on.

How light is affected by the presence of a lens depends on multiple factors. These are the refractive index of the glass the lens is made of, the lens thickness, the size of the lens, and the curvature and smoothness of the lens surfaces, i.e., the quality of the lens. Image defects produced by lenses are called *aberrations* and abound in practice. However, linear optics provides simple rules of thumb that hold quite accurately under the following conditions:

- Lenses are defect-free with perfectly curved surfaces onto which light undergoes specular reflection.
- Internal reflection of light within the lens is negligible. In practice, lenses within a microscope are treated with antireflection coatings.
- The lens thickness is small so that light rays hardly diffract when they traverse the lens.
- The light rays form a small angle with respect to the principal axis; they are said to be *paraxial*.

Apart from these assumptions, we need to follow a graphical convention. Generally *incoming* rays will be drawn from left to right and thus reach the left side of the lens. The *outgoing* rays will emerge from the right side of the lens and they will further travel to the left.

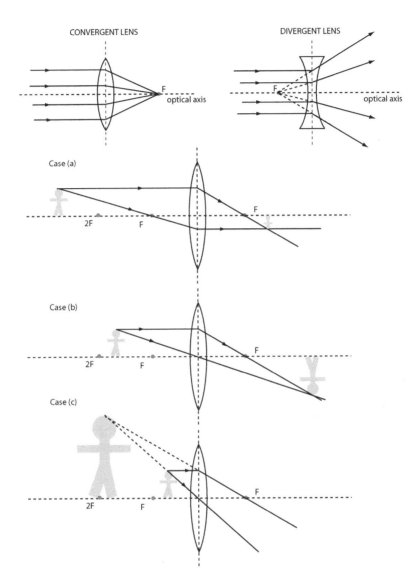

FIGURE 3.4 Light through a convergent and divergent lens. Convergent lenses bring light rays together to the focus, *F*, a point situated along the optical axis. Divergent lenses bring light rays apart as though they were all coming from the focus *F*, located on the same side as the incoming light. Image formation with a convergent lens: when the object is situated at a distance larger than 2*F* (Case a); when the object is situated between 2*F* and *F* (Case b); when the object is situated at a distance smaller than *F* (Case c).

- Incoming rays of light parallel to the optical axis of the lens deviate in such a way that the outgoing rays go through a point in the optical axis called *focus* on the other side of the lens. The distance between this point and the lens surface is called the *focal distance*, *F*. Likewise, incoming rays that go across the focus of the lens turn parallel upon traversing the lens.
- Rays traverse the lens unperturbed if their trajectories coincide with the optical axis of the lens or if they cross the center of the lens.

The image of a point object produced by a convergent lens depends on the distance between the object and the lens, the so-called *object distance*. We can thus differentiate three typical cases:

1. The object distance is larger or equal than twice the focal distance, $2F$.
2. The object distance is between $2F$ and F.
3. The object distance is smaller than F.

The three cases are depicted in Figure 3.4

In case 1, the lens produces a *real, inverted* image of the object, smaller than the real size. The distance of the image to the lens, the *image distance*, is between F and $2F$. A real image is produced by the deviated light rays on the other side of the lens, and it can be projected on a screen or captured by a camera. In case 2, the lens produces a *real, inverted* and *enlarged* image of the object. The object distance will be larger than $2F$. In case 3, the lens produces a *virtual, noninverted* and *enlarged* image of the object. The *virtual* image is produced by nonreal (extrapolated) rays of light that converge on the same side of the lens where the object is. The virtual image cannot be projected on a screen or captured by a camera.

WHEN LIGHT GOES THROUGH OBSTACLES

Any illuminated object comprises a series of obstacles and openings to the light that causes the latter to spread to some extent. This phenomenon is called *diffraction*. When projected on a screen, both the diffracted and nondiffracted light *interfere* either constructively or destructively, forming a more-or-less complicated pattern of bright and dark regions. This pattern has in appearance nothing to do with the shape and structure of the object; however, it contains all the information needed to generate the object's image. Here we will focus on diffraction patterns formed by simple objects: a single slit and a series of periodically arranged slits, which is called a diffraction grid.

LIGHT GOING THROUGH A NARROW SLIT

The diffraction pattern of a narrow slit of length *l* and width *D* is a row of alternate bright and dark regions perpendicularly oriented to the slit, as seen in

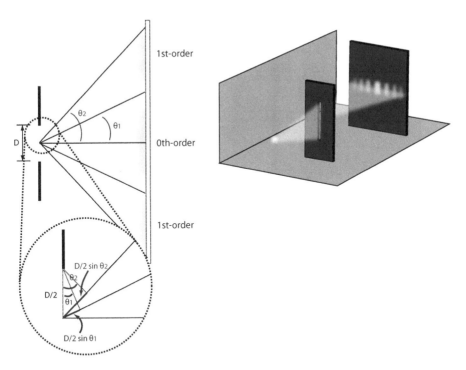

FIGURE 3.5 Light through a single, narrow slit of width D. The narrower is the slit, the larger is the diffraction pattern, composed of bright and dark regions, according to Equation (3.4).

Figure 3.5. Within the bright regions, constructive interference occurs, while in the dark regions, light rays cancel out due to destructive interference. As Figure 3.5 shows, destructive interference occurs when the light-path difference between the diffracted and nondiffracted rays is equal to an integer number of half the wavelength of light

$$\frac{D}{2}\sin\theta_m = m\frac{\lambda}{2}, \text{ with } m = \pm 1, \pm 2, \pm 3, \dots \quad (3.4)$$

where θ_m is the angle between the nondiffracted light and the diffracted light, which increases when the slit is narrower. The central region is the brightest, and it is called the zero-order diffraction; the lateral bands or fringes are symmetrically positioned on both sides of the central band, and they are called first-, second-, third-order diffraction maxima, according to the value of m.

LIGHT GOING THROUGH A GRID

When light goes through a series N of parallel slits separated by a distance d that is larger than the wavelength of light, the outgoing pattern is composed of

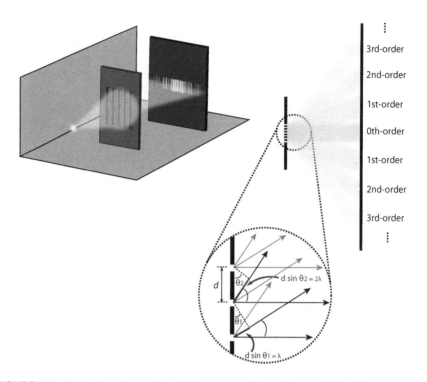

FIGURE 3.6 Light through a series of slits separated by a distance d. Providing d is larger than the wavelength of light, the diffraction pattern follows Equation (3.5).

a series of fringes, the primary maxima, which result from light diffracting at each particular slit, and the secondary maxima, the $N - 2$ less-bright fringes that appear between two consecutive primary fringes (Figure 3.6). When N is high, the primary fringes are positioned according to the *grating equation*

$$d \sin \theta_m = m\lambda, \text{ with } m = \pm 1, \pm 2, \pm 3, \ldots \qquad (3.5)$$

The smaller d is, the larger are the diffraction angle and the distance between two consecutive primary fringes.

ESSENTIAL PARTS OF AN OPTICAL MICROSCOPE:
THE IMAGE AND DIFFRACTION PLANES

Any optical microscope must contain the following parts: an *illuminator* or light source, a *condenser lens*, one or various *objective* lenses, a sample or specimen *stage*, and one or two *eyepieces* or *oculars*. The arrangement of these components

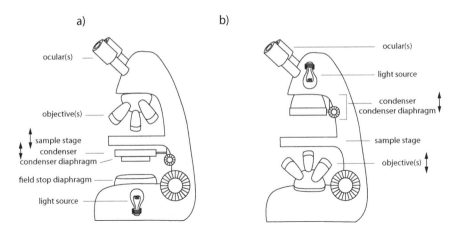

FIGURE 3.7 Microscope parts. Arrangement of basic components for an optical transmission microscope: (a) upright type, (b) inverted type. The double-pointed arrows indicate the components whose position can be adjusted.

within the optical transmission microscope is shown in Figure 3.7 for the two types of design: the upright (a) and the inverted (b) designs.

The light source, typically an incandescent lamp (Murphy 2001) or an optic fiber in the most modern versions, provides the illumination on the specimen. The illuminated area can be enlarged or reduced by the *field stop diaphragm*. The condenser focuses the light from the illuminator onto a small area of the sample that rests on the stage. The size of that area is defined by the *condenser aperture*, a hole or opening that can in turn be regulated by the *condenser diaphragm*. Each of the objective lenses collects the light diffracted by the specimen and forms an enlarged *intermediate* image close to the ocular or oculars. The intermediate image is a real image whose enlargement or magnification is determined by the objective type. Typical objective magnifications are 4×, 5×, 10×, 20×, 40×, 65×, and 100×. Each objective has a front focal plane that coincides with the specimen position and a rear focal plane, where the *back aperture* is. This aperture consists of a fixed circular hole that, like all apertures in a microscope, prevents stray light from entering the light path. The oculars are lenses that further enlarge the image created by the objective and, together with the eye retina, project the second and definite image that we perceive as a magnified, virtual image 25 cm in front of the eye. The process of such image formation is depicted in Figure 3.8a for an arrow-shaped object. Alternatively, an *image visualizing device*, such as a photo camera or a charge-coupled device (CCD), acquires the intermediate image for display.

An optical microscope has altogether eight focal planes arranged in two sets: a set of four *object* or *field* planes and a set of four *aperture* or *diffraction* planes. When the optical microscope is properly adjusted (Koehler 1993), all planes within a given set are *conjugate*, which means that they are all in focus when you look into the microscope. The planes have fixed, well-defined locations with respect to the components of the microscope, as seen in Figure 3.8b. The field planes are viewed through

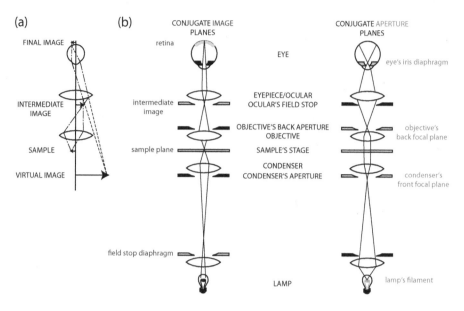

FIGURE 3.8 Conjugate planes. The image (a) and diffraction (b) sets of conjugate planes in a Koehler-adjusted microscope.

the oculars and comprise the field stop diagram, the specimen plane on the stage, the real intermediate image plane, and the retina or camera plane. The aperture planes, which can only be seen with the help of an eyepiece telescope or a Bertrand lens, are situated at the lamp filament; at the front aperture of the condenser, which is in turn located at the condenser diaphragm; at the back aperture of the objective lens; and at the exit pupil of the oculars, which should be comparable (or smaller) in size with the pupil of the eye, and it hangs a few millimeters above the oculars.

DIFFRACTION SETS THE LIMIT OF DETECTION, SPATIAL RESOLUTION, AND DEPTH OF FOCUS

As mentioned previously, objectives in a light microscope collect the *diffracted* light of an illuminated object and produce the image in the focal plane. Objectives possess apertures whose edges diffract light as well. The final image is thus generated by interference of the light rays spread by the sample itself and the light rays spread by the objective edges. Consequently, if the real object is a point with no dimensions, the point's image is not a point but a spot of light. This spot of light, or diffraction pattern, consists of a bright disk surrounded by a series of rings of lesser intensity, as seen in Figure 3.9a. The central disk as well as the rings around are separated by dark regions (minima) of nearly zero light intensity. The central spot is called the *Airy disk*, whose radius, r, is determined by the wavelength of light, λ, the refractive index, n, of the medium between the objective lens and the sample (the immersion medium), and the aperture angle of the lens, 2θ, as follows:

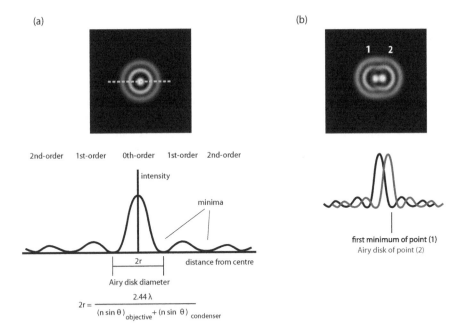

FIGURE 3.9 Diffraction spots—limit of detection. Image of a luminous point source and intensity profile along one of its axes (a). Rayleigh criterion that defines lateral resolution (b).

$$r = \frac{1.22\lambda}{2n\sin\theta} \tag{3.6}$$

As a result, the image of the object never represents the real object, and r determines the lower limit below which object sizes cannot be properly resolved. In other words, r defines the *limit of detection*. The expression $n\sin\theta$ is the so-called *numerical aperture*, and it is a characteristic parameter for objectives along with their magnification. According to Equation (3.6), the limit of detection decreases when the numerical aperture increases, and so does the sensitivity. For a constant aperture angle, the sensitivity increases with n. To increase n, it is necessary to fill the gap between the objective lens and the sample support with a fluid of higher refractive index than air; this is normally accomplished with *immersion fluids* such as oil or water (see following discussion). A small drop of fluid normally suffices to create a meniscus of fluid between the objective and the sample support.

In bright-field microscopy, the size of the Airy disk is not only affected by the numerical aperture of the objective, but also by the numerical aperture of the condenser. In the case of the latter being lower than the former, the expression for r results in

$$r = \frac{1.22\lambda}{(n\sin\theta)_{\text{objective}} + (n\sin\theta)_{\text{condenser}}} \tag{3.7}$$

Having introduced the numerical aperture and the shape of the point's image, we are now able to define the spatial resolution. The latter determines the minimal distance between proximal objects so that they are distinguishable in the final image; the smaller this distance is, the higher the spatial resolution, and vice versa. Taking into account the image of two proximal object points shown in Figure 3.9b, the points are resolved *when the Airy disk of one point coincides with the first diffraction minimum of the other point's image*. This is the so-called *Rayleigh criterion*, and hence the lateral resolution can be described by *r*.

Some Numbers: Lateral Resolution

Taking Equation (3.7) into account, let us estimate the maximal spatial resolution or, in other words, the minimum value for *r*. For a given wavelength, *r* decreases if the sum of the numerical apertures of both objective and condenser is high. This number cannot be larger than the sum of the refractive indexes of the immersion fluid of objective and condenser, since sin θ cannot be larger than 1. The condenser's immersion fluid is generally air ($n_{\text{air}} = 1.000$), water ($n_{\text{water}} = 1.33$), or oil ($n_{\text{oil}} = 1.51$–1.52; Cargille Labs 2012). Therefore,

$$r \geq \frac{1.22\lambda}{n_{\text{objective}} + n_{\text{condenser}}} \geq \frac{1.22\lambda}{1.52 + 1} = 0.48\lambda \approx \frac{\lambda}{2} \tag{3.8}$$

Rule of thumb: the maximal lateral resolution depends on the wavelength of light, and it can be approximated by λ/2.

Diffraction imposes a limitation not only on the lateral resolution, but also on the vertical resolution. Optical microscopy provides information of sample "slices" rather than sample planes. These "slices" or, more properly, optical sections are three-dimensional regions where the objects within the sample are in focus. The thickness of these optical sections is called *depth of field, Z*, and it is determined by the numerical aperture of the objective

$$Z = \frac{n\lambda}{(n\sin\theta)^2_{\text{objective}}} \tag{3.9}$$

Generally, the larger the numerical aperture of the objective, the thinner is the depth of field and thus the greater is the vertical resolution.

Some Numbers: Vertical Resolution

From Equation (3.9) it is possible to estimate the lower limit for the vertical resolution. Again, since sin θ cannot be larger than 1,

$$Z \geq \frac{n\lambda}{n^2} = \frac{\lambda}{n} \approx \lambda \qquad (3.10)$$

Rule of thumb: The minimum penetration depth is the wavelength of light. The thinnest optical section is thus as thick as one wavelength of the incident light.

INTERFERENCE SETS IMAGE FORMATION

In a way, proximal objects of a certain periodicity p can be simulated by a diffraction grating with slots equally spaced at a distance d. According to Abbe's theory for image formation in the light microscope, the slits are distinguishable in the image plane only if interference between zeroth and higher order diffracted rays occurs, generating image contrast. For this to happen, the objective should be able to collect at least two adjacent orders of diffracted light. The higher the number of orders of diffracted light the objective can detect, the better the image's quality, which, in principle, should scale with the numerical aperture of the objective.

In practice, however, polychromatic illumination, scattered light, and the reduction of the coherence of the waves that interfere in the image plane impair attaining optimal contrast. All these effects can be significantly reduced by decreasing the condenser aperture (Murphy 2001). However, according to Equation (3.7), this will increase the size of the light spots and reduce the spatial resolution. Therefore, high-resolution images can only be obtained at the expense of decreasing image contrast; likewise, an image with an optimal contrast can never be obtained at maximal resolution. Therefore a compromise should be taken.

OPTICAL CONTRASTS

For an object to become visible with an optical microscope, the wave characteristics of the light diffracted by the object should differ from those of the nondiffracted (zeroth-order) or background light, which is unaffected by the object's presence. If only the amplitudes of the waves associated with the diffracted and nondiffracted light are different, the objects absorb light, and in this case the optical contrast is directly detectable by the eye as changes in light intensity; the microscope operates in *bright field*. Dyes or pigments that absorb light of specific wavelengths make dyed specimens appear either in color when illuminated by white light or dark when illuminated by light of the complementary color. The objects are said to be *intensity objects*.

If the intensities of the diffracted and nondiffracted light are similar and only the wave phases differ, the object alters the *optical path* of the light, but it appears invisible to the eye. The optical path difference is defined as

$$\Delta = \left(n_2 - n_1\right) \cdot t \tag{3.11}$$

where n_2 and n_1 are the refractive index of the object and the surrounding medium, respectively, and t is the sample thickness. Therefore, in order to alter the optical path, the objects should have a distinct refractive index. Transparent objects like most biological specimens (e.g., living cells) are said to be *phase objects*. They diffract light and cause a phase shift of the rays of light that pass through them. Consequently, these objects may appear transparent when illuminated with white light. There are basically two ways to make them visible under the microscope. One way is to convert the phase difference into an intensity difference or, in other words, to convert the difference in optical path (refractive index and/or thickness) into optical contrast. That is what a *phase contrast microscope* does. Another way is to convert the *gradient in the optical path difference* into optical contrast. That is what a *differential interference contrast microscope* does.

Phase Contrast

In this kind of microscopy, the intensity of nondiffracted light is dimmed and its phase is either advanced, in the so-called *positive* phase contrast, or retarded, in the case of *negative* phase contrast, by $\pi/4$. To selectively vary the phase of the nondiffracted light without affecting the diffracted light, both beams are separated in such a way that they reach at different locations at the back aperture of the objective. To do that, a *condenser annulus* is required that is aligned with a *phase plate* at the back aperture of the objective. A phase plate is an etched ring of either reduced (positive) or increased (negative) thickness that lowers the light intensity and produces a phase shift of $\lambda/4$ or of $-\lambda/4$, respectively. Under Koehler illumination, the nondiffracted light is concentrated in a ring by the condenser annulus and focused on the phase plate with a phase shift of $+\lambda/4$ or $-\lambda/4$, in positive and negative phase contrast, respectively. The diffracted light with a phase shift of $-\lambda/4$ reaches the objective outside the phase plate. Consequently, the overall phase shift between the nondiffracted and diffracted light is either $\pi/2$ for positive phase contrast or zero for negative phase contrast, causing destructive or constructive interference, respectively, and thus making the object appear either dark or bright with respect to the background. In a positive-phase contrast microscope, the objects with higher refractive index than the surrounding medium will appear dark, whereas those objects with lower refractive index will appear bright. In a negative-phase contrast, the opposite occurs. The objects with higher refractive index will appear bright, whereas those with lower refractive index will appear dark.

Differential Interference Contrast

In differential interference contrast (DIC) microscopy, the sample is illuminated by pairs of closely spaced rays. If the components of the ray pair traverse a region in the sample where there is a gradient in the refractive index or sample thickness, the optical path for each ray will be different. Consequently, the rays will lose coherence, and hence a phase shift will be produced. A DIC microscope makes use of a polarizer that produces plane-polarized light and a condenser that contains a quartz prism, a *Wollaston* or *Nomarski prism*. This prism is made of quartz, a birefringent material that produces two plane-polarized light rays out of a plane-polarized one: an ordinary and an extraordinary ray. This is called the *ray pair generator* or *beam splitter*. The rays traverse the sample and are recombined by another Wollaston/Nomarski prism situated in the objective. If the components of the ray pair are coherent to one another, light remains plane-polarized and can be blocked by an analyzer, a cross polarizer located after the second prism. A cross analyzer will thus give a black background. However, if the components are phase shifted to one another, their recombination results in elliptically polarized light, which is partially transmitted by the polarizer.

Gradients in optical path length usually occur at the edges of sample objects, where either the refractive index, the object thickness, or both change more or less sharply. As a result, the images produced by a DIC microscope have a relief-like, shadow-cast appearance.

TAKE-HOME INFO

- Optimal performance relies on the proper positioning of the image and aperture planes, so that the planes within each set are conjugate. This is attained by Koehler illumination.
- The limit of detection is determined by the size of the Airy disk of a point source, which in turn is defined by the wavelength of light, the numerical aperture of the objective, and the numerical aperture of the condenser.
- Lateral resolution of two proximal point objects is defined by Rayleigh's criterion and increases with the numerical aperture of the objective and condenser. As a rule of thumb, it is comparable to half the wavelength of light.
- Vertical resolution increases together with the numerical aperture of the objective. As a rule of thumb, it is comparable to one wavelength of light.
- A must for image (contrast) formation: Abbe's theory—interference between zeroth and higher orders of diffracted light—must occur.
- Maximal lateral resolution and image contrast cannot be simultaneously achieved: Improving the image contrast is at the expense of losing lateral resolution and vice versa.

FLUORESCENCE MICROSCOPY: BESTOWING SPECIFICITY

Fluorescence microscopy is a type of optical microscopy that produces magnified images of *fluorescent* objects. In this case, the light source illuminates the sample and *excites* the fluorescent objects within, which *emit* visible light in turn. The image is formed by diffraction and interference of *just the light emitted* by those objects. Lateral and vertical resolutions in fluorescence microscopy are determined by the numerical aperture of the objective and the wavelength of the light emitted by the objects.

The discussion in the following background section will help the reader to understand the basics of fluorescence microscopy.

Background Information

BASICS ABOUT FLUORESCENCE

The basics of fluorescence presented here are based on the work of Valeur (2001) and Lakowicz (2006). Fluorescence is a type of *luminescence* that consists of the emission of visible light by electronically excited species. In fluorescence, the excitation is brought up by light as well, which implies that the species absorbs part of the light to get excited. Upon returning to the relaxed state, the species may emit light, although there are other ways of de-excitation that may compete with, or even hinder, fluorescence. They should be taken into account in order to understand the extent of fluorescence emission.

ABSORPTION AND EMISSION AND LIFETIME

Fluorescence is a molecular process that a few molecules called fluorophores exhibit, and it can be explained by quantum mechanics. In this regard, the graphical representation of Perrin-Jablonski is particularly useful, depicted in Figure 3.10. The phenomenon involves two processes: (a) *light absorption (excitation)* or promotion of electrons from a molecular orbital of lower energy to unoccupied orbitals of higher energy by absorption of photons of defined frequencies and (b) *light emission (de-excitation)*, where electrons return to their initial state of lower energy by emitting photons of defined frequencies. Both absorption and emission transitions occur between energy levels of equal multiplicity (typically between singlets, S) and hence they are allowed according to the selection rules of quantum mechanics. The *lifetime* defines the time scale for fluorescence to occur, and it is mainly determined by the time a molecule "lingers" in the excited state, which is on the order of 10^{-10} to 10^{-7} seconds. The absorption transition occurs in femtoseconds, and it is too fast to be considered. The emission photons are generally of lower energy than the excitation photons; hence fluorescence light has longer wavelengths than the excitation light.

The amount of light absorbed or emitted as a function of wavelength determines the fluorescence spectra, which consist of a series of maxima of defined width and height. These spectra can be considered as fingerprints of the fluorescence

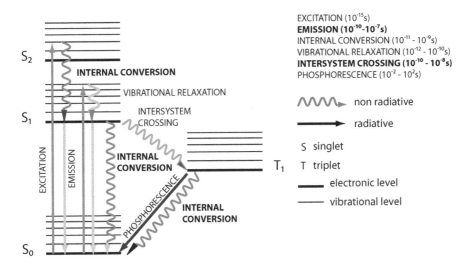

FIGURE 3.10 *(See color insert.)* Perrin-Jablonski diagram. Graphical representation of molecular energy levels and the transitions between them. Thick lines represent electronic energy levels: S denotes singlets; *T* denotes triplets. Thinner lines represent vibrational energy levels. Both radiative and nonradiative processes are transitions between energy levels and they are represented as arrows, pointing the direction where the transition occurs. Radiative processes, such as fluorescence and phosphorescence are represented as straight arrows; nonradiative processes, such as internal conversion or intersystem crossing, are represented as wavy arrows. The time scales of each process are shown between brackets.

activity of the *fluorophore with respect to its environment*. Figure 3.11 illustrates a simulation of the process and the corresponding excitation and emission spectra. Both have the same appearance; however, the emission spectrum is shifted to longer wavelengths. The emission spectrum is typically acquired using excitation light of a defined wavelength. One additional peculiarity of fluorescence is that the emission spectrum does not depend on the wavelength of the excitation light. Therefore, fluorophores are sometimes characterized by the wavelengths of the highest excitation and emission maxima, which is of practical use especially in fluorescence microscopy. In that context, we will refer to the wavelengths of highest excitation and emission maxima as simply excitation and emission wavelengths.

Quantum Yield and Fluorescence Quenching

As mentioned previously, fluorescence emission is just one of the many de-excitation pathways for a molecule to return to its relaxed state. Indeed, the maxima intensity (and hence the extent of fluorescence) can be altered or diminished by the occurrence of competitive processes in the close vicinity of the affected molecule, thus giving information about its environment. These processes can be radiative or nonradiative, depending upon whether or not they are accompanied by light emission, and they can occur within the molecule or through interaction with other molecules. Those processes that diminish fluorescence are the cause

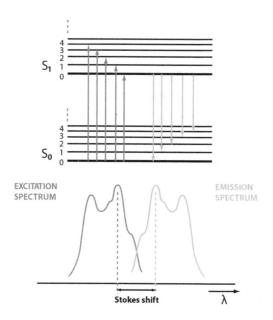

FIGURE 3.11 Excitation and emission spectra. The excitation peaks appear at shorter wavelengths than the emission peaks. The difference between the maximum of the excitation spectrum and the maximum of the emission spectrum is called the Stokes shift. Usually excitation and emission spectra mirror each other.

of *fluorescence quenching* and define the *quantum yield* of fluorescence. They include internal conversion, intersystem crossing, intramolecular charge transfer, and conformational changes.[*]

The quantum yield is the fraction of molecules that return to the relaxed state by fluorescence emission. It is a measure of the extent of fluorescence with respect to the rest of the competitive processes, expressed as a ratio of rate constants

$$\Theta = \frac{k_r}{k_r + k_{nr}} = k_r \tau \tag{3.12}$$

where k_r is the rate constant of the radiative process, in this case fluorescence; k_{nr} is the sum of the rate constants of the nonradiative processes; and τ is the lifetime of fluorescence emission, i.e., the lifetime of the excited singlet state.

Quantum yields take values between 0 and 1 for certain standard fluorophores, and they depend on temperature, the excitation wavelength, and the type of surrounding medium or solvent (Lakowicz 2006). They can be as low as 0.024 in the case of phenylalanine in water (260 nm, 23°C) and as high as 1.00 in the case of 9,10-diphenylanthracene in cyclohexane (366 nm). In everyday life, however, quantum yields can be much lower than 0.5, which indicates that just a small fraction of molecules will fluoresce. Fluorescence emission is weak

---

[*] For a detailed account of these phenomena, the reader is referred to Valeur (2001).

and thus requires optimal sensitivity of all optical components when applied to microscopy.

AN INTERESTING NONRADIATIVE PROCESS THAT PRODUCES FLUORESCENCE: RESONANCE ENERGY TRANSFER (RET)

We have seen that there are certain de-excitation processes that compete with fluorescence to bring the excited state of fluorophores back to their ground state. A relevant process involves the presence of another fluorophore that fulfills the following requirements:

- Its absorption spectrum partly overlaps the emission spectrum of the first fluorophore, as shown in Figure 3.12a.
- Both fluorophores are very close to each other (1–10 nm).

When these requirements are met, energy may be transferred from the first fluorophore, the *donor*, to the second fluorophore, the *acceptor*. As a result, the fluorescence emission of the acceptor is enhanced at the expense of the donor,

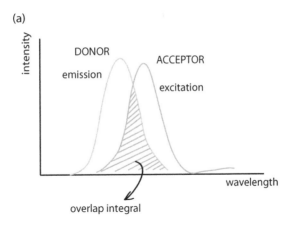

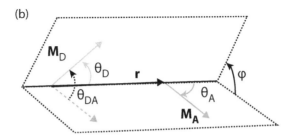

FIGURE 3.12 *(See color insert.)* Resonance energy transfer. Spectra (a) and orientations (b). (a) For energy transfer to occur, it is required that the donor's emission spectrum overlaps the acceptor's excitation spectrum. The striped area is the overlap integral, this magnitude appears in Equation (3.14). (b) Donor and acceptor have associated transition moments, M_D and M_A, respectively. Their spatial arrangement is defined by the angles θ_D, θ_A, θ_{DA}, and φ.

whose fluorescence is partly or totally quenched. The interaction mechanism of the energy transfer is not unique, and it can be either coulombic or caused by intermolecular orbital overlap (Valeur 2001). The efficiency of the energy transfer depends on the strength of the interaction (U) compared to the energy difference between the excited states of donor and acceptor (ΔE). This results in different RET scenarios, from very strong coupling ($U \gg \Delta E$) or to a very weak coupling ($U \ll \Delta E$). The energy balance is crucial to predict RET's dependence on the distance r between the donor and the acceptor: r^{-3} or r^{-6} for the cases of very strong-to-weak coupling and of very weak coupling, respectively (Valeur 2001). For the particular case of very weak coupling derived from long-range dipole–dipole interactions, the expression that relates the rate of energy transfer, k_T^{dd}, and the donor-acceptor distance, r, was formulated by Förster

$$k_T^{dd} = k_D \left(\frac{R_0}{r} \right)^6 = \left(\frac{1}{\tau_D} \right) \left(\frac{R_0}{r} \right)^6 \tag{3.13}$$

where k_D and τ_D are, respectively, the emission rate constant and the lifetime of the donor in the absence of the acceptor. R_0 is the critical distance or Förster radius, the distance at which transfer and spontaneous decay of the excited donor are equally probable ($k_T = k_D$). R_0 can be determined from spectroscopic data, and it is given by

$$R_0 = 0.2108 \cdot \left[\kappa^2 \Phi_D n^{-4} \int_0^\infty I_D(\lambda) \varepsilon_A(\lambda) \lambda^4 d\lambda \right]^{\frac{1}{6}} \tag{3.14}$$

where κ^2 is the orientational factor, Φ_D is the fluorescence quantum yield of the donor in the absence of acceptor, and n is the average refractive index of the medium in the wavelength range where spectral overlap is significant. The overlap integral is expressed in units of M^{-1} cm^{-1} nm^4 and depends on $I_D(\lambda)$, the normalized emission fluorescence spectrum of the donor

$$\left(\int_0^\infty I_D(\lambda) d\lambda = 1 \right),$$

and on ε_A, the molar absorption coefficient of the acceptor (in M^{-1} cm^{-1}). In this expression, R_0 is given in Å, and it is generally in the range of 15–60 Å.

The orientational factor κ^2 is determined by the spatial arrangement of the transition moments of both donor and acceptor molecules as follows:

$$\kappa^2 = \cos\theta_{DA} - 3\cos\theta_D \cos\theta_A = \sin\theta_D \sin\theta_A \cos\phi - 2\cos\theta_D \cos\theta_A \tag{3.15}$$

where θ_{DA} is the angle between the donor and acceptor transition moments, θ_D and θ_A are the angles between the latter and the separation vector, and φ is the angle between the planes that contain the transition moments (see Figure 3.12b). κ^2 can take values between zero (perpendicular transition moments) and 4 (collinear transition moments). If the molecules are free to rotate at a much faster rate than the de-excitation rate of the donor, κ^2 takes an average value equal to 2/3.

The determination of distances between donor and acceptor molecules is the main advantage of RET, and in this regard it is considered to be a *spectroscopic ruler*.

OPTICAL MICROSCOPY OF FLUORESCENT OBJECTS

Fluorescence microscopes are *reflection* optical microscopes that gather a few special features compared to conventional, bright-field optical microscopes, as seen in Figure 3.13. Essentially, they are equipped to generate an appropriate and stable excitation and to selectively collect the fluorescence emission so as to produce the fluorescence image.

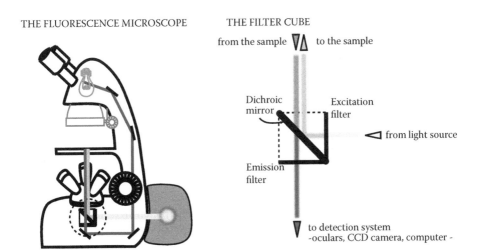

FIGURE 3.13 *(See color insert.)* The fluorescence microscope (left) and the filter cube (right). The fluorescence microscope is an optical microscope equipped with special illumination and optical filters compared with a standard bright-field optical microscope. The filter cube delivers the appropriate light to excite the fluorescent sample and just lets the fluorescence emission go through to the detection system.

Light Source

As mentioned previously, a fluorescent sample absorbs light of a defined wavelength to get excited and emit fluorescence. Illumination acts as an excitation source in fluorescence microscopy and hence should provide light of the "right" intensity and wavelength. The "right intensity" means that illumination should be intense enough to excite a sufficient amount of fluorophores and detect fluorescence. The "right wavelength" means that the light wavelength(s) of the excitation source should coincide with the excitation wavelength(s) of the fluorophore(s). This is the optimal case; however, this would require a different excitation source for each particular fluorophore, which makes it impractical. Instead, fluorescence microscopes usually have few excitation sources that provide light of wavelengths that are proximal to the excitation wavelengths of most fluorophores.

There are basically two ways to achieve this. One consists of filtering white light before it reaches the sample; the other consists of using monochromatic light. Using white light as an illumination source requires an *excitation filter.* This device blocks the light of wavelengths that greatly differ from those of the sample's excitation maxima and transmits the rest. In practice, fluorescence microscopes are equipped with arc lamps (mercury, argon) and a set of several filters, each one transmitting a particular range of wavelengths throughout the visible spectrum so that most fluorophores can be excited.

Monochromatic light sources are typically lasers or light-emitting diodes. These devices do not require any filters, since they already provide light with a specific wavelength. This makes them highly selective and effective excitation sources, especially when the wavelength of the light matches the excitation wavelength of the fluorophore. In addition, the use of lasers increases the lateral and vertical resolution by bringing the technique closer to the diffraction limit. The high-performance modes of fluorescence microscopy, such as laser scanning confocal microscopy or scanning near-field optical microscopy, make use of such illumination sources.

Collecting the Emitted Fluorescence: The Dichroic Mirror and the Emission Filter

When illuminating the sample with excitation light, part of it will be absorbed by the fluorescent objects within, which will in turn emit fluorescence. However, this is not the only phenomenon that takes place. Diffraction of excitation light by both fluorescent and nonfluorescent objects simultaneously occurs, and consequently the emitted fluorescence as well as the diffracted and the nondiffracted light reach the objective. Production of fluorescence images thus require that the fluorescence emission path should be separated from the excitation path so that only fluorescence emission reaches the detection system.

This is achieved by a *dichroic mirror.* A dichroic mirror is a device that totally reflects the light with wavelength below a certain transition value or threshold while it totally transmits the light with wavelength above that threshold. For an optimal performance, this transition wavelength should be intermediate between the excitation and emission wavelengths. If a dichroic mirror of such characteristics is positioned between the objective and the illumination source, the excitation light will be diverted toward the objective and sample, and the excitation light coming from the sample will be blocked off. Contrarily, the light emitted by the sample will be transmitted by the dichroic mirror and will reach the detection system.

To further block any remnants of excitation light and to gain selectivity in the detection, which is especially convenient when observing samples with different fluorophores, it is important to filter out the emitted light around the emission wavelength of the fluorophore of interest. This is done by the *emission filter*, which performs in a similar way to the excitation filter, however shifted to longer wavelengths.

In practice, the excitation filter, the emission filter, and the dichroic mirror comprise a set of filters spatially arranged in a cube, the *filter cube*. The appearance and performance of the cube are shown in Figure 3.13. Each filter cube is designed in such a way that the performance of the three devices is optimized for a particular set of fluorophores that are excited and emit in a certain range of wavelengths, e.g., within the range of blues, the range of greens, and the range of reds.

FLUORESCENCE FILTERS

There are three different types of filters according to their performance: *long pass*, *short pass*, and *band pass*. The long pass (LP) filters transmit light of wavelengths longer than a certain value, while they block the light of shorter wavelengths; in this regard, dichroic mirrors are long-pass filters. Short pass (SP) filters are those that transmit light of wavelength shorter than the threshold value and block the light of longer wavelengths. Band pass (BP) filters are said to be the most restrictive and hence the most selective, since they only transmit the light within a relatively narrow range of wavelengths. Hence BP filters have two transition wavelengths corresponding to the lower and the upper thresholds.

Like lenses, filters are not perfect, and the quality of their performance is determined by their transmittance to light. Ideally, filters should have a 100% transmittance to the wavelengths they let pass and 0% transmittance to the wavelengths they block. At the transition wavelength(s), the transmittance changes abruptly from 0% to 100%. Therefore, the transmittance as a function of the wavelength should be a step function for LP and HP filters and a rectangular function for BP filters. In practice, the more the transmittance function of a particular filter resembles the ideal behavior, the better its quality is. Further information about fluorescence filters can be found in the literature (Lakowicz 2006; Reichman 2000).

FLUORESCENT PROBES

Earlier in this section we explained the basics of molecular fluorescence. However, fluorescence microscopy produces images of fluorescent samples that are far bigger than molecules in most cases. With naturally occurring fluorescent samples, this imposes no problem. The constituting molecules are in large enough number to produce sufficient fluorescence. However, there are only a few such materials, which would make fluorescence microscopy of very low practical value. Fortunately, that is not the case, and fluorescence microscopy can also be applied to nonfluorescent samples, provided that they are labeled with a significant amount of fluorophore. *Fluorescence labeling* has become an essential step in fluorescence microscopy, and it comprises an extensive, ever-developing field in fluorophore synthesis, labeling strategies, and protocols (Lakowicz 2006; Haugland 2005).

DRAWBACKS OF FLUORESCENCE MICROSCOPY: THE EVER-PRESENT PHOTOBLEACHING

Photobleaching refers to the permanent loss of fluorescence by a dye due to photon-induced chemical damage or modification (Murphy 2001). It occurs when the fluorophore in its excited singlet state transits to a triplet state. In this state, the molecule is prone to react with other molecules, in particular with molecular oxygen, destroying the fluorophore and creating free radicals that can modify other molecules and produce in turn sample damage (*photodamage*).

Photobleaching is one of the main concerns in fluorescence microscopy; it cannot be completely avoided, but there are various ways to minimize its effect. Reducing the intensity of the excitation source is one of them, and care should always be taken not to overexpose the sample. To protect the latter from photodamage, it is always helpful to reduce the oxygen concentration whenever possible. High-performance modes of fluorescence microscopy based on confocal optics, total internal reflection, or near-field optics have their own mechanisms to minimize photobleaching, as will be described later. In this section we will focus on *two-photon excitation*.

Two-photon excitation occurs when the fluorophore promotes to its excited state by absorbing two photons simultaneously. The event is rather unlikely to happen, since two photons must synchronously interact with the fluorophore, and therefore it requires high illumination intensity. This is achieved by employing a pulsed laser that provides a high local instantaneous excitation intensity at the focal plane (Herman 1998). The probability for a fluorophore to absorb two photons simultaneously decreases with the fourth power of the distance along the optical axis, which makes it highly unlikely that excitation may occur out of the focal plane. This considerably reduces background fluorescence and out-of-focus interference, resulting in a depth of field that is highly similar to that produced by a confocal microscope, which is described in the next section.

Another significant advantage of using two-photon excitation is that the wavelength required for two-photon excitation is usually twice the wavelength required for one-photon excitation. For example, intense illumination with 800-nm light can induce fluorescence emission by a two-photon mechanism in a fluorophore that is normally excited by light of 400-nm wavelength in a single-photon mechanism (Murphy 2001). Under these circumstances, fluorophores that otherwise would be excited with UV light can be excited with visible light, even with infrared radiation of much less energy, thus reducing the risk of photodamage. Titanium:sapphire lasers with a peak efficiency around 800 nm and a working range of 750–1050 nm are usually employed as two-photon excitation sources. Since the selection rules are different in the case of multiphoton excitation compared to single-photon excitation, the two-photon excitation spectra may differ from those produced by single-photon absorption (Lakowicz 2006). It is thus always convenient to check beforehand the two-photon absorption spectrum of the fluorophore.

HIGH-PERFORMANCE MODES OF FLUORESCENCE MICROSCOPY

Under this title we include a series of techniques that bring fluorescence microscopy to the ultimate levels of temporal and spatial resolution and that have been successfully combined with scanning probe microscopy. We also mention two types of near-field optical microscopies that bring the technique unparalleled sensitivity when combined with fluorescence. Indeed, near-field optical microscopy bypasses the diffraction limit, as we will see in Chapter 4.

CONFOCAL LASER SCANNING MICROSCOPY (CLSM)

In standard fluorescence microscopy, emitted light from the illuminated sample is collected through the objective to generate the image. However, the objective not only collects the fluorescence emitted by the focused optical section of the sample, but also the light emitted by optical sections that are above and below the focal plane. The out-of-focus signal reduces contrast and spatial resolution, which impedes seeing the inner structure of the specimen. In particular, fluorescence microscopy alone cannot differentiate between a fluorescent solid and a fluorescent hollow particle.

The potential of CLSM lies in its capability of rejecting light coming from the planes above and below the focus plane and of producing images of thin optical sections within the sample. The microscope, depicted on the left-hand side of Figure 3.14, has one or several highly collimated laser light sources that excite a very small volume of the sample. Light emitted by the excited fluorophores is collected through the objective and the dichroic mirror, just as in conventional fluorescence microscopy. The confocality is brought about by the presence of a *pinhole* between the dichroic mirror and the detection system. The pinhole is an opening of adjustable size that is optically confocal with and conjugate to the specimen plane. As the right-hand side of Figure 3.14 shows, the pinhole aperture only lets light coming from the focal plane reach the detector.

This addresses the issue of producing a point within the sample. To produce the image of the whole sample, the laser is moved back and forth along the focal plane in a way called *raster scanning*. This is usually done by galvanometer-driven mirrors that deflect the excitatory laser beam along the *x*- and *y*-directions. The light from

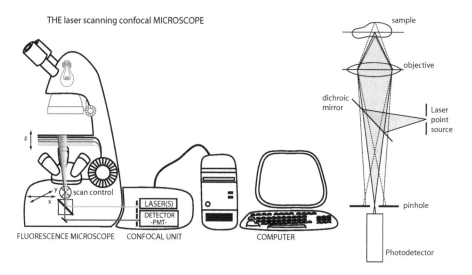

THE laser scanning confocal MICROSCOPE

sample

objective

dichroic mirror

Laser point source

z

y
x
scan control

LASER(S)

DETECTOR -PMT-

pinhole

FLUORESCENCE MICROSCOPE   CONFOCAL UNIT

COMPUTER

Photodetector

FIGURE 3.14 The confocal laser scanning microscope (left) and the optics involved (right). The CLSM is a fluorescence microscope equipped with one or several lasers, a scan control to direct the laser beam alongside the XY plane, and a motorized sample stage to generate the stack of images along the z-plane. Confocality is brought about by a pinhole that allows fluorescence coming only from the focal plane of the specimen to reach the detector.

each point within the focal plane is collected by a photomultiplier detector, and the image is electronically generated by integrating the photon signals point by point. Additionally, if the sample stage is moved alongside the z-direction, it is possible to generate a stack of confocal images from the bottom to the top of the sample (or vice versa). All these images can be electronically integrated to generate a magnified three-dimensional reconstruction of the sample.

There are thus several factors that contribute to the spatial resolution and image contrast in confocal microscopy apart from the numerical aperture of the objective: laser alignment and power, the size of the pinhole, the scanning rate, and the number of scanning points or pixels.

The laser spot in confocal microscopy should fill the back aperture of the objective (Valeur 2001); otherwise, the diffraction spot will be wide, which lowers the resolution. This is attained by a lens that expands the laser spot. Pinhole size is critical to both spatial resolution and contrast. The smaller the pinhole size, the thinner is the thickness of the focal plane along the z-axis, which improves the vertical resolution. Consequently, the exclusion of out-of-focal-plane signals will be more effective, which in turn improves image contrast. On the other hand, reducing the pinhole aperture decreases light intensity and hence the fluorescence that can reach the detector, which in turn reduces sensitivity and contrast. In practice, a compromise should be taken, and the optimum pinhole setting corresponds to an aperture giving 50% of the maximum light intensity (Valeur 2001). Scan rate defines the time the laser lingers on one specific point of the sample and hence determines the exposure time. Smaller scan rates give high-resolution images at the expense of missing dynamic processes that occur at higher speeds or of increasing the risk of photobleaching or photodamage. Faster scan rates give low-resolution images that capture dynamic events and minimize photobleaching and photodamage, although at the expense of reducing the signal-to-noise ratio. Scanning sampling directly affects the spatial resolution, since the image is electronically and not optically generated; however it should not compromise the spatial resolution given by the numerical aperture of the objective. In a confocal microscope, the minimum resolvable distance d between two points can be approximated by

$$d_{x,y} \approx \frac{0.4\lambda}{\left(n\sin\theta\right)_{\text{objective}}} \tag{3.16}$$

The optimal value for the pixel size is half of the value of d. Scanning samples that give out a pixel size smaller than $d/2$ produce oversampling and risk photobleaching; if the pixel size is larger than $d/2$, undersampling occurs together with loss of lateral resolution.

Fluorescence Lifetime Imaging Microscopy (FLIM)

Just as the pixels of an image acquired by standard fluorescence microscopy or CLSM represent the intensity of fluorescence emission, the pixels of an image captured by FLIM represent the fluorescence lifetime. The former gives information of the spatial distribution of fluorophores, while the latter gives information about the physical and chemical properties of their surroundings. When used in combination with confocal fluorescence microscopes, FLIM can thus provide a three-dimensional distribution of lifetimes.

When irradiated for a short period of time, a macroscopic sample containing a sufficient number of fluorescent molecules emits light with an intensity that decays exponentially with time. In most cases, the decay is a monoexponential function, and the decay time is the fluorescence lifetime.

$$I(t) = I_0 \exp\left(-\frac{t}{\tau}\right) \tag{3.17}$$

where $I(t)$ is the time-dependent fluorescence intensity, I_0 is the initial fluorescence intensity, and τ is the decay time. The time a molecule stays in the excited state before it returns to the relaxed state and emits fluorescence is a characteristic parameter that depends on the fluorophore's microenvironment. The great potential of FLIM resides in the fact that fluorescence lifetime depends on parameters that are characteristic of the fluorophore's close vicinity (Valeur 2001). These parameters can be chemical, such as pH, ion concentration (e.g., Ca^{2+}), or the presence of quenchers (e.g., heavy metals or oxygen). The presence of other fluorescent molecules that can in turn absorb light emitted by the fluorophore (if sufficiently close) can alter the fluorescence lifetime of the latter by resonance energy transfer (RET). On the other hand, lifetimes can also be affected by physical parameters (medium viscosity or temperature) or molecular interactions (dipole coupling or H-bonding).

There are mainly two methods for detecting fluorescence lifetimes: time-domain FLIM and frequency-domain FLIM. Both are depicted in Figure 3.15. Time-domain FLIM uses a pulsed source as excitation, which illuminates the sample for a relatively short time (a *gated pulse*) and fluorescence is detected at various delays. For detecting single exponential lifetimes, two delays are normally sufficient (t_1 and t_2). The fluorescence intensity during these two time intervals will be collected, and lifetime can be obtained by applying the following equation:

$$\tau = \frac{t_2 - t_1}{\ln\left(D_1 / D_2\right)} \tag{3.18}$$

These types of measurements demand rapid responses of the detection system (typically microchannel plate photomultipliers) due to the short decay times (10^{-7}–10^{-10} s). High-resolution images, however, demand extremely long computational times due to the fact that the excitation pulse and the fluorescence decay are overlapped. In this case, pixel-by-pixel signal deconvolution usually follows light acquisition in order to separate both signals and obtain the correct lifetime. Even with the most efficient detection systems (time resolution <10 ps), space resolution is on the order of 100 μm (Valeur 2001).

Frequency-domain FLIM uses a continuous-wave laser and an acousto-optical modulator to produce a sinusoidal excitation of high frequency, ω. Consequently, the fluorescence emission will be sinusoidal as well; however, since it has a defined lifetime, the fluorescence signal will be delayed with respect to the excitation signal. The delay of harmonic signals is expressed as a phase shift that is detected by the microscope. Together with the phase shift, the intensity of both excitation and fluorescence signals is detected, which is expressed as intensity ratio or modulation. Lifetime values can be computed from phase shifts (τ_Φ) and modulations (τ_M) according to

$$\tau_\Phi = \frac{1}{\omega} \tan^{-1} \Phi \tag{3.19}$$

and

$$\tau_M = \frac{1}{\omega}\left(\frac{1}{M^2} - 1\right)^{1/2} \tag{3.20}$$

If the lifetime values calculated from Equations (3.19 and 3.20) are identical, the fluorescence decay is monoexponential. If the decay is multiexponential, τ_Φ will be smaller than τ_M, and in order to extract the individual lifetimes, it is necessary to acquire images at various frequencies. The number of images should be at least the number of incognita; for an N-exponential decay, $2N - 1$ frequencies are needed (N lifetimes and $N - 1$ relative amplitudes or preexponential factors).

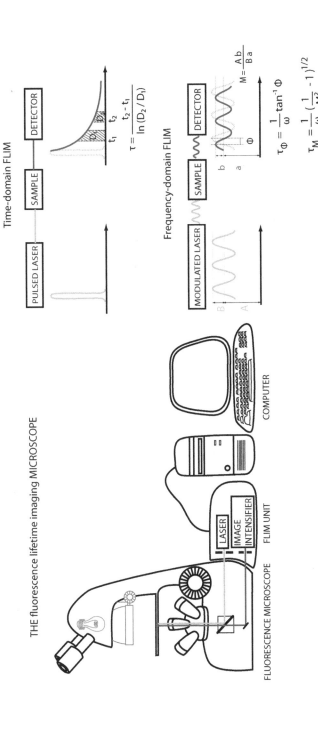

FIGURE 3.15 The fluorescence lifetime imaging microscope (FLIM) and the modes of obtaining lifetimes. The FLIM is a fluorescence microscope equipped with special ways of modulating the excitation light, either as a gated pulsed (time domain) or as modulated signal (frequency domain). The detector, an image intensifier is activated for a short period of time at various delays to collect the fluorescence in the time-domain mode. Lifetime is then calculated from the delay time and the intensity collected during the time intervals. Alternatively in the frequency-domain mode, the performance of the image intensifier is modulated at the same frequency (homodyne) or different frequency (heterodyne) as that of the incident light. In this case, lifetime is calculated either from the offset and amplitude ratios (modulation ratio, M) or from the relative delay (Φ) of the excitation and emission signals. These two values, τ_M and τ_Φ, are identical if the decay is monoexponential.

Total Internal Reflection Fluorescence (TIRF): A Near-Field Microscopy

We have already seen that the depth of focus Z in bright-field optical microscopy defines the vertical resolution, which typically is on the order of the wavelength of light (400–700 nm). However, it is possible to get images of optical sections as thin as 30–300 nm (Axelrod et al. 2010) and hence overcome the diffraction limit. TIRF is a kind of surface fluorescence microscopy, since it provides images of a restricted region within the sample that is immediately adjacent to the surface. In this case, "surface" refers to the interface between the sample and the supporting medium.

TIRF Is Based on the Optical Principles of Total Internal Reflection

This optical phenomenon occurs when light, coming from a medium of higher refractive index (such as glass), reaches a medium with lower refractive index (such as air or water). As we have already seen, part of the light is usually reflected and part of it refracted. However, at a certain angle of incidence, called the *critical angle*, practically all light is reflected and none is transmitted. Total reflection occurs at incidence angles that are *equal to or larger* than the critical angle. The critical angle $\theta_{critical}$ is defined as the incidence angle such that the angle of transmission (or angle of refraction) is equal to 90°. In other words, refraction occurs along the interface, and no light traverses the second medium. Applying the Snell-Descartes law of Equation (3.3) gives

$$\sin\theta_{critical} = \frac{n_2}{n_1} \text{, with } n_2 < n_1 \tag{3.21}$$

Some Numbers: Critical Angles between Media in Microscopy

Let's consider borosilicate glass as a typical support for samples and a watery specimen, such as living cells. Glass has a refractive index of 1.517, while living cells typically have a refractive index between 1.33 and 1.38 (Axelrod et al. 2010; Ross 2000). Thus, in this case, the critical angle would be between 61° and 65°.

An Evanescent Wave Acts as the Excitation Source in TIRF

When light of a specific wavelength strikes the surface at a higher angle than the critical angle, an *evanescent wave* is generated in the sample, very close to the boundary between the two media. The evanescent wave is a standing electromagnetic wave (contrary to a propagating wave such as light) of the same wavelength as that of the incident light. The intensity of the evanescent wave decays exponentially with the distance from the interface, typically being 40% of the initial intensity at 100 nm (Axelrod et al. 2010). Consequently, if an evanescent wave has the wavelength of the excitation light, only those sample fluorophores most proximal to the surface and contained in a section of thickness equal to the penetration depth of the evanescent field can be excited. The penetration depth Z_{TIRF} is a function of the wavelength of the incident light, the angle of incidence, and the refractive indices of the media that define the interface according to

$$Z_{TIRF} = \frac{\lambda}{4\pi\left(n_1^2 \sin^2\theta_i - n_2^2\right)^{1/2}} \qquad (3.22)$$

Some Numbers: Penetration Depth

If we consider the water–glass interface, it is possible to apply Equation (3.22) to estimate penetration depths. Figure 3.16 shows a plot of the penetration depth Z_{TIRF} versus the angle of incidence (at values higher than the critical angle) and at particular wavelengths: 422 nm (violet), 473 nm (blue), 534 nm (green), and 587 nm (yellow). Z_{TIRF} decreases with increasing angle of incidence and with the wavelength of incident light. Values are in the range of several tens of nanometers (30–100 nm).

As a type of fluorescence microscopy, the incoming light comes from the objective and reaches the sample at a higher angle than the critical angle. To avoid undesired reflections and diffractions, the gap between the objective and the sample support (normally a microscope slide or coverslip made of borosilicate glass of $n = 1.517$) is filled with immersion liquid of similar refractive index as that of the sample support.

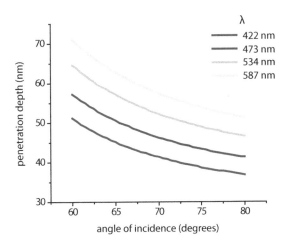

FIGURE 3.16 Penetration depths in TIRF (Z_{TIRF}). The graph shows the penetration depth as a function of the incident angle at various excitatory wavelengths. Z_{TIRF} decreases with the angle of incidence as well as with the wavelength of incident light.

TOTAL INTERNAL REFLECTION FLUORESCENCE MICROSCOPY

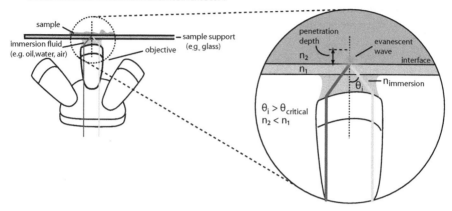

FIGURE 3.17 Total internal reflection microscopy. The figure illustrates how total internal reflection is attained using the objective lens. The angle that the objective sustains should be larger than the angles at which total internal reflection occurs. The interface where total reflection occurs is defined by the sample and the sample support. Preferably, the immersion fluid should have a similar refractive index to that of the sample support.

TIRF microscopes are standard fluorescence microscopes equipped with either white light or monochromatic illumination. Illumination at angles higher than the critical angle is mainly achieved in either of two ways: via an *illumination side prism* or via the *objective lens*.

In the first case, the illumination reaches the sample at the desired angle through a prism that is attached at the side of the sample support (i.e., a coverslip). However, this setup has the disadvantage of providing limited access to the specimen, which should be confined within the area occupied by the prism. In the second case, shown in Figure 3.17, the objective of the appropriate numerical aperture is used to deliver the excitation light at the desired angle. Since critical angles are in the range of 60°, objectives of high numerical apertures are needed. In particular, an appropriate objective for TIRF should be such that the angle sustained by the objective must be larger than the critical angle

$$\sin\theta_{objective} > \sin\theta_{critical} \Rightarrow NA_{objective} > \frac{n_{immersion} \cdot n_1}{n_2} \tag{3.23}$$

Some Numbers: Appropriate Objectives for TIRF

Taking again the glass–water interface as a reference, we can apply Equation (3.23) to estimate the minimal numerical aperture for an objective to be used in TIRF. Since $n_1 = 1.52$ and $n_2 = 1.33$, the numerical aperture of the objective should be higher than 1.1 if the immersion fluid is air, 1.52 if the immersion fluid is water, or higher than 1.7 if the immersion fluid is oil.

FLUORESCENCE SCANNING NEAR-FIELD OPTICAL MICROSCOPY (FLUORESCENCE SNOM)

The basic principles of SNOM are described in Chapter 4, and we refer the reader to it in order to fully understand this section. In a few words, SNOM is the technique that combines the topographical capabilities of a scanning probe technique with the features of near-field optical microscopy. Among the different optical contrasts available in SNOM, fluorescence is by far the most extensively used, and this is mainly due to the fact that contrast is easy to understand and less susceptible to coupling artifacts with the topological signal, which eases image interpretation. Moreover, fluorescence contrast can be attained with aperture and apertureless probes as well as with tip-on-aperture probes (Dunn 1999; Bouhelier 2006; Hartschuh 2008). Fluorescence SNOM is also advantageous if compared to other modes of fluorescence microscopy: Lateral resolution is at least tenfold improved, since it is not limited by light diffraction, which makes the technique unequaled among the rest of the optical microscopies; excitation is locally delivered to a tiny region (nanometer wide) within the sample, which significantly reduces fluorescence background; and the intensity of the excitation light is much weaker, which considerably lessens the risk of photobleaching and photodamage.

The applications of fluorescence SNOM abound in the most diverse research fields where detection of quantum effects or even of single molecules is required. For example, fluorescence SNOM has been employed to characterize the photoluminescence of single quantum wells that form part of semiconductor devices. In the field of soft-matter physics, the technique has been particularly useful in the study of conformations of single molecules in polymer films, not to mention in the field of biosciences, where the technique can be used to ascertain the distribution of single-molecule receptors, of protein clusters, or of lipids in both artificial and native cell membranes.

As a type of near-field microscopy, fluorescence SNOM relies on two main requirements

The presence of a small probe, generally a sharp and small tip with or without aperture, that either delivers excitation light to the sample or collects the sample fluorescence

The probe's close proximity to the sample, generally 5–10 nm apart

In fluorescence aperture SNOM, two operating modes are possible. One of them is the *illumination* mode, where the incident light goes through the aperture of the SNOM tip and the fluorescence is collected by the objective, usually in transmission. The other is the *collection* mode, where the sample is illuminated from the far field and the SNOM tip collects the fluorescence. Illumination mode is usually preferred, since the excitation volume is as small as the tip size, which significantly reduces fluorescence background and improves the signal-to-noise ratio. The SNOM setup is often combined with an inverted optical fluorescence microscope, as shown in Figure 3.18. The excitation beam propagates along the optical fiber that transmits the light to the sample through the subwavelength aperture, illuminating a very small volume of the sample, normally as large as the aperture size. The transmitted light

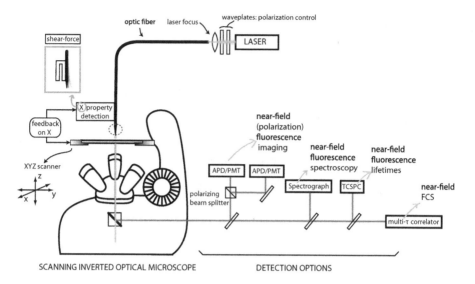

FIGURE 3.18 Setup for fluorescence aperture SNOM in illumination mode. The setup is mounted on a scanning inverted optical microscope in transmission; far-field illumination is replaced by an optic fiber with a subwavelength aperture at its end. The laser provides the light that propagates along the optic fiber, which through the aperture illuminates a tiny area of the sample. The polarization state of the incoming light can be set at any value. Tip-sample distance is kept constant via shear force-feedback in most cases. Fluorescence and residual excitation light are collected by an optical objective of high numerical aperture (typically 1.3). This setup can be combined with different detector systems for each fluorescence property of interest.

is collected by an optical objective of high magnification (usually 60×–100×) and high numerical aperture (0.6–1.3) and delivered to the detector. Dichroic mirrors and notch filters are used to filter out the excitation radiation before it reaches the detector, and the weak fluorescence intensity coming from such a small sample volume is amplified by photomultiplier tubes (PMTs) or avalanche photodiodes (APDs). A polarizing beam splitter can redirect the light that is x-polarized or y-polarized to a different photocounting detector in order to obtain the state of polarization of the emitted fluorescence. In single-molecule detection, this information is especially valuable, since it can provide the spatial orientation of molecular dipole moments.

In both aperture and apertureless fluorescence SNOM, the probe (or sample) is moved along the surface plane in a raster manner, and the emission fluorescence is collected point by point, which in turn produces the SNOM image. The distance between the probe and the sample is held constant during the scanning through shear-force feedback in most cases. This SPM feedback mechanism operates in a similar way to the alternating contact feedback described in Chapter 2. In this case, however, the probe is set to vibrate parallel instead of normal to the sample plane at its resonance frequency by means of a dither piezoelectric device or a tuning fork; the amplitude of such vibration changes due to shear forces between the tip and

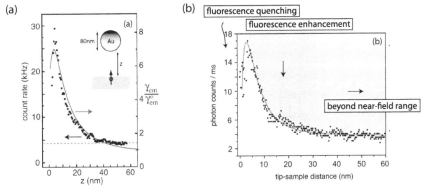

from Anger P et al. 2006, Phys. Rev. Lett. 96:113002   from Höppener C et al. 2009, Nano Lett. 9:903-908

FIGURE 3.19 Optical approach curves in fluorescence SNOM: finding the correct tip-sample distance. (a) The count rate and (b) the photon counts as a function of tip-sample separation. In both cases, the fluorescence activity increases due to field enhancement. At distances lower than 5 nm, fluorescence decreases due to metal-induced quenching. In both experiments, the optimal operating distance for SNOM measurements was set to 5 nm. (Figure (a) is reprinted from Anger, Bharadwaj, and Novotny 2006. With permission. Figure (b) is reprinted from Höppener, Beams, and Novotny 2009. With permission.)

sample that are distance dependent. The feedback ensures that minimal changes in the vibration amplitude occur, which results in minimal changes in the tip–sample distance. But how large should this distance be?

To determine the optimal tip–sample distance, it is very useful to perform *optical approach curves*, where the fluorescence emission or the fluorescence rate is collected as a function of the tip–sample distance. The curves look like those shown in Figure 3.19; in both cases, the apertureless probe is a gold nanoparticle of 80-nm diameter that acts as a nanoantenna. At large distances, the near-field fluorescence is not detected; when the distance gets smaller, the fluorescence intensity increases, but it may eventually decay at distances smaller than 5 nm. The drop in fluorescence at small distances can be caused by quenching induced by the metal tip or by contact between the tip and the sample. Therefore, the optimal distance should be small enough to get the maximum intensity of the near-field emission fluorescence but large enough so that the intensity is not affected by tip-induced quenching or tip–sample contact. The curves of Figure 3.19a and 3.19b exemplify the first case, where the distance for SNOM measurements was set to 5 nm (Anger, Bharadwaj, and Novotny 2006; Höppener and Novotny 2008).

One of the most typical configurations of the apertureless variant makes use of a gold nanoparticle or a cantilever tip that does not possess an aperture. In this case, the illumination of the tip is done by a focused laser placed either aside the probe at an angle of 40°–45° from the surface normal or along the tip axis, above or below the sample. Figure 3.20 shows the latter case, where a single optical objective illuminates and collects the emitted light from below the sample. For apertureless SNOM,

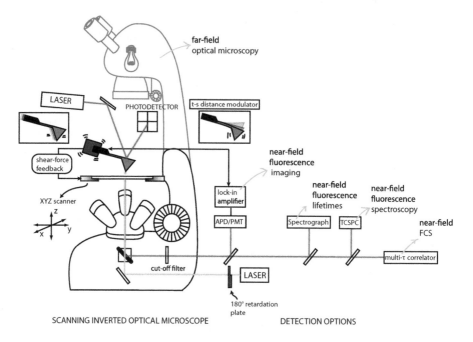

FIGURE 3.20 Setup for fluorescence apertureless SNOM. The configuration shows the SNOM in combination with a scanning inverted optical microscope and an atomic force microscope. The apertureless probe is a cantilever tip onto which a laser polarized along the tip axis is focused to produce the field enhancement. Laser focusing is done in this case through the optical objective placed below the sample; in this case, it is necessary to use a higher-order mode laser. The tip-sample distance is controlled by the AFM-driven feedback, usually based on shear forces. Both near-field and far-field background are collected by the optical objective toward the filter cube and a cutoff filter that removes the red laser radiation coming from the AFM and employed in the cantilever alignment. The background signal, usually large compared with the near-field fluorescence, the signal of interest, is filtered out in a lock-in amplifier, set at the resonance frequency of the cantilever or its higher-order harmonics. Just as in fluorescence aperture SNOM, the setup is compatible with several detection modes.

the incoming light should be polarized along the tip axis, and to achieve that in such a geometry, a higher-order mode laser is required (for further details, see Chapter 4). This setup is most convenient if far-field optical microscopy is to be used as well, since it makes room for top illumination through a standard condenser. In addition to the filter cube (dichroic mirror and notch filters), a cutoff filter is required to filter out the light coming from the laser that is employed to align the cantilever tip. However, the main drawback of such a setup is the large background signal. The laser spot is large compared to the size of the tip, and other areas around the probe may be illuminated and scatter light in turn, which interferes with the near-field signal. To circumvent this, the tip-to-sample distance is set to vary sinusoidally (in other words, it is set to modulate) by vertically oscillating the probe at its resonance frequency or at any of its harmonics. This in turn modulates the near-field signal, but not as much as the background signal. A lock-in amplifier is then used to filter out the near-field signal if set either at the resonance frequency of the cantilever or at any of its higher harmonics.

More than Near-Field Fluorescence Imaging

One of the great potentials of fluorescence SNOM is that it is also possible to obtain emission or excitation spectra from localized regions of the sample by placing a spectrograph as detector. Betzig suggested the need to acquire such spectra in both collection and illumination modes, since these operating modes are not symmetric and hence the spectra are not identical (Betzig and Trautman 1992). On the other hand, special care must be taken to ensure that the near-field spectra are free from artifacts caused by either the presence of the near-field probe, the metal coating, or the feedback conditions (Dunn 1999).

Apart from fluorescence spectroscopy, time-resolved fluorescence SNOM allows one to obtain fluorescence lifetimes as a function of the sample position. In this case, the setup is modified by replacing the illumination source by a continuous-wave laser that generates light pulses hundreds of picoseconds wide. Instead of notch filters and avalanche photocounting detectors, the pulsed emission goes through a monochromator that filters out the excitation light before reaching a time-correlated single-photon counting (TCSPC) detector. The TCSPC system is able to detect the time evolution of the fluorescence decay with a resolution of a few tens of picoseconds or less (Ambrose et al. 1994). However, it has been found that lifetimes may be affected if the fluorescent dyes are proximal to the metal edges of aperture probes. Indeed, there have been studies to investigate the change in fluorescence lifetime of molecules as they travel across the illuminated area of an aperture probe or as the latter scans across them. In some cases, lifetimes are found to decrease when the molecule is near the edge of the illuminated area, or in other words, when the molecule faces the metal edge of the aperture. Lifetimes are thus maximal when the molecules face the center of the aperture, and the effect at the edges can be explained in terms of fluorescence quenching induced by the metal edge. However, in other case studies, the opposite trend has been observed, which rules out fluorescence quenching as a plausible cause.

Recently the feasibility of combining fluorescence correlation spectroscopy (FCS) with SNOM to study the diffusion of single molecules in membranes has been demonstrated (Vobornik et al. 2009). FCS mainly profits from the near-field probes, since they are capable of illuminating very small volumes. In this case, the setup is equipped with a multi-tau correlator placed after the avalanche photodiode, as seen in Figures 3.18 and 3.20.

Photobleaching or Background—Despite Weak and Highly Localized Illumination

Weak and highly localized illumination may not be enough to prevent substantial photobleaching or background in SNOM. There have been some attempts to further diminish the risk of photobleaching and background noise that are based on two physical phenomena related to fluorescence: fluorescence resonance energy transfer (FRET) and two-photon excitation.

(F)RET-SNOM

As mentioned in the preceding section, RET and FRET rely on the presence of a donor–acceptor pair. The emission spectrum of the donor overlaps the excitation spectrum of the acceptor in such a way that the donor's radiative decay or fluorescence emission serves as excitation for the acceptor as long as the donor is properly excited. As a result of this energy transfer, the intensity of the donor fluorescence decreases, whereas that of the acceptor increases.

FRET only occurs if the donor and acceptor molecules are sufficiently closed, approximately 1–10 nm, which makes it especially convenient for near-field microscopy, where the probe and sample should be in close proximity. FRET-SNOM can be set in both aperture and apertureless probes (Sekatskii 2004).

If the probe is modified with donor molecules and the sample contains acceptor molecules as shown in Figure 3.21a, FRET occurs only when the SNOM probe is sufficiently close to an acceptor molecule. Under these circumstances, a small sample volume is illuminated with the emitted fluorescence of just a few molecules at the probe, which substantially reduces over-illumination, background noise, and photobleaching. FRET-SNOM is also possible if acceptor molecules are immobilized on the probe and donor molecules are in the sample, as depicted in Figure 3.21b. In this case, the excitation radiation is delivered through the probe to the sample donor,

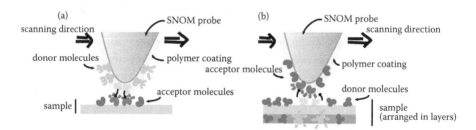

FIGURE 3.21   *(See color insert.)* FRET-SNOM. (a) Donors at the probe, acceptors at the sample; in this case, the donor-coated tip delivers the light of suitable wavelength to excite the donors, but not the acceptors. FRET occurs whenever the donor-coated tip encounters an acceptor molecule at the appropriate distance (1–2 nm), which in turn emits near-field fluorescence; those acceptor molecules that are not contained within the illumination volume or situated farther apart will not fluorescence. (b) Acceptors at the probe, donors at the sample; the acceptor-coated tip delivers the light that excites the donors, and not the acceptors. In this way, FRET occurs only when the tip meets a vicinal donor at the surface. Consequently the tip will emit light at the expense of the donor molecule. This configuration is especially convenient to study the distribution of donors within layer-structured samples from changes in their fluorescence intensity; donors closest to the tip will undergo FRET, which will undermine fluorescence emission and hence appear dimmed, while distant donors will appear brighter.

which may undergo FRET only if the probe containing the acceptor molecules is at the appropriate distance. This configuration is particularly advantageous to study the distribution of fluorescent markers in layer-structured samples. In this case, molecules in layers that are closer to the probe will weakly fluoresce, whereas molecules in the distant layers will fluoresce with higher intensity.

The technique has an important drawback though. There are not donor-acceptor pairs that are photostable enough to withstand the conditions of SNOM measurements (Sekatskii, Dietler, and Letokhov 2008). Continuous illumination at the probe or repetitive scanning at the sample may have an impact on either the donor-acceptor's integrity or the donor-acceptor's fluorescent properties.

Tip-Enhanced Two-Photon Excitation

Using a two-photon excitation source in an apertureless SNOM setup is expected to dramatically increase the signal-to-background ratio by a factor of 10^3 (Sánchez, Novotny, and Xie 1999). The source consists of a mode-locked laser, that is, a laser that produces a stream of pulses with a typical width of 100 femtoseconds and a frequency of about 80 MHz (Denk, Strickler, and Webb 1990). Under these circumstances, the probability that a molecule simultaneously absorbs two photons of different wavelengths becomes appreciable (Lakowicz 2006; Murphy 2001).

OPTICAL MICROSCOPIES: SUMMARY

Table 3.1 compiles the main features of some of the members of the optical microscopy family. We have included those techniques that, to our knowledge, have ever been used in combination with the scanning probe microscope, and we focused on both their performance characteristics and their limitations. Among them, fluorescence SNOM is the only technique that integrates both SPM and near-field optical microscopy.

COMBINED OM-SPM TECHNIQUES: EYESIGHT TO THE BLIND

The implementation of light microscopies to the scanning-probe technique has turned out to be a most profitable combination, especially in the field of biosciences. Indeed, far-field microscopy makes it possible to position the probe onto a particular micro-sized object—such as a particle or a cell—with great ease and precision. The tasks of identification and localization of specific sample constituents have been greatly eased by the recent developments in software that overlay the optical and scanning probe images.

A suitable instrumental configuration requires (a) the creation of an obstacle-free optical path to the SPM probe and (b) that the sample be analyzed without compromising the instruments' stability and quality of performance (Geisse 2009). Additionally, the visible red laser that is usually employed in most SPM-probe detection systems is a potential source of interference and cross-talk with optical data, and it is usually filtered out or replaced with lasers that do not emit in the visible

TABLE 3.1

Summary of Optical Microscopy Techniques Ever Used in Combination with a Scanning Probe Microscope

CHARACTERISTICS	OPTICAL MICROSCOPY		HIGH-PERFORMANCE FLUORESCENCE MODES			
	BRIGHT-FIELD	FLUORESCENCE	FLIM	CLSM	TIRF	SNOM
Lateral resolution		$\lambda/2$		$\lambda/3$		$\lambda/10 - \lambda/50$
Vertical resolution		λ			10^2 nm	50 nm
Sample requirements	refractive/diffracting	fluorescent				
Illumination volume	10^{18} nm^3			$10^8 - 10^7$ nm^3	$10^6 - 10^5$ nm^3	10^5 nm^3
Local/global technique	non local		local (lifetimes)	thin optical sections (1 μm)	thin optical sections (10^2 nm) immediately above the substrate	highly local, point-by-point illumination
Contrast	transmissivity, reflectivity, absorptivity; PhC: n_{object}·n_{medium}; DIC: dn/dr (r: lateral distance)	fluorescence intensity at λ or $\Delta\lambda$	fluorescence lifetime			
Other features	non destructive; subsurface imaging	high contrast (non-fluorescent background); chemically selective: identification of sample components	local physicochemical environment	Z stacks - 3D reconstructions	single molecule detection	single molecule detection & orientation; simultaneous topography & optical scanning; combinable with TIRF, FCS, spectroscopy, two-photon excitation
Limitations	contrast at the expense of resolution and viceversa; diffraction limited; stray light, lense aberrations; chemically non-selective	background fluorescence +; sample photodamage and sample photobleaching +			local information just above the substrate; not beyond	probe optimisation required; probe-induced artifact; poor S/N ratio

non-fluorescent samples: fluorescence dye labelling required range, such as infrared. The cylindrical piezo scanners, if ever used (Henderson and Sakaguchi 1993; Nagao and Dvorak 1998), are positioned above the sample, controlling the probe's movement and thus allowing reflected light or epifluorescence microscopy to be performed below the sample plane with an inverted optical microscope. In these cases, transmitted-light optical microscopy is simply not possible (Henderson and Sakaguchi 1993), but it can be integrated if one uses low-angle illumination with optic fibers (Nagao and Dvorak 1998) or with customized condensers (MFP, Asylum). There are, however, configurations commercially available that consider the use of microscope condensers in their designs and hence provide the required top-down optical access. Often flexural piezo scanners rather than piezoelectric tubes are employed. As for the convenience of a tip-moving or a sample-moving setup, there is no common agreement, since most established commercial setups are based either on one or the other concept. However, for steady-state or time-resolved optical microscopy or when simultaneous optical and scanning probe imaging is required, a tip-scannable configuration would be preferred. On the other hand, tip-enhanced fluorescence SNOM would mostly profit from a sample-moving setup.

Especially advantageous is the configuration based on an inverted optical microscope for transparent samples, as shown in Figure 3.22. As mentioned previously, the probe is mounted above the sample (so that probe scanning is done on this side), whereas the optical image is collected from below, which eases the access of the optical objective to the sample. This enables the use of high numerical aperture objectives with short working distances (below 1 mm). Both bright-field microscopy in transmission or epifluorescence are possible with such a setup. However, the former demands a considerable effort in scanning probe design to accommodate optical condensers of high numerical aperture (0.55 or higher) without compromising the quality of optical imaging or instrument stability. The combination of the scanning probe with an upright microscope is a more recent conception (Mangold et al. 2008), which has considerably extended the application of the combined techniques and epifluorescence to nontransparent substrates and hence to a wider range of materials.

Example A: Combining Scanning Probe Microscopy and Bright-Field Optical Microscopy

Nagao and Dvorak (1998) were among the first to report the combination of an atomic force microscope (AFM) with bright field and epifluorescence to investigate living cells. The work at that time represented a new approach for studies in cell biology and physiology. Samples consisting of COS (fibroblast-like) and BESM (bovine embryonic skin and muscle) cells were measured under physiological conditions by means of an enclosed, home-made cell culture chamber (coined the *Dvorak chamber*, after one of the authors). Cytoplasmic regions of living COS cells were successfully imaged in intermittent-contact mode with lateral and vertical resolutions of 70 and 3 nm, respectively. Figure 3.A1 shows the pioneering setup on an inverted light microscope. Figure 3.A2 shows the bright-field and AFM images of a small region within a single cell; the round-like protuberances are attributed to two *T. gondii* parasites infecting the cell.

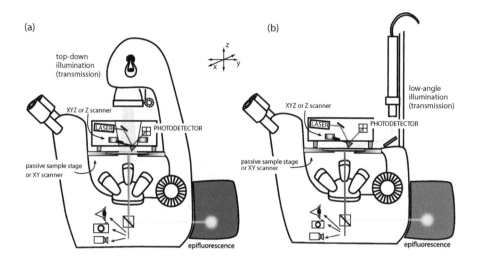

FIGURE 3.22 The tandem atomic force microscope with an optical-fluorescence micro-scope. The AFM is usually a compact module that rests on the microscope stage that holds the sample and can also work as XY scanner in some designs. The AFM module, or AFM head contains the cantilever, the laser position detecting system and either the Z or the XYZ scanner. (a) Design that allows top-down illumination for transmitted light microscopy. (b) Design that allows low-angle illumination for transmitted light microscopy. Epifluorescence is performed from below the sample, just as in conventional fluorescence microscopy. The optical images can be visualized by a camera, a photo camera, or through the oculars.

SPM and Optical Fluorescence Microscopy

From all the SPM-based techniques, the scanning force microscope, or AFM, has found a most convenient partner in fluorescence microscopy. The synergy between both techniques is evident. On the one hand, AFM provides topographical information of the sample surface at a resolution beyond the diffraction limit; however, it lacks the ability to identify similarly sized objects in a complex system or sense objects underneath the surface. On the other hand, fluorescence microscopy can selectively identify sample constituents below the sample surface. AFM probes objects at a cost of their integrity. Samples should be hard enough to resist continuous tip-mediated "plowing" or tapping. On the other hand, fluorescence microscopy can detect features of delicate samples without the risk of mechanical manipulation, but it can cause photodamage.

Some technical issues that are specific to the combination of AFM with fluorescence microscopy must be taken into account. One is the tip luminescence, and the other is synchronizing the illumination source with sample scanning.

Far from being a topic of no importance, one must consider the likely interference of the scanning probe into the fluorescence activity of the sample. A direct spatial correlation between the topographic and optical signals requires that the scanning probe be aligned within the excitation focus of the fluorescence microscope. This imposes a problem if the probe material can be excited and emits light in the same wavelength

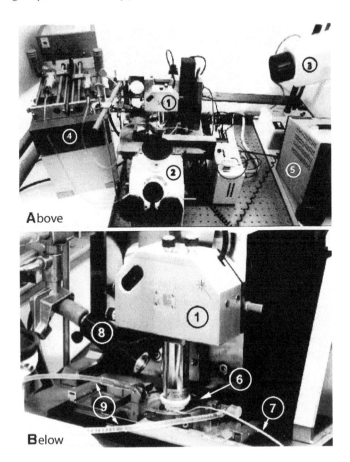

FIGURE 3.A1   Setup used by Nagao and Dvorak to study living cells. Above: overview of the integrated light and atomic force microscope. Below: close-up view of the environmental cell culture chamber. (1) AFM scanner head, (2) inverted light microscope, (3) air stream incubator, (4) infusion/withdrawal pump for perfusion measurements, (5) video monitor, (6) Dvorak-Stotler controlled-environment culture chamber, (7) temperature probe, (8) optic fiber illuminator for transmitted light microscopy, (9) perfusion tubing. (Adapted from Nagao and Dvorak 1998. With permission.)

range as the sample. This occurs mainly in the case of AFM cantilevers and tips made of amorphous Si_3N_4, where the luminescence covers a broad spectral region when illuminated with light of 800-nm wavelength, as seen in Figure 3.23. Contrarily, tips made of crystalline silicon are not luminescent and therefore are incapable of interfering with the sample's optical signal. Therefore it is convenient in these cases to use crystalline silicon probes or other low-luminescent materials.

Excitation must also be handled with care, especially in the case of those fluorescence techniques that require sample scanning, such as confocal microscopy. Confocal microscopy has lower lateral resolution than scanning-probe microscopy, and therefore the pixel number that comprises a confocal image is much lower than that of an SPM image. If simultaneous confocal and probe scanning were to be

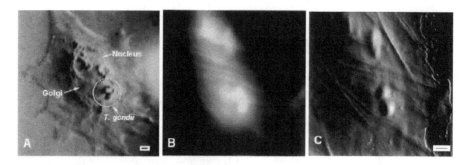

FIGURE 3.A2 (A) Bright field optical image; (B) AFM topography image; (C) amplitude (error signal) image of a *T. gondii*-infected BESM cell. The positions of two *T-gondii* parasites are depicted in circles on the image (a). The scale bars correspond to 10 mm. (Reprinted from Nagao and Dvorak 1998. With permission.)

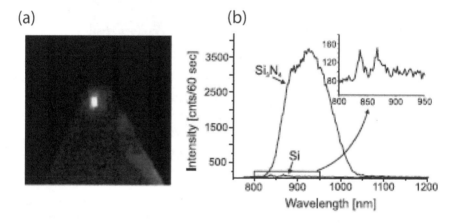

FIGURE 3.23 Emission luminescence of Si_3N_4 and Si tips. (a) Luminescence image of a cantilever made of Si_3N_4. (b) Emission luminescence spectrum of Si and Si_3N_4 tips. (Both figures reprinted from Kassies et al. 2005. With permission.)

done at the resolution of the SPM technique, the confocal image would be greatly oversampled at the expense of the sample's integrity, which could suffer from photobleaching due to overexposure. One way to minimize sample exposure to the excitation light consists of periodically illuminating the sample for a short period of time while it is being scanned. This is achieved through *gated excitation*, where the excitation source is switched off and on a defined number of times during a line scanning and before and after a complete frame scan. In this way, the illumination will be grossly distributed among a few pixels, binning the signal for the confocal image.

Ever since it was conceived in the early 1990s, the combined instrumentation has found its major applications in the field of bioscience, where it is used as an essential tool in the identification and characterization of cell organelles, cell membranes, and the distribution of cell receptors within (Frankel et al. 2006; Doak et al. 2008) or in the localization and characterization of ordered phases in mixed lipid layers (Shaw et al. 2006). These last examples account for *view&view* applications, so to say, where both AFM and fluorescence microscopy are used as imaging tools. *Change&view* applications refer to those where the AFM tip is used as a sample manipulator and fluorescence microscopy is used to visualize the result. Tip-induced injection of particles or molecules delivered by carbon nanotubes into living cells (Chen et al. 2007) and tip-induced stretching, dissection, or collection of fluorescent DNA strands (Hards et al. 2005; Peng et al. 2007) are among the most innovative applications in this category.

Example B: Combining Scanning Probe Microscopy and Fluorescence Microscopy

The morphology and structure of neuronal *Aplysia* bag cell growth cones has been studied using a combined atomic force and fluorescence microscope (Grzywa et al. 2007). The growth cones are fanlike shapes at the very tip of nerve cells that form part of axons and dendrites. They can be divided in different regions. The peripheral (P) domain consists of flat lamellipodia that are subdivided by radial filopodial F-actin bundles. The central (C) domain contains microtubules and vesicles of various sizes, and the transition (T) domain is the region between the P and C domains. The structure, which is shown in Figure 3.B1, develops to form synaptic contacts with other cells to constitute the neuronal network. Whereas AFM provides information about cell heights of the different areas of the growth cone, fluorescence microscopy is used to estimate the volume of cellular compartments beneath the surface.

Figure 3.B2 compares the bright-field, the scanning-probe, and the fluorescence images of a single neuronal cell to correlate volume distribution with topology. Though the fluorescence signal is noisier and is affected by an intensity gradient from the nearby, highly fluorescent cell nucleus, both topography and fluorescence signals fairly match each other.

Example C: Combining Scanning Probe Microscopy with Fluorescence Lifetime Imaging

Micic and coworkers (2004) studied cell polarity through the distribution of a fluorescent fusion protein, SO0584-YFP, in single *S. oneidensis* bacteria (Figure 3.C1). Using a combined confocal optical and atomic force microscope setup, they simultaneously obtained topography and fluorescence data that revealed the distribution of the fusion protein throughout the whole organism (Figure 3.C2). They found that fluorescence intensity and lifetime were enhanced at one or both ends of the bacteria body, where the protein tends to accumulate. Lifetime imaging was also performed to create a point-by-point map of the environment inside the bacterium.

Example D: Combining Scanning Probe Microscopy with Confocal Laser Scanning Optical Microscopy (CLSM)

Pelling and coworkers (2009) employed a combined atomic force microscope and a laser scanning confocal microscope to study the mechanical and morphological

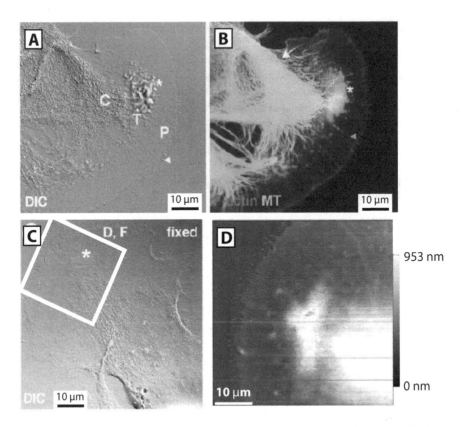

FIGURE 3.B1 *(See color insert.)* Structure of growth cones in an *Aplysia* bag cell shown in optical microscopy. (A) DIC optical micrograph and (B) the corresponding fluorescent image of the structure. Microtubules are dyed in green, whereas F-actin appears in red. Optical (C) and scanning probe (D) images of the same cell. (C) The DIC optical micrograph of a fixed cell; image (D) is the AFM image that has been produced by scanning the box area of the optical image. (Adapted from Grzywa et al. 2007. With permission.)

changes that occur in cells during apoptosis (programmed cell death). The cell targets in this study were human neonatal foreskin fibroblasts transiently expressing a series of fluorescent fusion proteins. The study focused on the transformations undergone by the cell cytoarchitecture (F-actin, microtubules [MT], and intermediate filaments [IF]) before and after triggering a caspase-dependent apoptosis with a chemical agent (staurosporine or STS). Additionally, they monitored the cytoskeleton reorganization in both apoptotic (i.e., STS-treated) and nonapoptotic (i.e., non-STS-treated) cells after the application of a localized deformation with an AFM cantilever tip.

It was found that, during the first stages of apoptosis, the actin fibers transform into actin monomers, partly disintegrating the F-actin network, which adopts a stellate distribution. Though both the MTs and IFs remain intact, their respective networks collapse around the nucleus, which in turn degrades as well. As a result, the cells become softer.

Z-stacks of confocal optical images of the cytoarchitecture were acquired before and after a localized compressive force was applied for 60 s (Figure 3.D1).

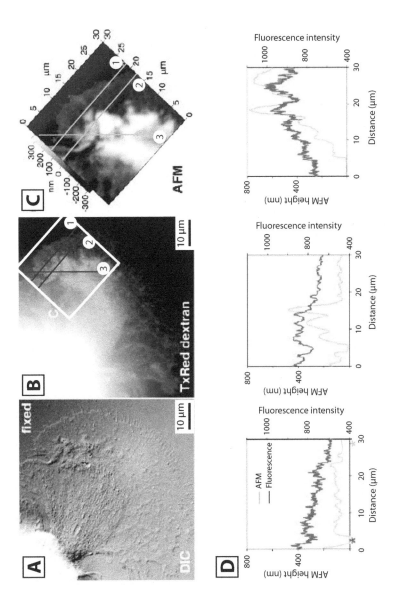

FIGURE 3.B2 Comparison of fluorescence and AFM imaging. (A) DIC image of a fixed neuron cell growth cone 1 h after the injection of fluorescent dextran, which acts here as a volume marker. (B) Fluorescence image of the same growth cone reveals an intensity distribution that corresponds to different volumes of cell compartments. The boxed area (C) shows the AFM-scanned region. In it, three scan lines are depicted that are compared with their corresponding fluorescence profiles in (D). (Adapted from Grzywa et al. 2007. With permission.)

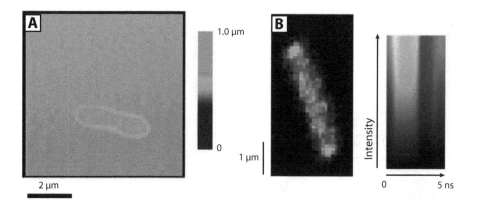

FIGURE 3.C1 *(See color insert.)* (A) Topography image of a single *S. oneidensis* bacterial cell on an agarose gel surface (the agarose keeps the bacteria alive in a dry environment). (B) Fluorescence lifetime image of the same kind of bacterium in solution. In both images, cell polarity is clearly shown, either as protuberance or as local increase of the fluorescence lifetime. (Reprinted from Micic et al. 2004. With permission.)

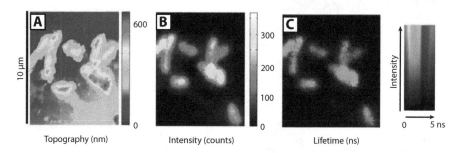

FIGURE 3.C2 *(See color insert.)* *S. oneidensis* bacterial cells on poly-l-lysine expressing the fluorescent fusion protein. Topography (A), fluorescence intensity (B), and fluorescence lifetime (C) images of the same group of cells. (Reprinted from Micic et al. 2004. With permission.)

The difference between these stacks produces subtraction images that illustrate the relative organization of the cytoskeleton due to the application of force. In nonapoptotic cells, the actin network does not deform, while MTs and IFs do displace—although in a nonisotropic manner—to and away from the point of force. In apoptotic cells, the reorganization is concentrated around the nuclear region where the tip is positioned. In this case, the monomeric actin displaces down and away from the tip, whereas the MTs and IFs deform isotropically out and away from the tip. These results are consistent with the modern mechanical picture of the cell in which a cortical elastic network made of F-actin surrounds a visco-elastic cytoplasm. The study illustrates that cell mechanics are determined by the different cytoskeletal components that, in turn, undergo substantial changes as apoptosis progresses.

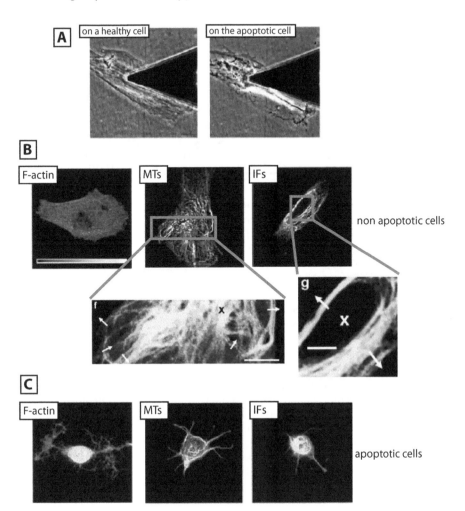

FIGURE 3.D1 *(See color insert.)* (A) Phase contrast micrographs showing an AFM cantilever exerting force on a healthy cell (left) and on the very same cell after 120-min treatment with the apoptotic agent STS (right). (B) Subtraction images of healthy cells on being deformed with an AFM cantilever. Image contrast originates from the relative displacement of the cytoskeleton fibers caused by the deformation. The F-actin image has low contrast, which means that they practically do not move in the process. However, they do displace MTs and IFs. The insets are overlapped images of the same optical plane before (in green) and after (in red) local deformation. Green-red stripes denote fiber displacements, and they are depicted by arrows. Not all the fibers move away from the point of force (marked as "X"). (C) Subtraction images of apoptotic cells after 120-min treatment with STS. The actin monomers concentrate around the nuclear region pressed by the AFM tip. MTs and IFs reveal maximum contrast around the nuclear region, which indicates that they move away from the point of force. (Reprinted from Pelling et al. 2009. With permission.)

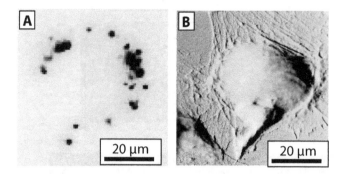

FIGURE 3.E1 Topography (A) and TIRF (B) images of an endothelial cell (HUVEC). Image (A) shows the cytoskeleton of the cell and the details of the apical membrane, whereas image (B) shows the distribution of the focal contacts of the basal membrane. (Reprinted from Mathur, Truskey, and Reichert 2000. With permission.)

Example E: Combining Scanning Probe Microscopy with Total Internal Reflection Fluorescence (TIRF)

The transmission of stress from the apical to the basal membrane in endothelial cells was studied with an integrated setup consisting of an atomic force microscope and a total internal reflection fluorescence microscope (Mathur, Truskey, and Reichert 2000). Localized forces in the range 0.3–0.5 nN were applied with the AFM cantilever tip on a particular region on the apical membrane of single human umbilical vein endothelial cells (HUVECs). The effect of these localized forces was studied using TIRF, which allowed the researchers to visualize the amount and distribution of focal adhesion points between the basal membrane of the cell and the substrate (Figure 3.E1).

The applied stress on cells, though localized, results in a global rearrangement of focal contacts at the basal membrane, as illustrated in Figure 3.E2. In all cases, a displacement of focal adhesion points was observed, with an overall increase of contact area after the force was released. However, this behavior was more reproducible when the force was applied above the nucleus than when the force was applied on the cell's periphery.

FLUORESCENCE SNOM AND SINGLE-MOLECULE DETECTION

Through the history of fluorescence SNOM, single-molecule detection has always served a dual purpose. On the one hand, single fluorescent molecules have constituted the ultimate test of novel instrumental configurations where important issues such as probe confinement, lateral resolution, and reliability need to be elucidated. On the other hand, and especially in the field of soft matter, they have been objects of study as tracers of other molecules or of few-molecule aggregates within complex supramolecular structures.

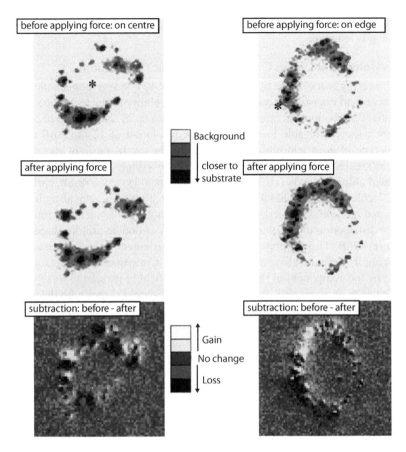

FIGURE 3.E2 Distribution of focal adhesion points before and after the application of a 0.35-nN force near the nucleus (left column) and on the cell's edge (right column). The first-row images are taken before the force was applied (the application point is marked with an asterisk). The second-row images are taken 5 min after the force was released. The last-row images are the corresponding subtraction images. Image contrast in this case is indicative of the presence (gain) and absence (loss) of the focal contacts and hence their lateral displacement. (Reprinted from Mathur, Truskey, and Reichert 2000. With permission.)

Single Fluorescent Molecules as Test Samples

Structurally simple and easy to make, the samples usually consist of highly dispersed fluorescent dyes in a matrix. The molecules are either embedded within or adsorbed onto polymer films, which act as adhesion promoters and, in some cases, quantum yield enhancers (Betzig and Chichester 1993). These types of samples have been used to show proofs-of-concept in SNOM, such as single-molecule diffusion in combination with fluorescence correlation spectroscopy (FCS, Vobornik et al. 2009) or with sequential SNOM imaging, single-molecule photoluminescence spectroscopy, or single-molecule fluorescence decay in nano- and mesostructured materials

(Dunn 1999). We will focus on single-molecule imaging and its potential to obtain molecular orientations.

Single-Molecule Imaging: Obtaining Molecular Orientations

The pioneering work of Betzig and Chichester (1993) on single-molecule detection using near-field microscopy established the applicability of the technique to unravel the spatial distribution and orientation of single molecules. Using tapered optic fibers as probes, they were able to attain a lateral resolution of 12 nm, still unequaled for this type of near-field probes. In this study, the excitation light was set at various polarization states, all oriented parallel to the sample plane: random (circular), orientated along one axis (x), or along the orthogonal (y). As a result, carbocyanine dye molecules were individually localized in optical images, appearing in different shapes and intensities. Indeed, the molecules did not show as identical round points, but as a distribution of circular, elliptical, arclike, or double-arclike shapes, as seen in Figure 3.24B. The variability of shapes was attributed to the fact that molecules were spatially oriented along different directions. Consequently, each molecule was excited by the component of the electromagnetic field in parallel to its dipole direction. To test their hypothesis, they compared the observed shapes with the theoretical predictions of their model based on dipole interactions. The calculations produced a collection of shapes that represented the distribution of the electromagnetic field along defined directions as a function of tip–sample distance. The results are graphically displayed in Figure 3.24A. Shape assignment to a particular dipole orientation was done by direct comparison of the experimental and the calculated data patterns and later used to reconstruct the molecules' orientations, which are shown in Figure 3.24C.

The approach of Betzig and Chichester (1993) has been followed by several groups. Molenda and coworkers (2005) calculated the dipole orientations of single terrylene diimide (TDI) molecules, imaged with a triangular aperture, as a function of the excitation polarization. The results are displayed in Figure 3.25. The shapes in the fluorescence SNOM images were compared with the theoretical predictions to assign dipole orientations with an accuracy better than 10° and lateral resolutions of 30 nm (Figure 3.25b). An improvement of the method was achieved by Frey and coworkers (2004), who employed scattering-type SNOM. In this work, tip-on-aperture probes were used to detect the orientation of cyanine dye (Cy-3) molecules attached at the termini of DNA strands with a lateral resolution of 10 nm. Applying the same model of previous works, symmetric double-lobed shapes were attributed to Cy-3 molecules lying parallel to the substrate plane, while asymmetric double-lobed shapes denoted certain molecular inclination. Molenda et al. additionally observed ring- and moonlike shapes that were not predicted by theory and attributed to effects caused by tip-induced fluorescence quenching.

Although this technique is quite valuable in terms of the information it already provides, truly molecular resolution remains a challenge in single-molecule imaging. Achieving lateral resolutions beyond 10 nm can only be done by producing sharper metal tips, which will in turn decrease the ever-present risk of quenching.

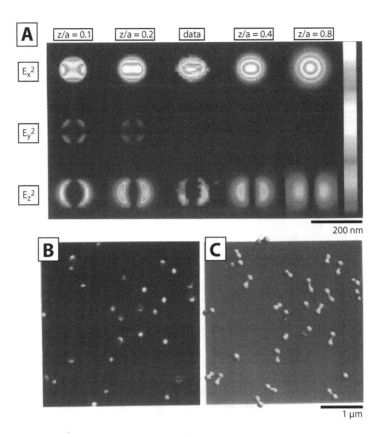

FIGURE 3.24 *(See color insert.)* Single-molecule detection with fluorescent SNOM. Assignment of molecular orientation. (A) Theoretical derivation of the distribution of electrical field components for a molecule at different distance between the latter and the SNOM probe (z normalized to the aperture diameter, a). Comparison with experimental data corresponding to a single molecule. (B) Experimental SNOM image with fluorescent spots corresponding to single molecules adopting different orientations. (C) Theoretically derived image where the molecular orientations are assigned to each spot of the measured image. (All figures reprinted from Betzig and Chichester 1993. With permission.)

Conformations of Single-Polymer Chains in Films

Fluorescence SNOM appears to be unequaled in ascertaining the conformation of single-polymer chains embedded in films. Other scanning probe microscopies such as AFM, though providing the right resolution, proved to be insufficient to selectively distinguish individual molecules out of many. Likewise, standard fluorescence microscopy, though specific as it may be, is restricted by the diffraction limit, and single-molecule detection is clearly out of reach.

The studies using fluorescence SNOM have mainly focused on the effect of two-dimensional confinement on the spatial conformation of single polymer chains; to

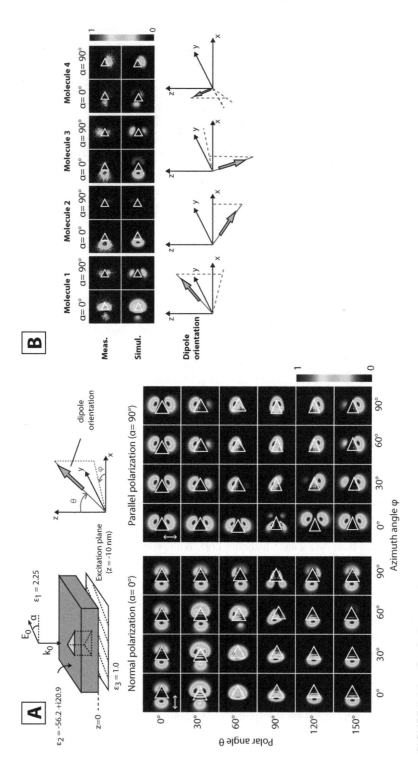

FIGURE 3.25 *(See facing page.)*

this aim, several ingenious approaches have been taken, all coming from a Japanese group in Kyoto University. Aoki and coworkers studied the conformations of single, fluorescently labeled poly(isobutyl-methacrylate) (PiBMA) chains in monomolecular thick films composed of the same polymer (Aoki, Anryu, and Ito 2005). Changing the molecular weight of the polymer had tremendous impact on the conformation of the fluorescent molecules. Ube and coworkers (2007) calculated the molecular strain of single poly(methyl methacrylate) (PMMA) chains in uniaxially stretched films by comparing the dimensions of the fluorescent spots in nonstretched and stretched samples. Phase-separated copolymer films, where the respective copolymer blocks are arranged in distinct nanodomains, offer a much more restrictive spatial confinement. This was readily exploited by Sekine and coworkers to study the chain conformation and chain-end spatial distribution of PiBMA blocks in PiBMA rich domains of just a few hundreds of nanometers width (Sekine, Aoki, and Ito 2009a, 2009b). Some of the results are displayed in Figures 3.26a–c.

Single-Molecule Diffusion

Low-diffusional processes, such as the translation and rotation of individual rhodamine-6G and carbocyanine (diC18) dyes in polymer films, can be tracked down by sequential SNOM imaging (Bopp et al. 1996). In these cases, the same area of a polymer film doped with dye molecules is repetitively scanned after a constant time interval of a few minutes. For translational diffusion, the trajectories of each molecule are obtained by registering the position of each molecule in all the frame images. From a considerable number of trajectories, the statistical treatment yields a mean square displacement and an average diffusion constant. In this way, both the directionality and the extent of diffusion of the process can be characterized. In the case of rhodamine-6G in films of polyvinylbutyral (PVB), the mean displacement suggested random-walk behavior (isotropic diffusion), although some of the molecules exhibited high directionality in their displacements. The diffusion constant was calculated to be 2.6×10^{-15} cm^2/s. The rotational diffusion is revealed by changes in the orientation of the transition dipole associated with the molecule, which can be tracked down by sequential polarization measurements at 0 and 90°. The time-dependent changes in intensity experienced by a molecule in both parallel- and cross-polarizer configurations are interpreted as changes in the orientation of the transition dipole and, hence, as molecular rotation.

For much faster diffusional processes, the combination of SNOM and fluorescence correlation spectroscopy (FCS) is most appropriate (Vobornik et al. 2009).

FIGURE 3.25 *(See color insert.)* Molecular orientations with a triangular aperture. (a) *Upper part*: Model that simulates the intensity patterns of a triangular aperture, illuminated at normal incidence with a plane wave that can be perpendicularly ($\alpha = 0°$) or vertically ($\alpha = 90°$) polarized. The spatial orientation of the absorption dipole moment can be characterized by two angles, the polar angle θ and the azimuth angle φ. *Lower part*: Orientation maps showing the distribution of fluorescent intensities experienced by a single molecule at different values of the pair (θ, φ). (b) Comparison between measured data and simulation data and assignment of the dipole orientation for four different molecules. (All figures adapted from Molenda et al. 2005. With permission.)

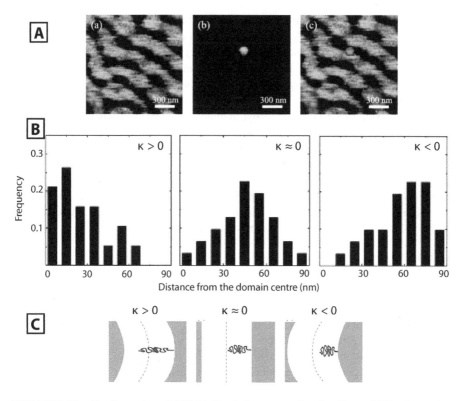

FIGURE 3.26 Conformation of PiBMA blocks in monomolecular films of diblock copolymers. (a) From left to right: topography, fluorescence, and overlay images of a film of diblock copolymer. The fluorescent spot represents a single copolymer block of PiBMA. The topography image shows a meander-like structure composed of higher and lower (PiBMA-rich) stripe domains. From the spatial distribution of fluorescent intensity, the authors calculated the position of the center of mass of the molecule in the PiBMA-rich domain. (b) Position of center of mass with respect to the domain center when the copolymer chain is located in linear stripes and in curved stripes (k is the curvature). (c) The results are interpreted as follows. The copolymer chain adopts a rather elongated conformation in stripes with positive curvature and a rather compressed one in stripes with negative curvature. (All figures reprinted from Sekine, Aoki, and Ito 2009. With permission.)

The configuration has recently been tested to obtain the diffusion coefficient of individual fluorescently labeled lipids in lipid bilayers.

Distribution of Molecules and Molecular Complexes of Biological Interest in Synthetic and Native Membranes

Life sciences, and especially cell biology, have particularly benefited from the imaging capabilities of fluorescence SNOM (de Lange et al. 2001; Dickenson et al. 2010). Indeed, the technique has greatly contributed in unveiling the localization and distribution of the molecules, molecular complexes, and small aggregates that play key roles in important cellular processes, such as signal transduction, cell adhesion, and molecular trafficking.

Colocalization studies of two of such aggregates thought to jointly regulate a certain biological process have also been frequently done to verify the link between spatial distribution and concerted function (Burgos et al. 2003; Dickenson et al. 2010). One of the most extensively studied model systems involves phase-separated lipid monolayers and bilayers as elementary constituents of cell membranes (Yuan and Johnston 2002; Ianoul et al. 2003). Of particular biological interest is the distribution of the raft marker glycolipid GM1 in supported phospholipid monolayers (Burgos et al. 2003; Abulrob et al. 2008), shown in Figures 3.27A–B. Lipid rafts are believed to be highly ordered lipid nano-aggregates that have been postulated to be crucial in early stages of cell signal transduction, though no experimental evidence of their existence in vitro has been found so far. The approach based on fluorescence SNOM to detect and identify lipid rafts is promising due to the high lateral resolution of the technique; however, its limited time resolution may be insufficient for use under in vitro conditions, since lipid rafts are believed to be highly dynamic.

Nuclear pore complexes (NPCs) are ordered protein clusters embedded in the double membrane of the cell nucleus that regulate molecular transport in and out of the nucleus. NPCs on native membranes have been particularly suitable to be investigated by fluorescence SNOM (Höppener et al. 2003; Dickenson et al. 2010), as shown in Figures 3.27C–E. Analogous studies on the distribution of membrane proteins in native membranes have involved cellular markers on T-cell membranes as an active part of: the immune response (Chen et al. 2008; Hu et al. 2009); c-type lecithin DC-SIGN on membranes of dendritic cells (Koopman 2004); calcium pumps integrated in the membranes of erythrocytes and cardiac myocytes (Höppener and Novotny 2008; Ianoul et al. 2004); and NADPH-oxidases in human hematopoietic stem cells (Frassanito et al. 2008).

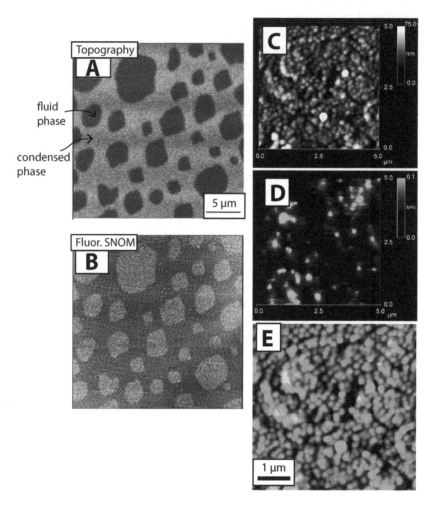

FIGURE 3.27 *(See color insert.)* Detection of lipid domains (A–B) and colocalization of nuclear pore complexes with regulating receptors (C–E). Figures (A) and (B) show topography and fluorescence SNOM images, respectively, of a phase-separated 1:1:1 sphingomyelin/cholesterol/dioleoylphosphatidylcholine (DOPC) lipid monolayer. The condensed phases appear higher in the topography image compared with the lower fluid phases. Fluorescence is brought about via a fluorescence marker (1,2-dipalmitoyl-sn-glycero-3-phosphoethanolamine-fluorescein, Fl-DPPE), which tends to accumulate in the fluid phase. Figures (C), (D), and (E) respectively show the topography, fluorescence, and overlay images of nuclear pore complexes (NPCs) and fluorescently marked IP_3 receptors. These receptors (appearing in (D) and as red spots in (E)) are said to regulate the activity of (NPCs) (appearing in (A) and as green spots in (E)). These studies were aimed to explore the connectivity between function regulation and spatial colocalization. (Figures (A and B) reprinted from Burgos et al. 2003. With permission. Figures (C, D, and E) reprinted from Dickenson et al. 2010. With permission.)

REFERENCES

Abulrob, A., Z. Lu, E. Brunette, D. Pulla, D. Stanimirovic, and L. J. Johnston. 2008. Near-field scanning optical microscopy detects nanoscale glycolipid domains in the plasma membrane. *J. Microsc.* 232:225–34.

Ambrose, W. P., P. M. Goodwin, J. C. Martin, and R. A. Keller. 1994. Alterations of single molecule fluorescence lifetimes in near-field optical microscopy. *Science* 265:364–67.

Anger, P., P. Bharadwaj, and L. Novotny. 2006. Enhancement and quenching of single-molecule fluorescence. *Phys. Rev. Lett.* 96:113002.

Aoki, H., M. Anryu, and S. Ito. 2005. Two-dimensional polymers investigated by scanning near-field optical microscopy: Conformation of single polymer chain in monolayer. *Polymer* 46:5896–902.

Axelrod, D., J. Steyer, W. Almers, R. Haugland, Y. Kwano, and R. G. Enders. 2010. Olympus application note. http://www.olympusmicro.com/primer/techniques/tirf/tirfinfo.html

Betzig, E., and R. J. Chichester. 1993. Single molecules observed by near-field scanning optical microscopy. *Science* 262:1422–25.

Betzig, E., and J. K. Trautman. 1992. Near-field optics: Microscopy, spectroscopy, and surface modification beyond the diffraction limit. *Science* 257:189–95.

Bopp, M. A., A. J. Meixner, G. Tarrach, I. Zschokke-Gränacher, and L. Novotny. 1996. Direct imaging single molecule diffusion in a solid polymer host. *Chem. Phys. Lett.* 263:721–26.

Bouhelier, A. 2006. Field-enhanced scanning near-field microscopy. *Microsc. Res. Tech.* 69:563–79.

Burgos, P., C. Yuan, M.-L. Viriot, and L. J. Johnston. 2003. Two-color near-field fluorescence microscopy studies of microdomains ("rafts") in model membranes. *Langmuir* 19:8002–8009.

Cargille Labs. 2012. Immersion Oil Specifications. http://www.cargille.com/immeroilspecs.shtml.

Chen, X., A. Kis, A. Zettl, and C. R. Bertozzi. 2007. A cell nanoinjector based on carbon nanotubes. *Proc. Natl. Acad. Sci. USA* 104:8218–22.

Chen, Y., L. Shao, Z. Ali, J. Cai, and Z. W. Chen. 2008. NSOM/QD-based nanoscale immunofluorescence imaging of antigen-specific T-cell receptor responses during an in vivo clonal Vγ2Vδ2 T-cell expansion. *Blood* 111:4220–32.

De Lange, F., A. Cambi, R. Huijbens, et al. 2001. Cell biology beyond the diffraction limit: Near-field scanning optical microscopy. *J. Cell Sci.* 114: 4153–60.

Denk, W., J. H. Strickler, and W. W. Webb. 1990. Two-photon laser scanning fluorescence microscopy. *Science* 248:73–76.

Dickenson, N. E., K. P. Armendariz, H. A. Huckabay, P. W. Livance, and R. C. Dunn. 2010. Near-field scanning optical microscopy: A tool for nanometric exploration of biological membranes. *Anal. Bioanal. Chem.* 396:31–43.

Doak, S. H., D. Rogers, B. Jones, L. Francis, R. S. Contan, and C. Wright. 2008. High-resolution imaging using a novel atomic force microscope and confocal laser scanning microscope hybrid instrument: Essential sample preparation aspects. *Histochem. Cell Biol.* 130:909–16.

Dunn, R. C. 1999. Near-field scanning optical microscopy. *Chem. Rev.* 99:2891–927.

Frankel, D. J., J. R. Pfeiffer, Z. Surviladze, et al. 2006. Revealing the topography of cellular membrane domains by combined atomic force microscopy/fluorescence imaging. *Biophys. J.* 90:2404–13.

Frassanito, M. C., C. Piccoli, V. Cappozzi, D. Boffoli, A. Tabilio, and N. Capitanio. 2008. Topological organization of NADPH-oxidase in haematopoietic stem cell membrane: Preliminary study by fluorescence near-field optical microscopy. *J. Microsc.* 229:517–24.

Frey, H. G., S. Witt, K. Felderer, and R. Guckenberger. 2004. High-resolution imaging of single fluorescent molecules with the optical near-field of a metal tip. *Phys. Rev. Lett.* 93:200801.

Geisse, N. A. 2009. AFM and combined optical techniques. *Mater. Today* 12:40–45.

Grzywa, E. L., A. C. Lee, G. U. Lee, and D. M. Suter. 2007. High-resolution analysis of neuronal growth cone morphology by comparative atomic force and optical microscopy. *J. Neurobiol.* 66:1529–43.

Hards, A., C. Zhou, M. Seitz, C. Bräuchle, and A. Zumbusch. 2005. Simultaneous AFM manipulation and fluorescence imaging of single DNA strands. *Chemphyschem* 6:534–40.

Hartschuh, A. 2008. Tip-enhanced near-field optical microscopy. *Angew. Chem. Int. Ed.* 47:8178–91.

Haugland, R. P. 2005. *The handbook: A guide to fluorescent probes and labelling technologies.* Eugene, OR: Invitrogen Corp.

Henderson, E., and D. S. Sakaguchi. 1993. Imaging F-actin in fixed glial cells with a combined optical fluorescence/atomic force microscope. *Neuroimage* 1:145–50.

Herman, B. 1998. *Fluorescence microscopy.* Oxford, U.K.: BIOS Scientific Publishers.

Höppener, C., R. Beams, and L. Novotny. 2009. Background suppression in near-field optical imaging. *Nano Lett.* 9:903–8.

Höppener, C., D. Molenda, H. Fuchs, and A. Naber. 2003. Scanning near-field optical microscopy of a cell membrane in liquid. *J. Microsc.* 210:288–93.

Höppener, C., and L. Novotny. 2008. Antenna-based optical imaging of single Ca^{2+} transmembrane proteins in liquids. *Nano Lett.* 8:642–46.

Hu, M., J. Chen, J. Wang, et al. 2009. AFM- and NSOM-based force spectroscopy and distribution analysis of CD69 molecules on human DC4+ T cell membrane. *J. Mol. Recognit.* 22:516–20.

Ianoul, A., P. Burgos, Z. Lu, R. S. Taylor, and J. L. Johnston. 2003. Phase separation in supported phospholipid bilayers visualized by near-field scanning optical microscopy in aqueous solution. *Langmuir* 19:9246–54.

Ianoul, A., M. Street, D. Grant, J. Pezacki, R. S. Taylor, and L. J. Johnston. 2004. Near-field scanning fluorescence microscopy study of ion channel clusters in cardiac myocyte membranes. *Biophys. J.* 87:3525–35.

Kassies, R., K. O. van der Werf, A. Lenferink, C. N. Hunter, J. D. Olsen, V. Subramaniam, and C. Otto. 2005. Combined AFM and confocal fluorescence microscope for applications in bio-nanotechnology. *J. Microsc.* 217:109–16.

Koehler, A. 1993. A new system of illumination for photomicrographic purposes. *Z. Wiss. Mikrosk.* 10:433–40.

Koopman, M. et al. 2004. Near-field scanning optical microscopy in liquid for high resolution single molecule detection on dendritic cells. *FEBS Letters* 573:6–10.

Lakowicz, J. R. 2006. *Principles of fluorescence spectroscopy.* Singapore: Springer.

Mangold S., K. Harneit, T. Rohwerder, G. Claus, and W. Sand. 2008. Novel combination of atomic force microscopy and epifluorescence microscopy for visualization of leaching bacteria on pyrite. *Appl. Environ. Microbiol.* 74:410–15.

Mathur, A. G., G. A. Truskey, and W. M. Reichert. 2000. Atomic force and total internal reflection fluorescence microscopy for the study of force transmission in endothelial cells. *Biophys. J.* 78:1725–35.

Micic, M., D. Hu, Y. D. Suh, G. Newton, M. Romine, and P. Lu. 2004. Correlated atomic force microscopy and fluorescence lifetime imaging of live bacterial cells. *Colloids Surf. B Biointerfaces* 34:205–12.

Molenda, D., G. Colas des Francs, U. C. Fischer, N. Rau, and A. Naber. 2005. High-resolution mapping of the optical near-field components at a triangular nano-aperture. *Opt. Express* 13:10688–96.

Murphy, D. B. 2001. *Fundamentals of light microscopy and electronic imaging.* New York: John Wiley & Sons.

Nagao, E., and J. A. Dvorak. 1998. An integrated approach to the study of living cells by atomic force microscopy. *J. Microsc.* 191:8–19.

Pelling, A. E., F. S. Veraitch, C. P.-K. Chu, C. Mason, and M. A. Horton. 2009. Mechanical dynamics of single cells during early apoptosis. *Cell Motil. Cytoskeleton* 66:409–22.

Peng, L., B. J. Stephens, K. Bonin, R. Cubicciotti, and M. Guthold. 2007. A combined atomic force/fluorescence microscopy technique to select aptamers in a single cycle from a small pool of random oligonucleotides. *Microsc. Res. Tech.* 70:372–81.

Reichman, J. 2000. *Handbook of optical filters for fluorescence microscopy*. Brattleboro: VT: Chroma Technology Corp.

Ross, S. T., S. Schwartz, T. J. Fellers, and M. W. Davidson. 2000. Total internal reflection fluorescence (TIRF) microscopy. http://www.microscopyu.com/articles/fluorescence/tirf/tirfintro.html.

Sánchez, E. J., L. Novotny, and X. S. Xie. 1999. Near-field fluorescence microscopy based on two-photon excitation with metal tips. *Phys. Rev. Lett.* 82:4014–17.

Sekatskii, S. K. 2004. Fluorescence resonance energy transfer scanning near-field optical microscopy. *Philos. Trans. R. Soc. Lond. A* 362:901–19.

Sekatskii, S. K., G. Dietler, and V. S. Letokhov. 2008. Single molecule fluorescence resonance energy transfer scanning near-field optical microscopy. *Chem. Phys. Lett.* 452:220–24.

Sekine, R., H. Aoki, and S. Ito. 2009a. Conformation of single block copolymer chain in two-dimensional microphase-separated structure studied by scanning near-field optical microscopy. *J. Phys. Chem. B* 113:7095–7100.

———. 2009b. Chain end distribution of block copolymer in two-dimensional microphase-separated structure studied by scanning near-field optical microscopy. *J. Phys. Chem. B* 113:12865–69.

Shaw, J. E., R. F. Epand, R. M. Epand, Z. Li, R. Bittman, and C. M. Yip. 2006. Correlated fluorescence-atomic force microscopy of membrane domains: Structure of fluorescence probes determines lipid localization. *Biophys. J.* 90:2170–78.

Ube, T., H. Aoki, S. Ito, J.-I. Horinaka, and T. Takigawa. 2007. Conformation of single PMMA chain in uniaxially stretched film studied by scanning near-field optical microscopy. *Polymer* 48:6221–25.

Valeur, B. 2001. *Molecular fluorescence: Principles and applications*. Weinheim, Germany: Wiley.

Vobornik, D., D. S. Banks, Z. Lu, C. Fradin, R. Taylor, and L. J. Johnston. 2009. Near-field probes provide subdiffraction-limited excitation areas for fluorescence correlation spectroscopy on membranes. *Pure Appl. Chem.* 81:1645–53.

Yuan, C., and L. J. Johnston. 2002. Phase evolution in cholesterol/DPPC monolayers: Atomic force microscopy and near-field scanning optical microscopy studies. *J. Microsc.* 205:136–46.

4 What Brings Scanning Near-Field Optical Microscopy

The Eyes at the Nanoscale

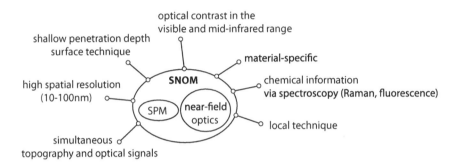

FUNDAMENTALS OF SCANNING NEAR-FIELD OPTICAL MICROSCOPY

We have seen that lateral resolution in conventional optical microscopy is limited by light diffraction. This means that objects smaller than approximately half the wavelength of light ($\lambda/2$) cannot be properly resolved. Near-field optical microscopy overcomes the diffraction limit by collecting the *near-field light*, which does not undergo diffraction, and thus produces optical images with significantly higher lateral resolution, typically in the range $\lambda/10$–$\lambda/50$ (Betzig and Chichester 1993; Hecht et al. 2000; Hartschuh 2008). As a scanning probe technique, scanning near-field optical microscopy (SNOM) can simultaneously provide topological images of objects. In essence, SNOM can in fact be considered a combined technique that encompasses the features of optical and scanning probe microscopies. According to the nature of the scanning probe, the illumination, and the type of optical contrast, SNOM can be classified into a series of derived subtechniques, as seen in Figure 4.1. Those techniques with spectroscopic capabilities, such as tip-enhanced Raman or infrared microscopy as well as fluorescence SNOM, are treated in detail in other chapters.

The history of SNOM starts off in the late 1920s, when E. H. Synge, an Irish scientist, envisioned the possibility of taking optical images of objects with nanometer resolution. His revolutionary ideas were supported by another eminent physicist, Albert Einstein, but they remained unrealizable until the 1970s, when Ash and

Scanning near-field optical microscopy/near-field scanning optical microscopy (SNOM/NSOM)	
Aperture SNOM	Apertureless SNOM

Polarization contrast SNOM
Fluorescence SNOM
 	
    Near-field fluorescence spectroscopy
    Fluorescence resonance energy transfer (FRET) SNOM
	Fluorescence microscopy
	Scanning tunneling optical microscopy
	(STOM)/photon scanning tunneling microscopy (PSTM)
	Near-field Infrared microscopy and spectroscopy
	Tip-enhanced Raman spectroscopy

FIGURE 4.1 The SNOM family. Continuous research in near-field optics has led to a number of techniques derived from both aperture and apertureless SNOM. The techniques share the common principles of near-field optical microscopy and are based on different mechanisms of near-field optical contrast. Fluorescence contrast can be generated in both aperture and apertureless SNOMs, while polarization contrast is only possible in the aperture mode. Raman and infrared contrasts are exploited in the apertureless modes.

Nicholls demonstrated the principle employing microwave radiation. One decade later, two research groups working in parallel extended the application of scanning near-field microscopy to visible wavelengths, borrowing the concepts and parts of the recently developed scanning tunneling microscope. Their work ignited extensive research in both technical development and application of SNOM that continues today. A more detailed historical background is depicted as a timeline in Figure 4.2.

The discussion in the following background section will help the reader to understand the basics of near-field optics.

Background Information

REVISITING THE INTERACTION OF LIGHT AND MATTER

In the previous chapter, we mentioned that light may undergo reflection, refraction, and diffraction in its interaction with matter and that all those physical phenomena together contribute to image formation in optical microscopy. The light in this way transformed by the object is collected by a *distant* objective and transmitted to a visualizing device. That is how *far-field optics* works.

However, this is only one side of the history, and in order to introduce the purpose of near-field optics and its raison d'être, it is necessary to introduce the whole picture of the interaction of light and matter.

Especially clarifying is the description of Courjon and Bainier (1994) that we try to reproduce here. We start by saying that an object can be described as a huge ensemble of oscillating charges and currents distributed in space in a certain manner, which can in turn create or perturb electromagnetic fields. That means that any object has an associated *light field* around it. An object's image is just a two-dimensional distribution of light intensity that represents the light field of the object and *not the object itself*. And that intensity comes from the light that can propagate from the object to the free space and reach the optical detector.

1928 -- E. H. Synge describes an experimental setup to get optical resolution beyond the diffraction limit using a very small light source that illuminates the sample at a very short distance. He proposes the use of a strong light source illuminating the back side of a thin, opaque metal film with a hole of approximately 100 nm diameter. The metal film should be placed no further away from the sample than the hole diameter. He never demonstrates his idea due to the difficulties of its realization.
(Synge, E. 1928. A suggested method for extending the microscopic resolution into the ultramicroscopy region, *Phil. Mag.* 6:356-362)

1956 -- J.A. O'Keefe proposes the principles of near-field imaging, without knowing the publications of Synge.
(O'Keefe, J.A. 1956. Resolving power of visible light, *J. Opt. Soc. Am.* 46:359-359)

A.V. Baez demonstrates near-field imaging at sound frequencies using his finger as object ,that is smaller than the wavelength of sound.
(Baez, A.V. 1956. Is resolving power independent of wavelength possible? An experiment with a sonic "microscope". *J. Opt. Soc. Am.* 46:901-901)

1972 -- E.A. Ash and G. Nicholls demonstrate the concept of near-field microscopy with microwave radiation (wavelength = 3 cm). Resolution of $\lambda/60$ for 1D objects and of $\lambda/20$ for 2D objects
(Ash, E.A., Nicholls. G. 1972. Super-resolution aperture Scanning Microscope. *Nature* 237:510-512)

1984 -- Pohl and Lewis simultaneously report the demonstration of near-field microscopy at visible wavelengths. As sub-wavelegnth light source they use a small aperture at the apex of a sharply pointed probe tip coated with metal. A feedback loop was implemented to keep the probe at a distance of few nanometers while it raster scans the sample.
(Pohl, D.W., Denk, W., Lanz, M. 1984. Optical stethoscopy: Imaging recording with resolution $\lambda/20$. *Appl. Phys. Lett.* 44:651-653)
(Lewis, A., Isaacson, M., Harootunian, A., Muray, A. Development of a 500 Å spatial resolution light microscope. *Ultramicroscopy* 13:227- 231)

1985 -- Wessel sets the basis for tip-enhanced apertureless SNOM. He suggested using nanometric metal particles to locally enhance an electromagnetic field.
(Wessel, J. 1985. Surface-enhanced optical microscopy. *J. Opt. Soc. Am. B* 2:1538-1541)

1993 -- Betzig and Chichester report single molecule detection using aperture SNOM
(Betzig, E., Chichester, R.J. 1993. Single molecules observed by near-field scanning optical microscopy. Science 262:1422-1425)

1994/1995 -- First realizations of SNOM using apertureless probes and interferometric detectors - coined aperture-less SNOM and scanning interferometric apertureless microscope (SIAM), respectively
(Zerhnhausern, F., O'Boyle, M.P., Wickramasinghe, H.K. 1994. Apertureless near-field optical microscopy. *Appl. Phys. Lett.* 65:1623-1625)
(Zernhausern F., Martin, Y., Wickramasinghe, H.K. 1995. Scanning interferometric apertureless microscopy: optical imaging at 10 Angstrom resolution. *Science* 269: 1083-1085)

1999 -- Scattering SNOM in the mid-infrared range
(Knoll, B., Keilmann, F. 1999. Near-field probing of vibrational absorption for chemical microscopy. *Nature* 399:134-137)

FIGURE 4.2   The history of scanning near-field optical microscopy.

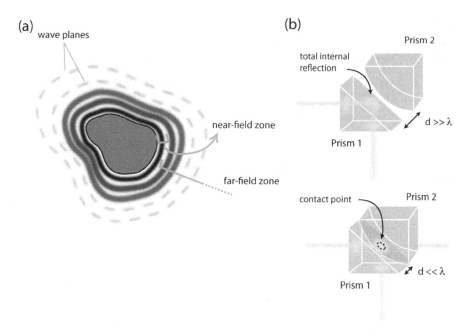

FIGURE 4.3   (a) The distribution of light fields in an object. Two zones can be distinguished: a near-field zone right above the surface of the object that extends a few nanometers away and a far-field zone that extends from the near-field boundary to infinity. The far-field zone mainly consists of propagating fields, whereas the near-field zone comprises evanescent waves, confined in a very limited space. (b) Newton's experiment, which evidences the optical tunneling effect. Light that undergoes total internal reflection on a side of a prism (prism 1) is transmitted to a second prism (prism 2) when the latter is situated sufficiently close (the distance d is small compared to the wavelength λ).

Figure 4.3a sketches the light distribution of an object, where two field regions are distinguished: a region very close to the object's surface, usually extending a few nanometers away, the *near-field* zone, and a *far-field* zone that extends from the near-field zone to infinity. The structure of the near-field is not simple, though, and it is composed of intermingled propagating and nonpropagating components. The latter are intimately connected to the object's material and are confined above its surface; they exist because of the presence of matter, and they rapidly vanish in free space. They are named nonpropagating or *evanescent waves*.

The emission of both propagating and evanescent waves by objects is a universal phenomenon that can be theoretically derived from the electromagnetic theory, regardless of the fact that the object is self-luminous or not. This is expressed in the Wolf-Nieto theorem, which states that light impinging on an object can be converted into propagating *and* evanescent waves. The former are associated with low spatial frequencies of the object (i.e., the coarse structure) and the latter to the high frequencies, the fine structure of the object.

Therefore, to observe the fine structure of an object, it is required to detect the near-field waves, and the way to do it is by applying the *optical tunneling effect*.

THE OPTICAL TUNNELING EFFECT

Three centuries ago, Isaac Newton realized the curious experiment sketched in Figure 4.3b. He observed that a beam illuminating the side of a prism can be totally reflected at an appropriate incidence angle. Starting from this observation, he approached the concave side of a second prism to the first one, expecting to observe light transmitted to the second prism once the prism sides were in contact and thus frustrate total reflection. He did observe light transmission at the point of contact, but to his surprise, he saw that light was also transmitted through those regions where the prism sides were not touching but were slightly separated. Though Newton did not come up with the right explanation at the time, the phenomenon can be explained by stating that, in the immediate vicinity above the surface of the prism where total reflection occurs, there is an evanescent field that can be transformed into a propagating field (and hence into a light beam) through the second prism following the continuity conditions at the interface.

The phenomenon is an example of the optical tunneling effect. A dielectric material (a second prism) immersed in an evanescent field can convert the latter into a propagating field. It is not a quantum mechanical effect, and it can be explained from Maxwell equations of classical electromagnetic theory.

Near-field optics makes use of the optical tunneling effect to detect the evanescent field of objects and observe their fine structure at a resolution beyond the diffraction limit:

- When a light beam impinges on an object, the latter will convert the light into propagating and evanescent fields.
- The evanescent fields do not obey the Rayleigh criterion; they are confined on the object's surface; and they are intimately connected with the fine structure of the object.
- If a light collector is placed a nanometric distance far from the object without touching it, it will in turn transform the evanescent field of the object into a propagating field that can reach the detector.
- Reciprocally, if a light emitter is used instead, the evanescent field created by the emitter will reach the vicinal object, which in turn will convert it into a propagating field.

A crucial part in the near-field microscope is therefore the light collector or the light emitter. In practice, to place such an element at a nanometric distance of the object requires the former to be very small. Standard devices for light detection and focusing (i.e., objective lenses) are not suitable, since they are too big, and the smallest focal distances are far too large compared to the nanometric dimensions demanded here. In the following we will describe the ways available to detect the evanescent fields in near-field microscopy.

OPTICAL FIBERS

An optical fiber is a cylindrical structure composed of a transparent core and a transparent coating, the so-called cladding. Both the core and the cladding are made of dielectric materials. The index of refraction of the cladding is lower than that of the core, so that light transmits along the core of the fiber by total internal reflection.

The nature (i.e., propagating or evanescent) of the near-field light is complex, and it is likely to differ from one SNOM mode to another, which itself constitutes a motive of relevant importance to generate discussion and theoretical modeling (Novotny, Pohl, and Hecht 1995; Alvarez and Xiao 2006a, 2006b). However, it is out of the question that the intensity of the near-field light is weak and rapidly decays with distance from the object, typically within a few hundreds of nanometers. In a conventional optical microscope, both the illumination source and the light collector are typically placed millimeters away from the sample and thus the near-field light goes undetected. At such distances, only the far-field light is detected, and light diffraction governs image formation.

In order to skip diffraction and attain high resolution, it is thus necessary to collect *only* the near-field light coming out of the sample. To do this, three main requirements should be fulfilled that are shared by all scanning near-field microscopes and depicted in Figure 4.4: (1) either the light source or the light collector is a probe of subwavelength dimensions; (2) either the illumination source or the light collector should be placed very close to the object; (3) either the probe or the sample can move back and forth in a *raster* manner to collect the light as a function of sample position. As in confocal laser-scanning optical microscopy, the image is acquired point by point.

Since the intensity of the near-field light varies with the distance to the object, care must be taken in keeping this distance short as well as constant during sample scanning. This is usually attained in a similar manner as in an atomic force microscope; the distance can be precisely set by a piezoelectric actuator whose performance is controlled by a *feedback mechanism*. Typical feedbacks in SNOM are the shear force, tapping mode, ion conductivity, or tunneling. These feedbacks will be briefly explained in a later section.

As mentioned previously, the near-field light is weak, and therefore near-field optical microscopes need to be equipped with highly sensitive detectors. Though optical objectives with high numerical apertures (e.g., NA 1.4) are normally sufficient for the least-demanding applications, single-molecule detection in fluorescence SNOM requires ultrasensitive devices, such as avalanche photodiodes (APDs) and photomultiplier tubes (PMTs).

BACKGROUND SUPPRESSION

On top of this, near-field signals are largely affected by considerable background radiation. Far-field backscattered light coming from the tip itself or from distant regions of the sample outside the measuring volume concomitantly accompanies the weak

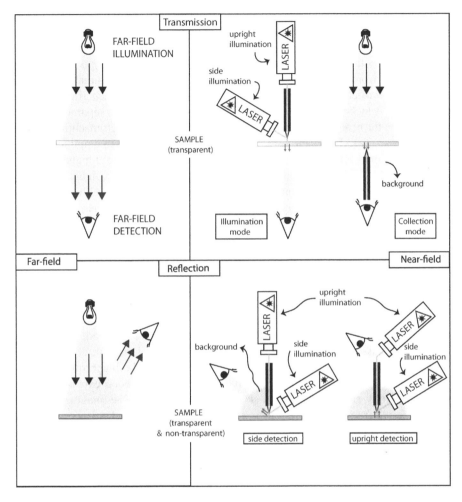

FIGURE 4.4 The far-field and the near-field setups. In far-field microscopy (left-hand side) both illumination and light detection are distant (i.e., millimeters away) from the sample. At such distances, light diffraction occurs, which governs image formation. In near-field microscopy (right-hand side) either the illumination source or the light collector is very close (within nanometers) to the sample. Light diffraction is thus skipped which leads to a significant improvement in lateral resolution. Near-field microscopy can be operated in either transmission or reflection modes. Light can be delivered to the sample through a very small aperture or through a small apertureless probe and collected by an optical objective in the far field in the so-called *illumination mode*. Alternatively, the sample can be illuminated from a distant light source and light can be collected through a nearby aperture in the *collection mode*. Apertureless probes need to be illuminated from an external light source. Focusing a large light spot on a tiny probe often results in large background signals.

near-field signal. The effect is especially acute in a certain mode of SNOM, called apertureless mode, as we will show later. Therefore near-field optical microscopes are equipped with the means to suppress background radiation and to increase the signal-to-noise ratio (Bouhelier 2006; Höppener, Beams, and Novotny 2009 and references within). One of the most common ways consists of varying the tip-to-sample distance in a sinusoidal fashion (modulation). Since the near-field intensity decays rapidly with distance from the sample, a modulation in the tip-to-sample distance produces a modulated near-field signal and a background signal that is not much affected by the modulation. A lock-in amplifier positioned before the detector can separate the background from the near-field signal. In the particular case where micromachined atomic force microscopy (AFM) cantilevers are used as probes, the lock-in amplifier is set to capture the signal at the fundamental resonance frequency (ω) and at higher harmonics (2ω, 3ω, 4ω, ...) of the cantilever. Signal demodulation through higher harmonics results in significant background suppression (Brehm et al. 2005). Other ways have recently been developed, including feedback modulation (Höppener, Beams, and Novotny 2009) and laser optical feedback imaging (LOFI, Blaize et al. 2008).

THE NATURE OF THE SNOM PROBE: APERTURE AND APERTURELESS SNOM

Depending on the nature of the probe, SNOM can be subdivided into two categories: *aperture SNOM*, if the probe consists of an aperture or opening of subwavelength dimensions where light goes through, and *apertureless SNOM*, if the probe lacks an aperture. Aperture SNOM was the first to be conceived and developed, but apertureless SNOM has been gaining relevance in recent years (Hartschuh 2008) and has caught up with its sister technique. In the last two decades, both aperture and apertureless SNOM and their respective imaging modes have been extensively used in diverse fields, ranging from solid-state physics to biology. The imaging modes in SNOM derive from the different near-field optical contrast mechanisms, which are summarized in Figure 4.1.

Aperture SNOM

Figure 4.5 illustrates the most popular mode of aperture SNOM, where the incoming light goes through a small aperture in an *opaque mask* and the near-field light is collected by an objective of high NA. This is the *illumination mode*. The aperture is positioned nanometers away from the sample surface, thereby illuminating a very small region within the sample; the objective, however, is placed millimeters away from the sample, thereby collecting the light in the far field. If the sample

---

FIGURE 4.5 *(See color insert.)* Transmission aperture SNOM. The figure on the left shows a typical SNOM setup mounted on an inverted microscope. Light is delivered through an optic fiber whose open end is positioned in close vicinity to the sample. Light polarization can be set at any value through polarizers at the output of the laser cavity. Transmitted light is collected by an objective of high numerical aperture and delivered to the detection system, usually consisting of photomultiplier tubes (PMTs) of avalanche photodiodes (APDs). In fluorescence SNOM both excitation and emission radiation can be collected by a dichroic mirror and two PMTs; in polarization contrast SNOM an analyzer or a polarizing beam splitter can be placed before the detectors to collect light at different polarization directions. *(continued)*

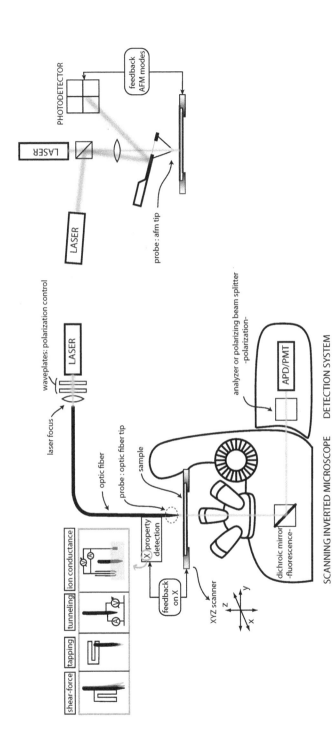

FIGURE 4.5 (*continued*) In the depicted case, the sample is mounted on an *xyz* scanner that keeps the sample at constant distance from the probe at all times and thus assists in acquiring the topological image. This is done by a feedback mechanism, usually focused on a property X, which is distance dependent and is set to a constant value. The left inset illustrates the most popular feedback-based properties. Instead of an optic fiber, it is possible to use a micromachined cantilever with an integrated tip in a typical AFM setup. The green laser is focused on the tip's aperture to create the near-field radiation, while the red laser is focused on the cantilever and reflected to the photodetector. This in turn delivers the signal that actuates the feedback.

lies between the probe and the objective, the SNOM is operating in *transmission*. Operation in *reflection* requires both probe and objective to be positioned on the same side of the sample plane; to avoid interferences caused by the presence of the probe, illumination is set along the normal to the sample plane, and light is collected from one side (see Figure 4.4).

The lateral resolution in aperture SNOM is determined by the size of the aperture; this means that the smaller the aperture, the higher is the resolution. On the other hand, the aperture should let sufficient light through to detect any signal and thus cannot be as small as one would desire. In particular, light transmission through a hole scales with $(a/\lambda)^n$, with $n = 3\text{--}4$ and a representing the aperture diameter (Burgos et al. 2002; Bethe 1944). Therefore a compromise must be taken, and theoretical calculations as well as experiments point to an optimal value of $0.35\ \lambda$ for the aperture diameter.

Apertureless SNOM

With aperture-free probes, the illumination source does not directly act on the sample, but it focuses onto the probe tip, which in turn delivers light onto the sample. "Light" in this context has a wider meaning, comprising any electromagnetic radiation within the visible and infrared range.

The probe in apertureless SNOM is not only a light carrier, but also a light *modifier*. It can increase the intensity of the light coming from the source, but it also can scatter it. The first case relates to the tip-enhanced apertureless SNOM techniques; the second case relates to scattering-type apertureless SNOM.

The probe can be illuminated in several ways, as seen in Figure 4.6. For example, illumination can be carried out from a side, at a certain angle from the surface normal where the probe is positioned. The detector in this case can be placed either beside the probe, as a kind of reflection mode convenient for opaque samples, or opposite the probe tip at the other side of a transparent sample, in transmission. Focusing a large laser spot on a tiny tip is not optimal, though, and this results in a large background signal. An alternative is to illuminate the sample by an evanescent

---

FIGURE 4.6 *(See facing page.)* Setups for apertureless SNOM. Different configurations are possible in combination with an AFM setup where the probe is a cantilever with a sharp tip. In all the depicted cases, the XYZ scanner is attached to the sample. Side illumination is done by focusing the laser on the probe tip with an optical objective placed aside; the near-field signal is collected by the same objective in reflection (as shown) or placed below the sample in transmission (not shown). Alternatively, the tip probe can be illuminated by an evanescent field; in this case, the detection system has to be placed on the side of the probe, either along the tip axis (on axis) or aside. The third option consists of illuminating the tip along the tip axis and detecting the backscattered light by means of focused high-order mode lasers. Detection in apertureless SNOM requires a means to suppress a large background from weak near-field signal. This is normally done by modulating the tip–sample distance with the dither piezo and inserting a lock-in amplifier after a photomultiplier tube or avalanche photodetector. On the other hand, both amplitude and phase of the near-field signal can be detected through an interferometer and a dual channel lock-in amplifier providing a reference signal is available. The reference signal comes from the illuminating source and does not interact with the sample.

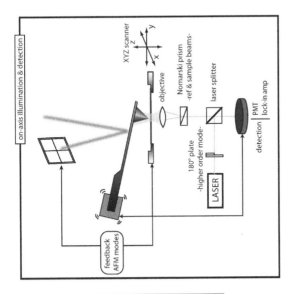

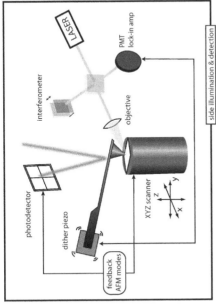

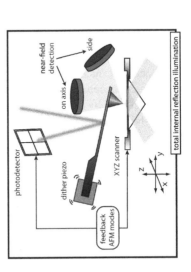

FIGURE 4.6 *(See facing page.)*

field. In this case, the illumination source is placed below the sample at an angle higher than the critical angle so that total internal reflection occurs; the detector can be placed either aside the probe or on top of it. This configuration is often referred to as *scanning tunneling optical microscopy* (STOM). A third option consists of illuminating the probe either from above or below along the probe axis. This on-axis configuration requires the use of highly focused, higher-order laser beams, as we will explain in the following discussion (Bouhelier 2006).

The great advantage of apertureless SNOM is that it can support spectroscopic capabilities, such as vibrational (IR) and Raman spectroscopy. Probe fabrication in this case is considerably easier, since apertures are not needed. In addition, topographical resolution is not compromised as in aperture SNOM, since in this case both topographical and optical resolution depend on tip size, both improving with tip sharpness. Nowadays, micromachined probe technology produces pointed probes with radii of curvature as small as 10–20 nm, and consequently apertureless SNOM can provide better resolution than the apertured counterpart.

Field Enhancement

When an electromagnetic field, E_0, is focused onto a sphere or an ellipsoid whose dimensions are small compared to the field's wavelength, field intensification occurs in the immediate vicinity of the small object, resulting in a field strength, E_s, larger than E_0. This local field enhancement, denoted as $\Gamma = E_s/E_0$, can be caused by the interplay of different mechanisms. One of these mechanisms is *dielectric* (also called *electronic*) *resonance* (Brehm et al. 2005; Bouhelier 2006), which occurs when the material the object is made of has a complex dielectric constant ($\varepsilon = \varepsilon' + i\varepsilon''$) whose real part is negative (–1 to –10) and whose imaginary part is negligible ($\varepsilon'' \approx 0$), i.e., a metal. The effect is especially high in metals that exhibit plasmon resonances in the visible region—like Au or Ag—although dielectric resonance can also occur on metals where no plasmon resonance can occur, such as tungsten (Bouhelier 2006). In the infrared region, the effect is especially high with polar crystals that exhibit phonon resonances, such as SiC or GaAs (Bouhelier 2006).

The other enhancement mechanism is the *lightning-rod effect*, an electrostatic phenomenon caused by confined electric charges in a conductor material with a special shape. The lightning-rod effect has an everyday application in the sharp poles on the building tops that accumulate electric charges and protect buildings from lightning. If the conductor's shape is a sharp nanosized tip, the effect leads to an abrupt increase of the surface charge density in that region, which brings about a localized field enhancement. The distribution of charges on the surface of the tip depends on the polarization of the impinging field, and it has been found that the field strength is maximum and localized at the apex of the tip when the polarization is along the main axis of the tip (Bouhelier 2006).

Antenna resonance comes as the third enhancement mechanism that occurs in nonresonant materials, such as silicon (Bouhelier 2006). The effect is especially relevant when the probe height is an odd multiple of half the wavelength of light (Bouhelier 2006; Hartschuh 2008). The geometry of the tip plays an equally important role in the capacity of the antenna to enhance fields; for example, in the case of an ellipsoid-shaped antenna, the critical parameter is the ratio of the semimajor axis to the radius of curvature. The antenna shape greatly influences the antenna's performance, and it has been found that antenna geometries other than spherical—such as bow-tie, half-wave, or quarter-wave antennas—improve lateral resolution (Höppener, Beams, and Novotny 2009).

Field-enhancement can thus be applied to SNOM if the illuminating electromagnetic field consists of an IR or visible laser beam and the sphere or ellipsoid is, for example, the sharp tip of an AFM cantilever. The technique is called apertureless tip-enhanced SNOM, and it is polarization dependent; optimal field enhancements are achieved only when the polarization of the light used to illuminate the tip is along the tip axis. Side illumination of *p*-polarized light hence provides a straightforward way to achieve the best performance. For upright illumination, however, light propagates along the tip axis, and an ordinary laser beam cannot induce field enhancement, since the fundamental Gaussian mode does not provide any polarization component along the axis of propagation. However, it is still possible to get field enhancement if highly focused, higher-order laser modes are used (Hartschuh 2008). In particular, lasers with Hermite-Gaussian (1,0) modes do possess a nonnegligible electrical field component along the axis of propagation (Novotny, Sánchez, and Xie 1998); they can be easily generated from ordinary laser beams by inserting 180° phase plates in the laser path (Novotny, Sánchez, and Xie 1998).

Scattering-Type SNOM

Scattering-type SNOM has been identified with tip-enhanced SNOM by some authors (Brehm et al. 2005) and clearly distinguished by others (Hartschuh 2008). Here we have chosen to differentiate both modes of apertureless SNOM, since it allows us to introduce a new physical phenomenon that causes near-field contrast: *near-field scattering*. This, in combination with interferometric detection, makes it possible to image the real and imaginary part of an object's refractive index.

Upon illumination, a vibrating tip scatters radiation when it interacts with a surface, and this radiation is detected in the far field with a distant detector. The scattered electric field, E_s, is combined with a reference signal, E_r, in an interferometer, which extracts the amplitude and the phase difference (i.e., the in-phase and out-of-phase components of the scattered light relative to the reference signal). Tip and sample act as coupled dipoles, and the scattered light is related to their respective susceptibilities. Acquisition of the in-phase and out-of-phase near-field scattered light thus makes it possible to simultaneously obtain the real and imaginary parts of the sample's susceptibility and hence the complex refractive index.

THE HISTORY OF SNOM IS THE HISTORY OF ITS PROBES

How are SNOM probes done in practice? Actually, probe manufacture can be considered as the most critical step in SNOM performance. Any breakthrough in probe fabrication has been accompanied by a significant improvement in SNOM lateral resolution and image quality. The requirements for SNOM probes are highly demanding. The probes must have a well-defined geometry, high durability, and reliability; in addition, aperture probes must have high light transmission. Last but not least, the method of fabrication should produce probes with a high degree of reproducibility.

Consequently, attempts to produce SNOM probes that comply with those requirements have been numerous; however, all have their advantages and disadvantages, and research on this topic is ongoing (Chibani et al. 2010; Zhang, Dhawan, and Vo-Dinh 2011).

Aperture Probes

Straight or bent single-mode optical fibers have been mostly used as aperture probes for SNOM (Betzig et al. 1991; Betzig and Trautman 1992). The very end of the optical fiber is turned into a fine tip by heating and pulling (Betzig et al. 1986) or by chemical etching (Jiang et al. 1991). Heating and pulling is a method borrowed from the micropipette technology and consists of heating the fiber with a CO_2 laser while the fiber is being pulled. Chemical etching is performed in either one or two steps and involves the immersion of the fiber in a hydrofluoric acid (HF) solution. The resulting *tapered fiber probes* are then coated with a metal of small skin depth to produce the opaque coating, so that light can only go through the aperture. Aluminum has been the metal of choice due to its small skin depth (about 7 nm at $\lambda = 500$ nm); coating thicknesses lie within 60–100 nm. The aperture at the fiber tip can be done during or after the metallization, and it should be as circular as possible for polarization measurements, as we will see later. In the first case, the aperture is produced if metal deposition is done at an angle relative to the fiber tip. However, this method lacks reproducibility. In addition, the metal edge around the aperture is usually grainy as a result of the evaporation, and it can produce in turn image artifacts, since the distance between the tip and the sample is not constant throughout the aperture area.

A better alternative consists of creating the aperture after metal deposition by focused ion beam (FIB) milling (Figure 4.7). This technology produces smooth and reproducible apertures by directing a focused beam of ions to the probe tip at a right angle to the probe axis. The tip of the fiber is sliced off, leaving an edge-free aperture of usually 50–100 nm in diameter. Adding up the aluminum coating all around the aperture, the overall probe diameter amounts to 250–300 nm. Having a flat, grain-free aperture of such dimensions results in loss of resolution in the topography images, since the SNOM probe acts as a blunt AFM probe. So gain in optical resolution is achieved at the expense of topographical resolution, which constitutes an important drawback for aperture probes.

Other procedures to create nanoapertures with flat rims involve electrolytic demetallization, which consists of immersing the very end of the probe, previously

---

FIGURE 4.7 *(See facing page.)* Aperture SNOM probes. Most typical aperture probes are the tapered optic fibers (left) and the cantilevered AFM tips. Fiber probes have a sharp end produced either by heat and pulling or by chemical etching. The fibers are coated with a metal of small skin depth, such as aluminum, so that light is trapped within the fiber. FIB slicing is used to create a flat-edge circular aperture. (SEM photographs of fiber probes reprinted from Burgos et al. 2002. With permission.) AFM cantilevers with integrated sharp tips are mostly made of SiN or SiO_2 to produce either hollow (shown) or solid tips (not shown), respectively. Aperture is produced by FIB slicing or drilling. Depending on the geometry of the tip (conical, triangular, or rectangular pyramid), the FIB process can create circular, triangular, or rectangular apertures. (SEM photographs of AFM cantilevers reprinted from Jin and Xu 2008. With permission.)

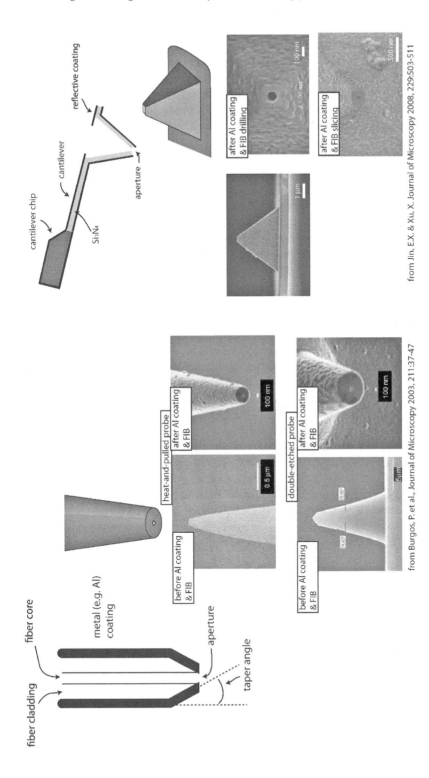

from Jin, E.X. & Xu, X. Journal of Microscopy 2008, 229:503–511

from Burgos, P. et al., Journal of Microscopy 2003, 211:37–47

FIGURE 4.7 (*See facing page.*)

coated with a layer of silver, in either an electrolytic solution or in a solid electrolyte. The probe end is dissolved off, and care must be taken to stop the process at the right time to avoid over-erosion or under-erosion. The process can be precisely controlled if it is integrated in a SNOM device; the probe is maintained in contact with the electrolyte and gradually retracted while the metal dissolution proceeds. The extent of the process can be monitored through the time evolution of the light flux and the current intensity between the probe and a counter electrode immersed in the electrolyte (Bouhelier et al. 2001).

As mentioned previously, light transmission in tapered probes depends on the aperture size. In the case of pulled tapered probes, light transmission increases to the fourth power of the aperture diameter (a^4), while it goes to the third power (a^3) in the case of double-etched probes (Burgos et al. 2002). Light propagates in one or several transverse modes along the fiber as long as the fiber core diameter is large compared to the light wavelength. The taper at the tip of an optical fiber can be depicted as a truncated cone, whose cross section decreases toward the aperture, as shown in Figure 4.7. If the fiber cross section decreases below a certain cutoff value, light of a particular mode either attenuates or becomes evanescent, and it does not propagate any further. For example, the propagating HE_{11} mode in single-mode optical fibers runs into cutoff when the inner diameter of the fiber is 160 nm ($\lambda/3$ at $\lambda =$ 488 nm, Hecht et al. 2000). For a SNOM probe to be efficient, light should reach the aperture, and hence the cut-off diameter (if ever attained) should be as close as possible to the aperture plane. It is thus to be expected that fiber probes with large taper angles will exhibit higher light transmissibility.

A simple way of improving light transmission at the aperture would consist of increasing the laser power coupled to the probe. However, tip heating is a concomitant drawback in aperture probes, which leads to tip damage and hence limits the maximum power that can be delivered to the probe. For aluminum-coated probes, deformation occurs upon elongation of the coating around the tapered end of the probe, which presumably induces thermal stress. The elongation increases with increasing laser power and depends on the light wavelength as much as 20 nm/mW at 780 nm and 100 nm/mW at 1300 nm (Lineau, Richter, and Elsaesser 1996). This greatly alters probe size and causes image artifacts, which makes the probe useless. On the other hand, heating has been found to depend on the taper geometry; Stähelin and coworkers (1996) reported heating coefficients for aluminum-coated probes of various taper angles. The temperature was measured 25 µm away from the probe tip, where maximum energy transfer occurs (La Rosa, Yakobson, and Hallen 1995), as a function of the input laser power. The study revealed that probes with small taper angles were more sensitive to heating (60K/mW) than those with larger taper angles (20K/mW). Laser power on the order of 9.5 mW was enough to induce tip damage. Other works report tip damage with laser input powers of 4.5 mW (La Rosa, Yakobson, and Hallen 1995).

Micromachined probes made of Si or SiN are a convenient alternative to tapered fibers. The fabrication technology allows the probes to be batch produced with high reproducibility and at low cost. In essence, the probes are atomic force microscopy cantilevers with either hollow tips or solid quartz tips with an opening at their apex (Mitsuoka et al. 2000; Kim 2001). The opening is made either by FIB slicing or by simply pressing the tip against a hard surface. On square or triangular pyramid tips, this usually leaves square and triangular apertures, respectively. Circular apertures can be produced on these types of probes or on solid tips if the ion beam is directed along the tip axis; in this case, the ion beam drills the probe to produce a circular hole, as seen in Figure 4.7 (Jin and Xu 2008). The use of these probes allows the implementation of the SNOM technique on an atomic force microscope with minimal modifications in the setup.

As a final note, it is worth mentioning that a circular aperture is not an essential requirement to produce well-performing aperture probes or to make polarization experiments possible. Other aperture geometries have been tried out in aperture SNOM with promising results: A tip with triangular aperture has been used to map dipole orientations of single molecules (Molenda et al. 2005), while large rectangular apertures have been used to scan small objects in a new SNOM variant called differential SNOM, where the resolution is not dictated by the size of the aperture but by the sharpness of the aperture edges (Ozcan et al. 2006).

Apertureless Probes

As mentioned previously, fabrication of apertureless probes is far simpler than aperture probes, since no aperture is required. One convenient approach involves the use of AFM cantilevers with well-defined, integrated sharp tips coated with gold or silver (Bouhelier 2006); chemical etching protocols used in scanning tunneling microscopy may also be "borrowed" to produce metal tips as apertureless probes (Formanek et al. 2006), as shown in Figure 4.8 (left). Alternatively, gold nanoparticles have also been used and found to be efficient field enhancers, either attached to cantilevers (Kim and Leone 2006) or to optical fibers (Kalkbrenner et al. 2001).

Tip-on-Aperture Probes

The combination of an aperture and an apertureless probe, the so-called tip-on-aperture probe, constitutes a great advance in SNOM performance (Frey et al. 2002; Hartschuh 2008). Figure 4.8 (right) shows the appearance of such a probe. The tip, located at the edge of the aperture, efficiently receives the electromagnetic radiation delivered by the optical fiber, producing in turn the field enhancement required for SNOM imaging. The tip-on-aperture probes do not make use of far-field illumination and thus do not suffer from impaired tip focus and large background signals; on the other hand, the presence of the small-sized tip makes high resolution possible. Tips in such probes are regarded as supports for surface plasmons (Frey et al. 2004) or for antenna resonances (Taminiau et al. 2007).

Tables 4.1 and 4.2 summarize the types of probes used in SNOM together with their characteristics, advantages, and limitations. Both aperture and apertureless probes are extensively used in modern SNOM setups.

FEEDBACK MODES IN SNOM

As mentioned previously, one of the basic requirements to attain high resolution in SNOM is that the probe should be kept very close to the sample, either at a fixed position above the sample or at a constant distance (gap width). The tip position or the

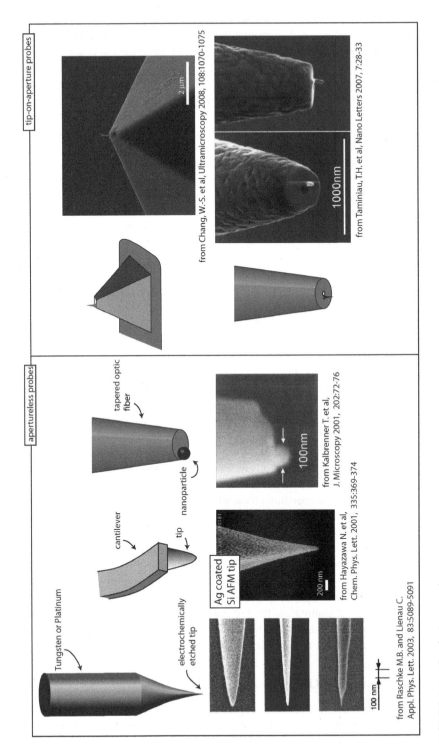

FIGURE 4.8 *(See facing page.)*

tip–sample distance should thus be precisely controlled within nanometers through-out the whole scanning process, including the initial tip-to-sample (sample-to-tip) approach. To accomplish that, several feedback mechanisms can be applied, just as in other scanning-probe microscopies. The most common feedbacks used in SNOM are the shear-force and tapping-mode feedbacks.

In shear-force feedback, the SNOM probe is dithered laterally at its resonance frequency so that it vibrates with a small amplitude, typically less than 10 nm. Upon approaching the sample, shear forces dampen the amplitude of the probe oscilla-tion. The variations in amplitude are then used by the feedback mechanism to read-just the tip–sample distance until the vibration amplitude reverts to a predefined value. The tapping-mode feedback works in a similar way to the shear-force feed-back. In this case, however, the probe is set to oscillate along the surface normal. In both feedback modes, vibration is usually driven to the probe by a piezoelectric tuning fork: The probe is glued to one of the prongs, which are either oriented per-pendicular (in shear-force feedback) or parallel (in tapping-mode feedback) to the surface plane (Höppener et al. 2002).

High resolution and optical contrast are intimately related to the feedback effi-ciency in SNOM, which is determined by the capacity of the probe to detect small shear forces. This capacity increases with the quality factor, Q, of the probe vibra-tion. Indeed, one of the major challenges in SNOM is to achieve full applicabil-ity in a liquid environment. Optical contrast in liquids is seriously hampered, since vibration dampening due to viscous dissipation severely interferes with shear-force dampening. There have been some attempts to circumvent this by using alterna-tive feedback methods such as (a) ion conductivity (Rothery et al. 2003) by using thin and sharp tips in the case of electrochemically etched probes, since dampening strongly depends on tip shape (Fragola et al. 2004), (b) keeping the tuning fork in air while the tip is immersed in liquid (Höppener et al. 2002; Koopman et al. 2004), or (c) using hydrophobic tips in an attempt to reduce the interaction between the tip and the surrounding (aqueous) fluid, thereby reducing the liquid-induced dampening (Sommer and Franke 2002).

CONTRAST MECHANISMS IN SNOM

Near-field optical microscopy shares with conventional microscopy most of its contrast mechanisms: optical density, luminescence, birefringence. Here we focused our attention on three contrast mechanisms not contemplated in the far-field section that are especially relevant in near-field optics: polarization contrast, refractive index, and

FIGURE 4.8 *(See facing page.)* Typical apertureless SNOM probes (left) and tip-on-aperture probes (right). *(Left)*: Electrochemically etched STM tips and AFM cantilever-integrated tips are among the most popular apertureless probes. (*Left* SEM photograph reprinted from Raschke and Lienau 2003. With permission. *Middle* SEM photograph reprinted from Hayazawa et al. 2001. With permission. *Right* SEM photograph reprinted from Kalkbrenner et al. 2001. With permission.) *(Right)*: Gold nanoparticles can be attached at the end of either optic fibers or cantilever tips to produce antenna-mediated field enhancements. (Top image reprinted from Chang, Bauerdick, and Jeong 2008. With permission. Bottom images reprinted from Taminiau et al. 2007. With permission.)

TABLE 4.1

Classification of Aperture Probes and Their Characteristics

Aperture Probes	Aperture Radius (a)	Taper Angle (α)	Lateral Resolution (≈aperture size)	Transmission Coefficient (Γ) $f(a, \alpha)$	Advantages	Limitations
Straight or bent optical fibers	50–200 nm (typically ≥ 30 nm)	20°–90°	50–100 nm (routine)	100 nm 30° 10^{-6}–10^{-5} 100 nm 42° 10^{-3}	Low background High control over illumination polarization	Low light throughput when aperture is smaller than 20 nm Tip damage at input power higher than 10 mW Cut-off effect: reduced output via propagating-to-evanescent wave conversion Poor reproducibility Stiff probes (100 N/m)
Heated-and-pulled	100 nm < 100 nm (60 nm) >20–500 nm	Low values (20°)	$\lambda/50$ 150 nm $\lambda/43$	$(a/\lambda)^4$ (Bethe/Bouwkamp) 10^{-4}–10^{-6} (a = 100–200 nm) 10^{-4}–10^{-6}		Less stable Lower lateral resolution Damage when input power is increased from 0.6 to 6 mW
Double-etched		Higher values (90°)		$(a/\lambda)^3$ 10^{-3}–0.05 (a = 50–100 nm)	More stable Higher lateral resolution	Damage when input power is 9.5 mW, decreasing with decreasing taper angle
Micromachined AFM probes	50–150 nm	30°	50–150 nm	10^{-2}–10^{-3}	Wider range of stiffness (1–100 N/m), higher applicability	Higher reproducibility, batch processable

TABLE 4.2
Classification of Apertureless Probes and Their Characteristics

Apertureless Probes	Tip Radius (r)	Lateral Resolution (≈tip size)	Enhancement Factor (Γ) f (r, opening angle, enhancement mech.)	Advantages	Limitations
Gold, silver sharp probes	<20 (10–30 nm)	10–30 nm	Approx. 2000 (SP) Up to 10^4 (Raman)	Better resolution than aperture probes Easier to manufacture	Poor light focusing—large background
Gold nanoparticles	100 nm	100 nm	Better for prolate spheroids than spheres		Lateral resolution limited by the particle size
Tungsten probe	20–100 nm	20–100 nm	Up to 10^2	Easier to etch than gold Stiffer—limits tip damage	

infrared vibrational contrast. An important mechanism of contrast caused by fluorescence has been profusely applied in numerous SNOM studies. Fluorescence SNOM has been treated in detail in Chapter 3 as the near-field option of fluorescence microscopy, since it exhibits extensive capabilities with respect to its far-field counterpart.

Polarization Contrast

Nontransparent objects that in theory cannot transmit light may appear bright in the images acquired in transmission-mode aperture SNOM. Likewise, images of homogeneous metal objects may appear distorted or may show areas with enhanced brightness. These findings may be considered counterintuitive, but they can be understood if polarization effects are taken into account. Betzig and coworkers (1992) made a systematic study on the effect of polarized light on images of micrometer-sized aluminum rings on glass. The aperture probe consisted of a heat-and-pull tapered fiber through which light was transmitted at a defined polarization plane, called a *probe state*. An additional polarizer was placed before the detector, defining another polarization plane, the detector state. Figure 4.9 shows SNOM images of the rings at different states of polarization at the probe and at the detector. When both probe and detector states are parallel, the rings appear bright just where the aluminum structure is perpendicular to the polarization plane. This appears to fulfill the boundary condition of near-field optics, where the electric field should be perpendicular to the object structure to get maximum signal. On the other hand, if the probe and detector states are mutually perpendicular, the rings exhibit bright sections oriented at 45° to the polarization directions. The outcome in this case was totally analogous to the effect of a third polarizer between two crossed polarizers. When circularly polarized light was used instead, the rings appeared homogeneously dark regardless of the polarization state of the detector.

These results make us exercise caution when interpreting contrast in SNOM images of metal objects or corrugated samples, especially when polarized light is used. The same effect can be encountered if the aperture is noncircular, where the transmission coefficient may vary with the polarization plane. In addition, corrugated samples exhibit facets that may be differently oriented to the polarization direction and hence bring about polarization artifacts. This effect can be even more pronounced if accompanied by vertical movement of the tip that is required to acquire the topological image (Hecht et al. 2000). Consequently, cross talk between optical and topological signals may occur, and care must be taken to discriminate a topography artifact from a real near-field signal. Measurements at constant height, where the probe is held at a fixed distance from the sample during scanning, are preferred to measurements at constant gap width (tip-to-sample distance), where the probe should follow the contour of the sample by moving up and down.

Alternatively, images of the sample at different illumination polarizations may be helpful to identify topography artifacts. However, the fact that metals can change the polarization state of light and enhance or hinder light transmissibility can result in several applications. One of these applications consists of using the *Faraday effect* exhibited by certain magnetic materials to infer their distribution of magnetization states. The Faraday effect rotates the polarization direction of the incident light that

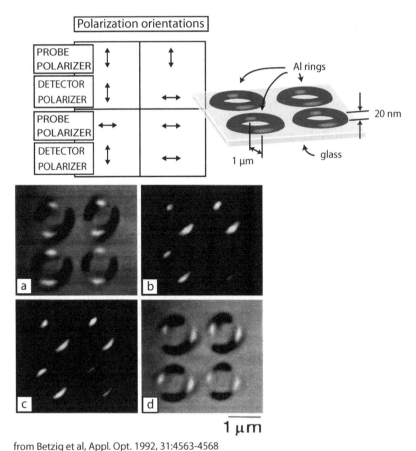

from Betzig et al, Appl. Opt. 1992, 31:4563-4568

FIGURE 4.9 Polarization contrast in SNOM. A sample consisting of four aluminum rings deposited on glass is imaged with polarized light through an aperture probe. The polarization direction at the probe and at the detector can be set either to vertical ($\updownarrow$) or to horizontal ($\leftrightarrow$) via two polarizers. The images correspond to a specific combination of both probe and detector polarizers as defined in the polarization matrix (above the images). Maxima of near-field intensity are detected alongside the polarization directions when both polarizers are parallel, either vertically (a) or horizontally (d). When the polarizers are perpendicular to each other—cases (b) and (c)—intensity reaches a maximum along the diagonal (45°). (SNOM images reprinted from Betzig, Trautmann, Weiner, et al. 1992. With permission.)

impinges on a material. The changes typically amount to 1°, and the direction (clockwise or counterclockwise) depends on the material's magnetization (up or down). If a polarizer is placed just before the detector at an angle of 1° from being orthogonal to the illumination polarization, it is possible to differentiate the different magnetization domains from the SNOM contrast. Domains with opposed magnetization will appear dark or bright if they bring complete light extinction or partially attenuated transmission, respectively.

Refractive Index Contrast

A convenient mechanism that can be applied to dielectrics or low-contrast biological samples is based on detecting changes in the real part of refractive index. The discontinuity in the refractive index results in noticeable variations of intensity of the scattered light due to probe–sample coupling (Trautman et al. 1992; Betzig and Trautman 1992). The scattered light shows a strong dependence on refractive index, which makes the near-field contrast especially sharp if compared to that of the far-field counterpart (Trautman et al. 1992): Regions with higher refractive index appear brighter than regions with lower refractive index. Grooves engraved on PMMA films thus appeared dark due to the smaller refractive index of air (1.000) compared to that of PMMA (1.489) (Trautman et al. 1992).

In scattering-type SNOM, where interferometric detection is employed and background is properly suppressed, it is possible to simultaneously map the amplitude and phase of the near-field scattered light and, hence, to obtain both the real and imaginary parts of the refractive index or the dielectric permittivity (Brehm et al. 2005).

Infrared-Vibrational Contrast (Infrared Apertureless SNOM)

An interesting, material-specific contrast arises when infrared wavelengths are employed as illuminating sources in apertureless SNOM. Apart from its scattering ability, any material that has distinct vibrational bands can be excited with infrared light of defined wavelength. Consequently, the material will absorb radiation corresponding to that specific wavelength and will appear dark in the resulting SNOM image.

The illumination sources for infrared SNOM are typically CO_2 (Knoll and Keilmann 1999) or CO laser beams (Taubner, Hillenbrand, and Keilmann 2004), which provide infrared radiation in the wavelength ranges of 9–10 μm and 5.5–6 μm, respectively. Under these conditions, samples composed of materials that exhibit maximum infrared absorption at different wavelengths can be resolved. For example, binary mixtures of vibrationally active polymers, such as those consisting of polystyrene and polymethylmethacrylate, can exhibit vibrational contrast if the infrared illumination is tuned to the vibrational resonance of either material, and the image contrast is correspondingly reversed (Knoll and Keilmann 1999).

SUMMARY

In the last 20 years, SNOM has evolved into a family of techniques based on the aperture and apertureless variants. Each one has its advantages and limitations, as Table 4.3 shows, and therefore it is not entirely correct to consider that one technique surpasses the other or vice versa. Apertureless SNOM provides better lateral resolution at the expense of a diminished signal-to-noise ratio due to large background signals that should be suppressed in order to get meaningful results. On the other hand, aperture SNOM is less affected by background, but the resolution is hampered by the low light transmissibility of the apertures, which cannot be smaller than 30 nm in diameter.

TABLE 4.3

The Characteristics and Limitations of Both Aperture and Apertureless SNOM

	Aperture SNOM	Apertureless SNOM
Lateral resolution	Lateral resolution = aperture size	Lateral resolution = tip size
Illuminating source	Light through an sub-λ aperture	Tip-induced field enhancement
	1–100 nm	
Vertical resolution	1–100 nm	
Illumination efficiency	Light transmission increases with taper angles, aperture size at the expense of image resolution	Maximum enhancement with p-polarized light
Chemical information	Indirect, through fluorescent markers	Direct
Other features	Applicable to wavelengths in the visible range	Applicable to wavelengths in the visible and mid-infrared range
	Illumination and detection can be set at any polarization state	Subsurface imaging as deep as 50 nm is possible with infrared light
	Supports polarization, fluorescence spectroscopy	Supports Raman, infrared, and fluorescence spectroscopies
Limitations	Reduced light transmissibility that imposes a limit in the aperture size and in the lateral resolution	Large background
	Probe optimization required	
	Limited applicability in fluid conditions	
	Confined on the surface of materials	
	Cross talk between topography and optical signals	

As the high-resolution version of optical microscopy, SNOM's great potential lies in the nondestructive detection of single molecules or few-molecule aggregates. However, being a local, raster-based technique, its versatility is limited compared with the far-field optical techniques. Tip–sample interaction in SNOM and the sources of field-enhancement are either not clear or not fully understood, which impedes understanding of optical contrast in some cases or judging the extent of coupling between the topographical and optical signals (cross talk). Despite its great potential, the application of SNOM as a monitoring tool of functional processes in living cells is still far from being routine and remains a challenge.

APPLICATIONS OF SNOM

The applications of SNOM have greatly expanded as a tool that simultaneously provides topological and chemical information of a sample's constituents. In particular, apertureless SNOM has gained great importance in solid-state chemistry, since it can address important mechanistic questions on chemical transformations (i.e., oxidation, diazotization, photodimerization, surface hydration, hydrolysis). In structural biology, it has been employed to identify submicroscopic cell organelles in a nondestructive, staining-free way, and it has been used in oncology to differentiate cancerous tissue from healthy tissue in the early stages of the disease. The technique has also found numerous applications in industry, such as assessing the chemical resistance of dental alloys as well as the processes of paper glazing and textile dyeing and determining the quality of blood-storage bags (Kaupp 2006). In this regard, the SNOM variants based on fluorescence contrast and tip-enhanced Raman spectroscopy are ubiquitous among the research works in bioscience and materials sciences. These SNOM modes and their applications are described in detail in the chapters dedicated to optical microscopy (Chapter 3) and chemical spectroscopy (Chapter 5), and they will not be further mentioned here.

In this section, we will exemplify the importance of SNOM in three application lines: that of biology, to characterize the morphology of living cells in their physiological environment; that of materials science, to chemically identify sample constituents in composite layers of metals, semiconductors, and dielectrics; and that of plasmonic devices, to map field distributions of plasmonic resonances in gold nanostructures.

Example A: A Combined SNOM-SICM Setup

The first works reporting the use of SNOM on living cells under physiological conditions were done by Korchev and coworkers (2000) and Rothery and coworkers (2003). Respectively, they managed to image living cardiac myocytes and A6 epithelial cells by combining SNOM with scanning ion conductance microscopy (SICM).

Their approach was twofold innovative: On the one hand, they used ion conductance as the feedback magnitude to keep the tip–sample distance constant and generate the topography images (Korchev et al. 2000). The experimental setup is schematized in Figure 4.A1, and exemplifying images of living cells are shown in Figure 4.A2. The SNOM tip, a glass micropipette apex of diameter 100–500 nm, was filled with the same electrolyte solution that was used to maintain the cells. A dc electric potential was applied between an Ag/AgCl electrode inside the micropipette, and a distant electrode was immersed in the sample medium. The flow current between these electrodes, a magnitude that strongly depends on the tip–sample separation, was converted into voltage and used as the input signal for the electronic feedback. In this way, the flow current and hence the tip–sample distance were kept constant throughout the whole scanning. On the other hand, Rothery and coworkers (2003) proposed a novel method of generating near-field light at the end of the micropipette that does not require probe coating. The micropipette contains a fluorescent dye, fluo-3, that emits light when associated with calcium ions (Ca^{2+}). Inside the calcium-free medium of the micropipette, the dye does not fluoresce. However, when immersed in a calcium-containing solution as the sample's medium, the fluo-3 diffuses out of the micropipette and combines with the Ca^{2+} ions. The resulting complex is fluorescent, exhibiting a maximum intensity at 526 nm when illuminated at 488 nm. This complex readily forms at the edge of the micropipette tip and thus constitutes a highly local light source that does not require micropipette coating.

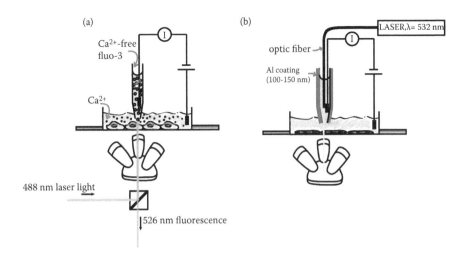

FIGURE 4.A1 Experimental setup used by Rothery (a) and Korchev (b). SICM controls the tip–sample separation and generates the topography data; the micropipette hosts one of the electrodes and a multimode fiber (Korchev et al. 2000) or a solution containing a calcium indicator, fluo-3 (Rothery 2003) as near-field light sources. A dc potential is established between two electrodes and the flow current is used by the feedback control.

Although their approach has contributed in extending the application of SNOM to living cells and liquid environments, important technical issues, such as the improvement of lateral resolution or the effect of cross talk graphical signals, are still pending.

Example B: Visible and Mid-Infrared SNOM and Subsurface Imaging

Material identification of sample constituents is a prominent capability of SNOM that arises from a distinct near-field interaction between the tip and the particular component. This interaction mainly depends on the dielectric constant of the latter, which in turn is responsible for the near-field optical contrast displayed in the image.

Taubner and coworkers exploited this feature to map metal/semiconductor/dielectric composite samples and identify structures lying below the sample's surface (Taubner, Keilmann, and Hillenbrand 2005). The samples consisted of hexagonally arranged gold islands deposited on silicon wafers and partially buried in a polystyrene layer. Gold, silicon, and polystyrene are clearly distinguishable from their optical amplitude signals when irradiated with visible light ($\lambda = 633$ nm)—from bright (gold) to dark (polystyrene)—as seen in Figure 4.B1. The polystyrene layer, which appears as a high diagonal stripe in the topography image, hides a few gold islands that are, however, visible in the optical image. Although the intensity of the buried gold islands is much reduced compared to that of the unburied ones (50%), it still shows that optical imaging of a subsurface feature is possible. Using visible light, it is thus possible to resolve features 5 nm below the sample

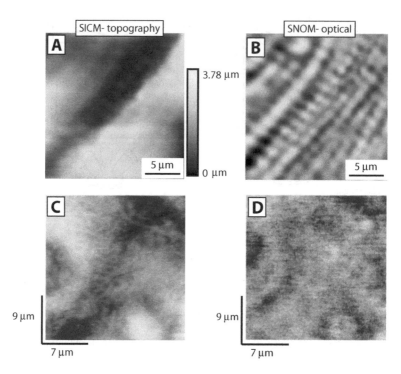

FIGURE 4.A2 (A) Topographical (SICM image) and (B) optical images of living cardiac myocytes showing the striated pattern of the sarcomeres. (C) Topographical and (D) optical images of living A6 cells showing the cell boundaries and the small microvilli in the topography. The topography and optical images do not show exactly the same features; however, the authors could not rule out a possible contribution of cross talk between both signals. The vertical scale for (c) is 1.82 mm. (Images (A) and (B) reprinted from Korchev et al. 2000. With permission. Images (C) and (D) reprinted from Rothery et al. 2003. With permission.)

surface. Infrared light ($\lambda = 10.7\ \mu m$) has more penetration depth than visible light, which allows distinguishing subsurface features 10 times deeper, as illustrated in Figure 4.B2.

Example C: Mapping Plasmonic Resonances in Gold Nanostructures

Plasmon resonances of gold nanostructures can be locally excited and mapped by a SNOM probe small enough to minimally disturb the sample's resonance modes. If scattered light coming both from tip and sample is detected, the resulting image is difficult to interpret, since both signals may be strongly coupled. Vogelgesang and coworkers found a way to circumvent this problem by using s-polarized radiation to selectively excite the sample without altering the tip, which is polarizable along the surface normal (p-polarizable) (Vogelgesang et al. 2008). If the localized plasmonic mode of the sample exhibits strong vertical components along the surface normal, the tip will be excited and consequently will scatter light off to the detector. The acquired signal will thus be related to the local field components of the sample's resonance mode that are parallel to the surface normal.

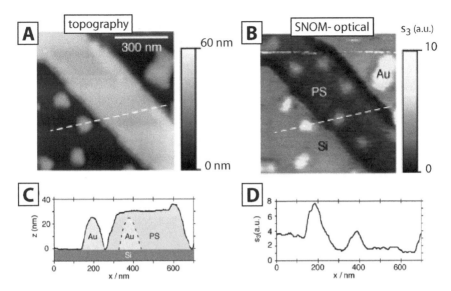

FIGURE 4.B1 Topography (A) and visible-SNOM (B) images of a sample consisting of gold (Au) islands on a silicon (Si) surface, partly covered by a polystyrene (PS) layer with their respective line profiles in (C) and (D). The Au islands buried under a 5-nm-thick PS layer are still distinguishable in the optical image, though their intensity is approximately 50% that of the uncovered ones. (Images (A–D) reprinted from Taubner, Keilmann, and Hillenbrand 2000. With permission.)

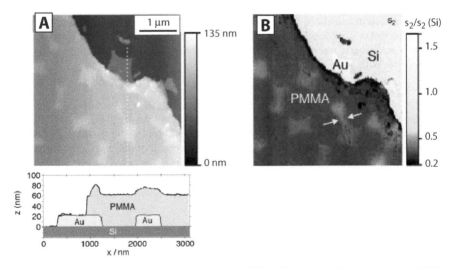

FIGURE 4.B2 Topography (A) and mid-infrared SNOM (B) images of a sample where PS is replaced by a 62-nm-thick poly(methyl methacrylate) (PMMA) as the covering layer. In this case, the gold islands are buried under a 60-nm-thick PMMA, but they are still distinguishable in the optical image. (Reprinted from Taubner, Keilmann, and Hillenbrand 2005. With permission.)

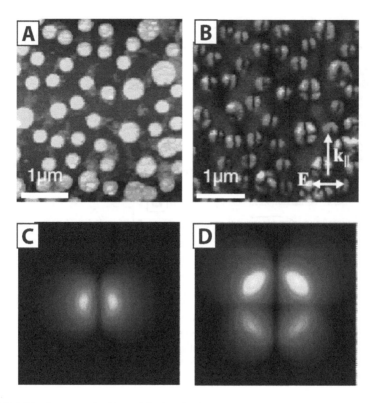

FIGURE 4.C1 Topography (A) and SNOM (B) images of a sample containing gold disks of 140 and 300 nm diameter. Images (C) and (D) show the field strength distribution along the vertical direction calculated for disk-shaped metal structures exhibiting dipolar (two lobes) and quadripolar (four lobes) resonances. The excitation radiation was s-polarized and of 800-nm wavelength, resembling the experimental conditions. (Reprinted from Vogelgesang et al. 2008. With permission.)

This approach was tested in gold disks of two different sizes, 140 and 300 nm diameter, which in the SNOM image exhibit two-lobe and four-lobe shapes, respectively (Figure 4.C1). These shapes exactly resembled the theoretical predictions for the field distribution along the surface normal of dipolar and quadrupolar plasmonic eigenmodes in metal disks, respectively (Figures 4.C1C and D). After validating their experimental approach, they applied it to characterize the plasmonic response of long gold wires and bow-tie nanoantennas (Figure 4.C2).

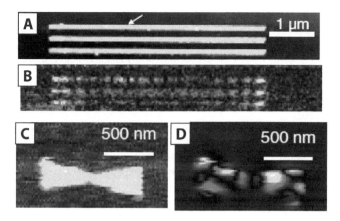

FIGURE 4.C2 Topography (A and B) and SNOM (C and D) images of the plasmon distribution of gold wires and a bow-tie antenna. (Reprinted from Vogelgesang et al. 2008. With permission.)

REFERENCES

Alvarez, L., and M. Xiao. 2006a. The role of propagating and evanescent waves in scanning near-field optical microscopy. *Opt. Commun.* 260:727–32.

———. 2006b. Theoretical analysis on the resolution of collection mode scanning near-field optical microscopy. *Opt. Rev.* 13:254–61.

Bethe, H. A. 1944. Theory of diffraction by small holes. *Phys. Rev.* 66:163–82.

Betzig, E., and R. J. Chichester. 1993. Single molecules observed by near-field scanning optical microscopy. *Science* 262:1422–25.

Betzig, E., A. Lewis, A. Harootunian, M. Isaacson, and E. Kratschmer. 1986. Near-field scanning optical microscopy (NSOM): Development and biophysical applications. *Biophys. J.* 49:269–79.

Betzig, E., and J. K. Trautman. 1992. Near-field optics: Microscopy, spectroscopy, and surface modification beyond the diffraction limit. *Science* 257:189–95.

Betzig, E., J. K. Trautman, T. D. Harris, J. S. Weiner, and R. L. Kostelak. 1991. Breaking the diffraction barrier: Optical microscopy on a nanometric scale. *Science* 251:1468–70.

Betzig, E., J. K. Trautman, J. S. Weiner, T. D. Harris, and R. Wolfe. 1992. Polarization contrast in near-field scanning optical microscopy. *Appl. Opt.* 31:4563–68.

Blaize, S., B. Bérenguier, I. Stéfanon, et al. 2008. Phase sensitive optical near-field mapping using frequency-shifted laser optical feedback interferometry. *Opt. Express* 16:11718–26.

Bouhelier, A. 2006. Field-enhanced scanning near-field optical microscopy. *Microsc. Res. Tech.* 69:563–79.

Bouhelier, A., J. Toquant, H. Tamaru, H.-J. Güntherodt, and D. W. Pohl. 2001. Electrolytic formation of nanoapertures for scanning near-field optical microscopy. *Appl. Phys. Lett.* 79:683–85.

Brehm. M., H. G. Frey, R. Guckenberger, et al. 2005. Consolidating apertureless SNOM. *J. Korean Phys. Soc.* 47:S80–S85.

Burgos, P., Z. Lu, A. Ianoul, et al. 2002. Near-field scanning optical microscopy probes: A comparison of pulled and double-etched bent NSOM probes for fluorescence imaging of biological samples. *J. Microsc.* 211:37–47.

Chibani, H., K. Dukenbayev, M. Mensi, S. K. Sekatskii, and G. Dietler. 2010. Near-field scanning optical microscopy using polymethylmethacrylate optical fiber probes. *Ultramicroscopy* 110:211–15.

Courjon, D., and C. Bainier. 1994. Near-field microscopy and near-field optics. *Rep. Prog. Phys.* 57:989–1028.

Formanek, F., Y. de Wilde, G. S. Luengo, and B. Querleux. 2006. Investigation of dyed human hair fibres using apertureless near-field scanning optical microscopy. *J. Microsc.* 224:197–202.

Fragola, A., L. Aigouy, P. Y. Mignotte, F. Formanek, and Y. De Wilde. 2004. Apertureless scanning near-field fluorescence microscopy in liquids. *Ultramicroscopy* 101:47–54.

Frey, H. G., F. Keilmann, A. Kriele, and R. Guckenberger. 2002. Enhancing the resolution of scanning near-field optical microscopy by a metal tip grown on an aperture probe. *Appl. Phys. Lett.* 81:5030–32.

Frey, H. G., S. Witt, K. Felderer, and R. Guckenberger. 2004. High-resolution imaging of single fluorescent molecules with the optical near-field of a metal tip. *Phys. Rev. Lett.* 93:200801/1–4.

Hartschuh, A. 2008. Tip-enhanced near-field optical microscopy. *Angew. Chem. Int. Ed.* 47:8178–91.

Hecht, B., B. Sick, U. P. Wild, et al. 2000. Scanning near-field optical microscopy with aperture probes: Fundamentals and applications. *J. Chem. Phys.* 112:7761–74.

Höppener, C., R. Beams, and L. Novotny. 2009. Background suppression in near-field optical imaging. *Nano Lett.* 9:903–8.

Höppener, C., D. Molenda, H. Fuchs, and A. Naber. 2002. Scanning near-field optical microscopy of a cell membrane in liquid. *J. Microsc.* 210:288–93.

Jiang, S., N. Tomita, H. Ohsawa, and M. Ohtsu. 1991. A photon scanning tunnelling microscope using an AlGaAs laser. *J. Appl. Phys.* 30:2107–11.

Jin, E. X., and X. Xu. 2008. Focussed ion beam machined cantilever aperture probes for near-field optical imaging. *J. Microsc.* 229:503–11.

Kalkbrenner, T., M. Ramstein, J. Mlynek, and V. Sandoghdar. 2001. A single gold particle as a probe for apertureless scanning near-field optical microscopy. *J. Microsc.* 202:72–76.

Kaupp, G. 2006. Scanning near-field optical microscopy on rough surfaces: Applications in chemistry, biology and medicine. *Int. J. Photoenergy* 2006:1–22.

Kim, B. J., J. W. Flamma, E. S. Ten Have, M. F. Garcia-Parajo, N. F. Van Hulst, and J. Brugger. 2001. Moulded photoplastic probes for near-field optical applications. *J. Microsc.* 202:16–21.

Kim, Z. H., and S. R. Leone. 2006. High-resolution apertureless near-field optical imaging using gold nanosphere probes. *J. Phys. Chem. B* 110:19804–9.

Knoll, B., and F. Keilmann. 1999. Near-field probing of vibrational absorption for chemical microscopy. *Science* 399:134–37.

Koopman, M., A. Cambi, B. I. de Bakker, et al. 2004. Near-field scanning optical microscopy in liquid for high resolution single molecule detection on dendritic cells. *FEBS Lett.* 573:6–10.

Korchev, Y. E., M. Raval, M. J. Lab, et al. 2000. Hybrid scanning ion conductance and scanning near-field optical microscopy for the study of living cells. *Biophys. J.* 78:2675–79.

La Rosa, A. H., B. I. Yakobson, and H. D. Hallen. 1995. Origins and effects of thermal processes on near-field optical probes. *Appl. Phys. Lett.* 67:2597–99.

Lineau, Ch., A. Richter, and T. Elsaesser. 1996. Light-induced expansion of fiber tips in near-field scanning optical microscopy. *Appl. Phys. Lett.* 69:325–27.

Mitsuoka, Y., T. Niwa, S. Ichihara, et al. 2000. Microfabricated silicon dioxide cantilever with subwavelength aperture. *J. Microsc.* 202:12–15.

Molenda, D., G. Colas des Francs, U. C. Fischer, N. Rau, and A. Naber. 2005. High-resolution mapping of the optical near-field components at a triangular nano-aperture. *Opt. Express* 13:10688–96.

Novotny, L., D. W. Pohl, and B. Hecht. 1995. Light confinement in scanning near-field optical microscopy. *Ultramicroscopy* 61:1–9.

Novotny, L., E. J. Sánchez, and X. S. Xie. 1998. Near-field optical imaging using metal tips illuminated by higher-order Hermite-Gaussian beams. *Ultramicroscopy* 71:21–29.

Ozcan, A., E. Cubukcu, K. B. Crozier, B. E. Bouma, F. Capasso, and G. J. Teamey. 2006. Differential near-field scanning optical microscopy. *Nano Lett.* 6:2609–16.

Rothery, A. M., J. Gorelik, A. Bruckbauer, W. Yu, Y. E. Korchev, and D. Klenerman. 2003. A novel light source for SICM-SNOM of living cells. *J. Microsc.* 209:94–101.

Sommer, A. P., and R.-P. Franke. 2002. Near-field optical analysis of living cells in vitro. *J. Proteome Res.* 1:111–114.

Stähelin, M., M. A. Bopp, G. Tarrach, A. J. Meixner, and I. Zschokke-Gränacher. 1996. Temperature profile of fiber tips used in scanning near-field optical microscopy. *Appl. Phys. Lett.* 68:2603–5.

Taminiau, T. H., R. J. Moerland, F. B. Segerink, L. Kuipers, and N. F. van Hulst. 2007. $\lambda/4$ resonance of an optical monopole antenna probed by single molecule fluorescence. *Nano Lett.* 7:28–33.

Taubner, T., R. Hillenbrand, and F. Keilmann. 2004. Nanoscale polymer recognition by spectral signature in scattering infrared near-field microscopy. *Appl. Phys. Lett.* 85:5064–66.

Taubner, T., F. Keilmann, and R. Hillenbrand. 2005. Nanoscale-resolved subsurface imaging by scattering-type near-field optical microscopy. *Opt. Express* 13:8893–99.

Trautman, J. K., E. Betzig, J. S. Weiner, et al. 1992. Image contrast in near-field optics. *J. Appl. Phys.* 71:4659–63.

Vogelgesang, R., J. Dorfmüller, R. Esteban, R. T. Weitz, A. Dmitriev, and K. Kern. 2008. Plasmonic nanostructures in aperture-less scanning near-field optical microscopy (aSNOM). *Phys. Stat. Sol. B* 245:2255–60.

Zhang, Y., A. Dhawan, and T. Vo-Dinh. 2011. Design and fabrication of fiber-optic nanoprobes for optical sensing. *Nanoscale Res. Lett.* 6:18–24.

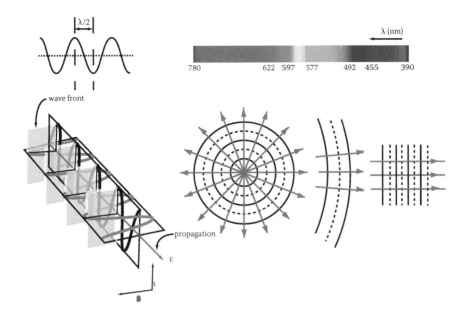

COLOR FIGURE 3.2 Light as an electromagnetic wave (above). The wavelengths of visible light range from 390 nm (blue light) to 780 nm (red light). Wavefronts and light rays (below). Both electric and magnetic fields oscillate perpendicularly to the direction of propagation (transversal wave), defining the wave fronts as surfaces of constant phase. Light emerging from a punctual source propagates in all directions and the wave fronts are spherical. When the source is sufficiently far away, the wave front turns into planar and the light rays are parallel from one another.

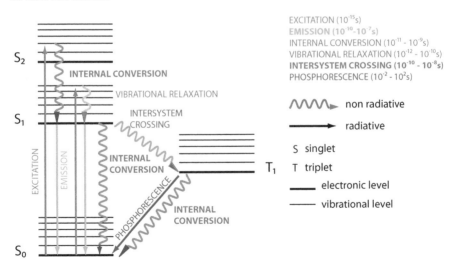

COLOR FIGURE 3.10 Perrin-Jablonski diagram. Graphical representation of molecular energy levels and the transitions between them. Thick lines represent electronic energy levels: S denotes singlets; T denotes triplets. Thinner lines represent vibrational energy levels. Both radiative and nonradiative processes are transitions between energy levels and they are represented as arrows, pointing the direction where the transition occurs. Radiative processes, such as fluorescence and phosphorescence are represented as straight arrows; nonradiative processes, such as internal conversion or intersystem crossing, are represented as wavy arrows. The time scales of each process are shown between brackets.

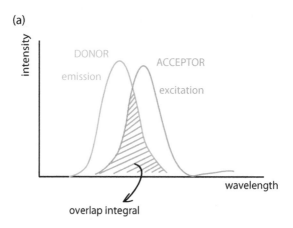

(a)

intensity

DONOR

emission

ACCEPTOR

excitation

wavelength

overlap integral

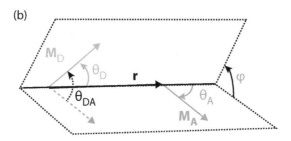

(b)

M_D

θ_D

r

φ

θ_{DA}

θ_A

M_A

COLOR FIGURE 3.12 Resonance energy transfer. Spectra (a) and orientations (b). (a) For energy transfer to occur, it is required that the donor's emission spectrum overlaps the acceptor's excitation spectrum. The striped area is the overlap integral, this magnitude appears in Equation (3.14). (b) Donor and acceptor have associated transition moments, M_D and M_A, respectively. Their spatial arrangement is defined by the angles θ_D, θ_A, θ_{DA}, and φ.

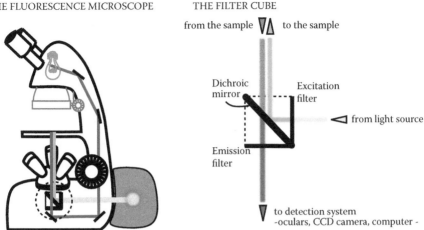

THE FLUORESCENCE MICROSCOPE   THE FILTER CUBE

from the sample   to the sample

Dichroic
mirror   Excitation
filter

from light source

Emission
filter

to detection system
-oculars, CCD camera, computer -

COLOR FIGURE 3.13 The fluorescence microscope (left) and the filter cube (right). The fluorescence microscope is an optical microscope equipped with special illumination and optical filters compared with a standard bright-field optical microscope. The filter cube delivers the appropriate light to excite the fluorescent sample and just lets the fluorescence emission go through to the detection system.

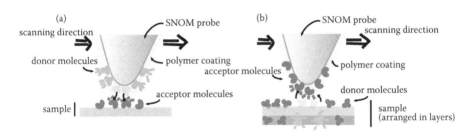

(a)   SNOM probe   (b)   SNOM probe
scanning direction   scanning direction
donor molecules   polymer coating   polymer coating
                         acceptor molecules   acceptor molecules
                         acceptor molecules   donor molecules
sample   sample
                                                    (arranged in layers)

COLOR FIGURE 3.21 FRET-SNOM. (a) Donors at the probe, acceptors at the sample; in this case, the donor-coated tip delivers the light of suitable wavelength to excite the donors, but not the acceptors. FRET occurs whenever the donor-coated tip encounters an acceptor molecule at the appropriate distance (1–2 nm), which in turn emits near-field fluorescence; those acceptor molecules that are not contained within the illumination volume or situated farther apart will not fluorescence. (b) Acceptors at the probe, donors at the sample; the acceptor-coated tip delivers the light that excites the donors, and not the acceptors. In this way, FRET occurs only when the tip meets a vicinal donor at the surface. Consequently the tip will emit light at the expense of the donor molecule. This configuration is especially convenient to study the distribution of donors within layer-structured samples from changes in their fluorescence intensity; donors closest to the tip will undergo FRET, which will undermine fluorescence emission and hence appear dimmed, while distant donors will appear brighter.

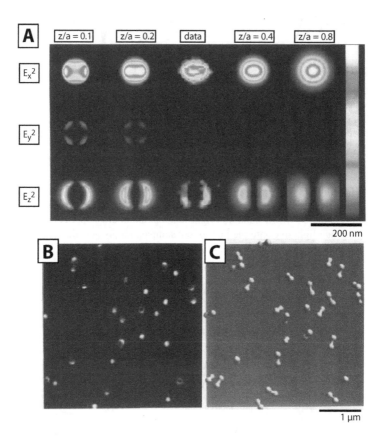

COLOR FIGURE 3.24 Single-molecule detection with fluorescent SNOM. Assignment of molecular orientation. (A) Theoretical derivation of the distribution of electrical field components for a molecule at different distance between the latter and the SNOM probe (z normalized to the aperture diameter, a). Comparison with experimental data corresponding to a single molecule. (B) Experimental SNOM image with fluorescent spots corresponding to single molecules adopting different orientations. (C) Theoretically derived image where the molecular orientations are assigned to each spot of the measured image. (All figures reprinted from Betzig and Chichester 1993. With permission.)

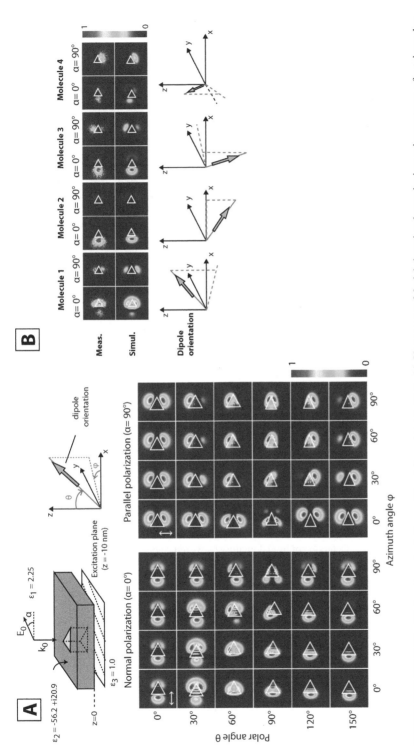

COLOR FIGURE 3.25 Molecular orientations with a triangular aperture. (a) *Upper part:* Model that simulates the intensity patterns of a triangular aperture, illuminated at normal incidence with a plane wave that can be perpendicularly ($\alpha = 0°$) or vertically ($\alpha = 90°$) polarized. The spatial orientation of the absorption dipole moment can be characterized by two angles, the polar angle θ and the azimuth angle φ. *Lower part:* Orientation maps showing the distribution of fluorescent intensities experienced by a single molecule at different values of the pair (θ, φ). (b) Comparison between measured data and simulation data and assignment of the dipole orientation for four different molecules. (All figures adapted from Molenda et al. 2005. With permission.)

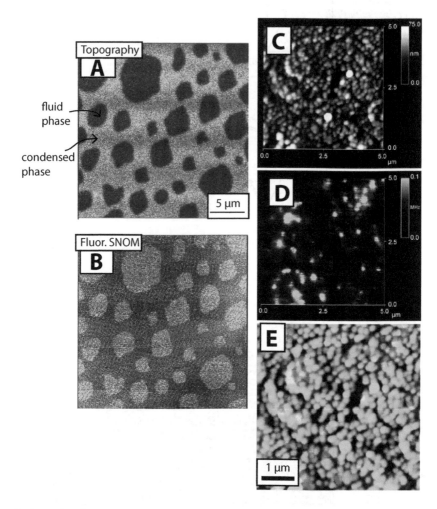

COLOR FIGURE 3.27   Detection of lipid domains (A–B) and colocalization of nuclear pore complexes with regulating receptors (C–E). Figures (A) and (B) show topography and fluorescence SNOM images, respectively, of a phase-separated 1:1:1 sphingomyelin/cholesterol/dioleoylphosphatidylcholine (DOPC) lipid monolayer. The condensed phases appear higher in the topography image compared with the lower fluid phases. Fluorescence is brought about via a fluorescence marker (1,2-dipalmitoyl-sn-glycero-3-phosphoethanolamine-fluorescein, Fl-DPPE), which tends to accumulate in the fluid phase. Figures (C), (D), and (E) respectively show the topography, fluorescence, and overlay images of nuclear pore complexes (NPCs) and fluorescently marked IP$_3$ receptors. These receptors (appearing in (D) and as red spots in (E)) are said to regulate the activity of (NPCs) (appearing in (A) and as green spots in (E)). These studies were aimed to explore the connectivity between function regulation and spatial colocalization. (Figures (A and B) reprinted from Burgos et al. 2003. With permission. Figures (C, D, and E) reprinted from Dickenson et al. 2010. With permission.)

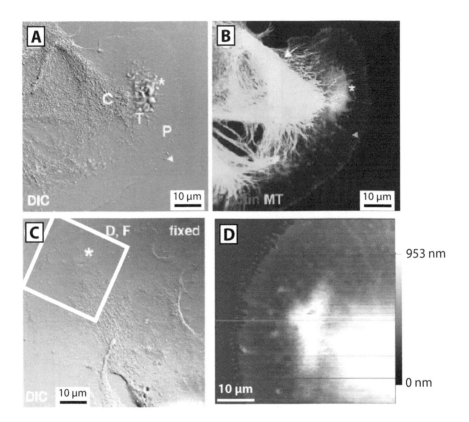

COLOR FIGURE 3.B1 Structure of growth cones in an *Aplysia* bag cell shown in optical microscopy. (A) DIC optical micrograph and (B) the corresponding fluorescent image of the structure. Microtubules are dyed in green, whereas F-actin appears in red. Optical (C) and scanning probe (D) images of the same cell. (C) The DIC optical micrograph of a fixed cell; image (D) is the AFM image that has been produced by scanning the box area of the optical image. (Adapted from Grzywa et al. 2007. With permission.)

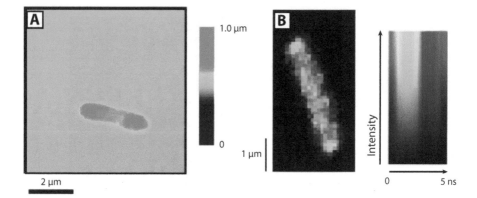

COLOR FIGURE 3.C1 (A) Topography image of a single *S. oneidensis* bacterial cell on an agarose gel surface (the agarose keeps the bacteria alive in a dry environment). (B) Fluorescence lifetime image of the same kind of bacterium in solution. In both images, cell polarity is clearly shown, either as protuberance or as local increase of the fluorescence lifetime. (Reprinted from Micic et al. 2004. With permission.)

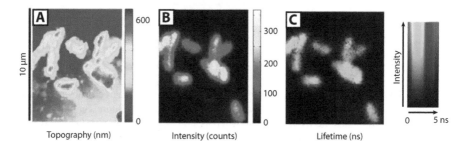

COLOR FIGURE 3.C2 *S. oneidensis* bacterial cells on poly-l-lysine expressing the fluorescent fusion protein. Topography (A), fluorescence intensity (B), and fluorescence lifetime (C) images of the same group of cells. (Reprinted from Micic et al. 2004. With permission.)

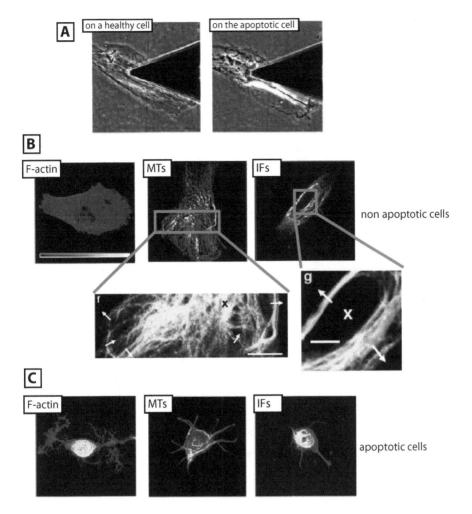

COLOR FIGURE 3.D1 (A) Phase contrast micrographs showing an AFM cantilever exerting force on a healthy cell (left) and on the very same cell after 120-min treatment with the apoptotic agent STS (right). (B) Subtraction images of healthy cells on being deformed with an AFM cantilever. Image contrast originates from the relative displacement of the cytoskeleton fibers caused by the deformation. The F-actin image has low contrast, which means that they practically do not move in the process. However, they do displace MTs and IFs. The insets are overlapped images of the same optical plane before (in green) and after (in red) local deformation. Green-red stripes denote fiber displacements, and they are depicted by arrows. Not all the fibers move away from the point of force (marked as "X"). (C) Subtraction images of apoptotic cells after 120-min treatment with STS. The actin monomers concentrate around the nuclear region pressed by the AFM tip. MTs and IFs reveal maximum contrast around the nuclear region, which indicates that they move away from the point of force. (Reprinted from Pelling et al. 2009. With permission.)

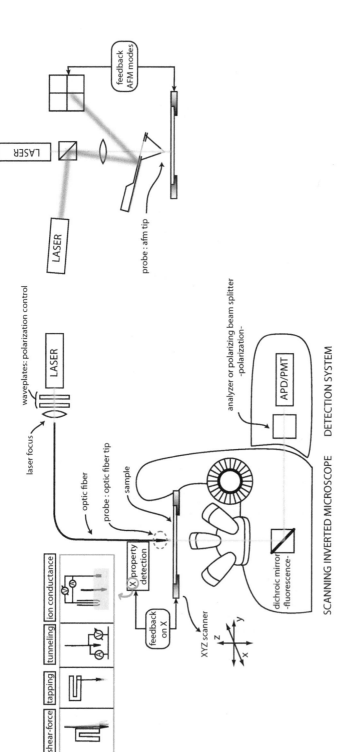

COLOR FIGURE 4.5 Transmission aperture SNOM. The figure on the left shows a typical SNOM setup mounted on an inverted microscope. Light is delivered through an optic fiber whose open end is positioned in close vicinity to the sample. Light polarization can be set at any value through polarizers at the output of the laser cavity. Transmitted light is collected by an objective of high numerical aperture and delivered to the detection system, usually consisting of photomultiplier tubes (PMTs) of avalanche photodiodes (APDs). In fluorescence SNOM both excitation and emission radiation can be collected by a dichroic mirror and two PMTs; in polarization contrast SNOM an analyzer or a polarizing beam splitter can be placed before the detectors to collect light at different polarization directions. In the depicted case, the sample is mounted on an xyz scanner that keeps the sample at constant distance from the probe at all times and thus assists in acquiring the topological image. This is done by a feedback mechanism, usually focused on a property X, which is distance dependent and is set to a constant value. The left inset illustrates the most popular feedback-based properties. Instead of an optic fiber, it is possible to use a micromachined cantilever with an integrated tip in a typical AFM setup. The green laser is focused on the tip's aperture to create the near-field radiation, while the red laser is focused on the cantilever and reflected to the photodetector. This in turn delivers the signal that actuates the feedback.

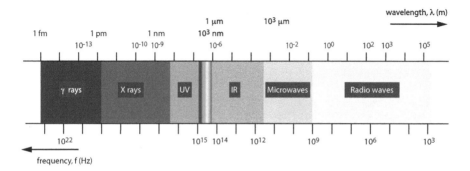

COLOR FIGURE 5.3 The electromagnetic spectrum beyond the visible. The electromagnetic spectrum encompasses most diverse types of radiation; from the highly energetic γ-rays of very short wavelengths (from femtometers to picometers) to the widespread radiowaves of long wavelengths (from few centimeters to hundreds of kilometres).

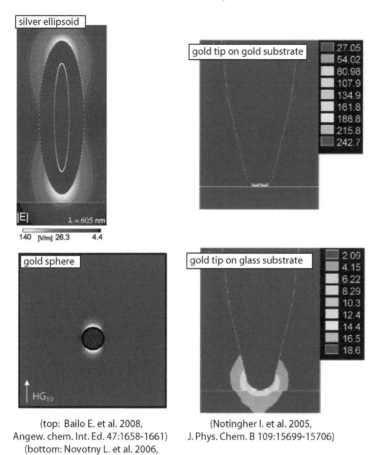

(top: Bailo E. et al. 2008,
Angew. chem. Int. Ed. 47:1658-1661)
(bottom: Novotny L. et al. 2006,
Annu. Rev. Phys. Chem. 57:303-331)

(Notingher I. et al. 2005,
J. Phys. Chem. B 109:15699-15706)

COLOR FIGURE 5.11 Tip-enhancement on curved surfaces. Calculated field distribution around a silver ellipsoid, a gold sphere and a gold tip on a gold and on a glass substrate. Colour scale of the left-hand graphics: the electric field increases from dark purple to yellow. (*Top*: reprinted from E. Bailo, V. Deckert, *Chem. Soc. Rev.* 37:921-930, 2008, with permission *of the Royal Society of Chemistry. Bottom*: reprinted with permission from L. Novotny, S.J. Stranik, *Annu. Rev. Phys. Chem.* 57:303-331, 2006. Copyright 2006 Annual Reviews). Colour scale of the right-hand graphics: the electric field increases from purple to red (Reprinted with permission from I. Notingher, A. Elfick, *J. Phys. Chem. B* 109:15699–15706, 2005. Copyright 2005 American Chemical Society).

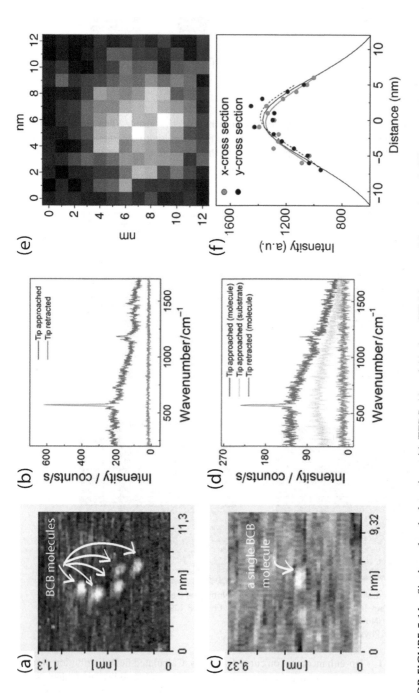

COLOR FIGURE 5.14 Single molecule detection with TERS. (a) and (c) show STM images of five and one brilliant cresyl blue (BCB) molecules, respectively, and their corresponding Raman spectra in (b) and (d). Tip enhancement is clearly shown if one compares the Raman spectra taken when the tip is close to the molecule(s) (tip approached, red curve) and when the tip is far away (tip retracted, grey curve). (e) and (f) show the TERS image of a single BCB molecule on flat gold and the respective intensity profile along the vertical and the horizontal (x and y) axes. (Reprinted with permission of J. Steidtner, B. Pettinger, *Phys. Rev. Lett.* 100:236101, 2008. Copyright 2008 by the American Physical Society, URL http://link.aps. org/abstract/PRL/v100/e236101.)

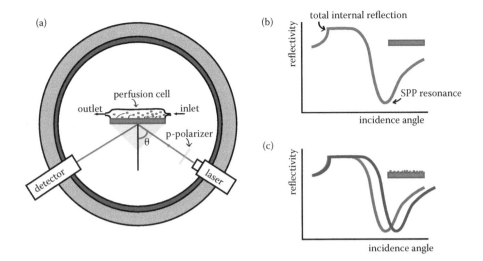

COLOR FIGURE 6.12 SPR setup. (a) The film sample rests on a prism according to the Kretschmann configuration. The interface of interest is exposed to a fluid carrying an adsorbate that is delivered through a perfusion cell. The light beam coming from a laser at an incidence angle θ is p-polarized by the polarizer, traverses the prism, reaches the sample and is reflected back towards the detector. Both laser and detector arms can be rotated by means of a computer controlled goniometer and hence θ can be varied. In this way it is possible to monitor the reflectivity as a function of the incidence angle. (b) Typical reflectivity curve on a bare surface, showing the angles at which total internal reflection and the SPP resonance occur. (c) Reflectivity curves of a bare surface (in red) and on the same surface after material has been deposited (in brown). Whereas the angle of total internal reflection does not vary, the resonance angle shifts to a higher value.

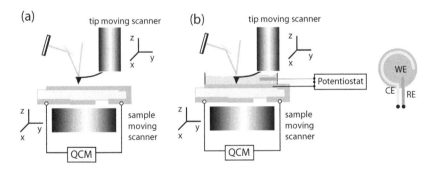

COLOR FIGURE 6.22 The combined AFM-QCM technique. (a) The two-in-one version makes use of the exposed side of the QCM sensor as the substrate for AFM studies. The AFM can be either a tip-moving or a sample-moving setup. (b) The three-in-one version includes an electrochemical cell to perform voltammetry. The exposed side of the QCM sensor acts as AFM substrate and additionally as working electrode. The counter electrode (yellow) is a metal wire looped at its end in a three-quarter-of-a-circle shape. Together with the reference electrode (green), they both are immersed in the solution, hovering over the working electrode. The electrodes are connected to a potentiostat that applies the potential in a controlled fashion and measures the current as a function of time or of the applied potential.

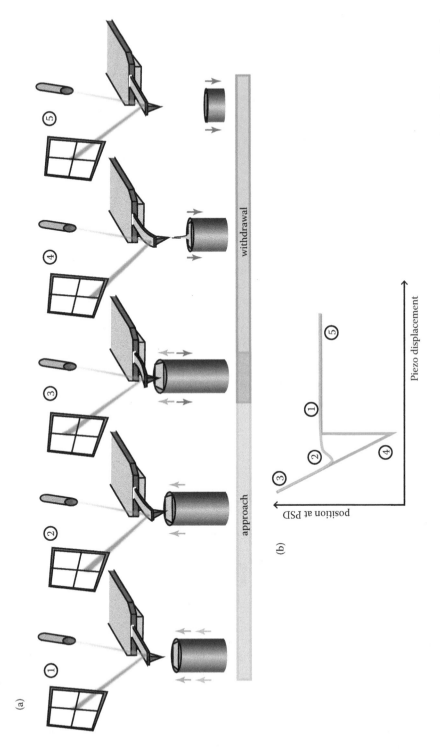

COLOR FIGURE 7.4 Force curves measured with the AFM. (a) The position of the laser beam that is reflected at the backside of the cantilever is registered at the PSD at the same time as the sample is moved toward (1–3) and away from (3–5) it. In this case, the sample rests on the top of a tubular piezo actuator, or simply *piezo*, that moves the sample up and down. (b) The position at the PSD is thus plotted as a function of the piezo displacement during approach (light orange curve) and during withdrawal (green curve).

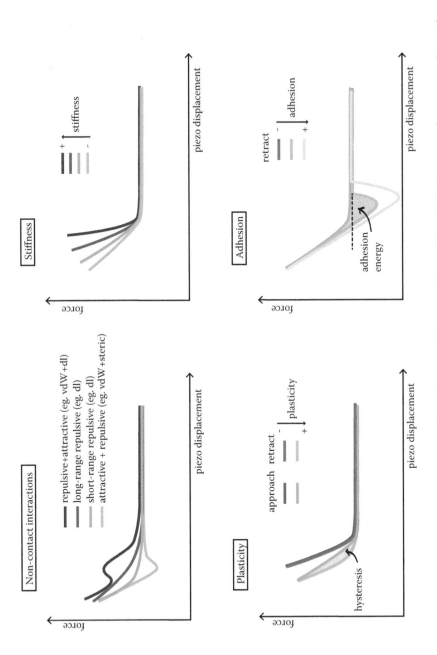

COLOR FIGURE 7.9 The information in a force curve. From the approach curves, it is possible to infer the nature of non-contact interactions (top-left graph) or qualitatively address the sample stiffness (top-right graph). The hysteresis between the approach and retract curves in the contact regime is indicative of the degree of plasticity in the sample (bottom-left graph), whereas the adhesion is a measure of the affinity between the sample and the probe (bottom-right graph).

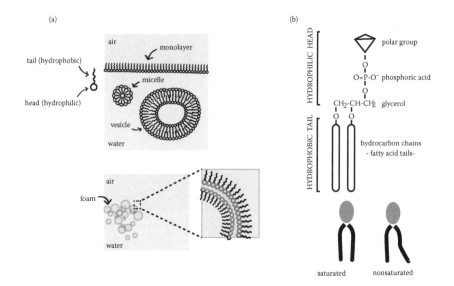

COLOR FIGURE 8.10 Amphiphilic molecules. (a) A surfactant is an amphiphilic molecule composed of a hydrophilic head and a hydrophobic tail. Surfactants can form stable structures in both hydrophobic and hydrophilic media, such as micelles, vesicles or foams. At the interface between a hydrophobic and a hydrophilic fluid, such as air and water, they readily form monolayers. (b) Phospholipids are amphiphilic molecules of particular biological value. The chemical structure shows that their head is composed of glycerol, a phosphoric group and a polar (or ionic) group. The hydrocarbon chains are the tails of fatty acids linked to the glycerol liked through ester bonds. If the hydrocarbon chains do not contain double bonds, the lipids are said to be *saturated*. Otherwise, they are nonsaturated. The double bond impairs the free rotation of the carbon–carbon bond, and hence reduces the flexibility of the hydrophobic tail. The presence of the double bond is depicted as a kink in the tail.

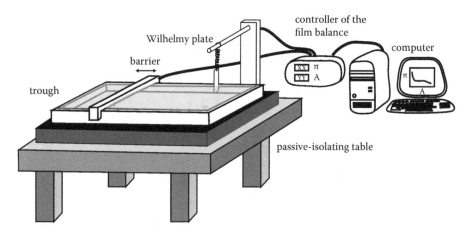

COLOR FIGURE 8.11 Setup of a film balance. A trough usually made of polytetrafluoroethylene (PTFE) is filled with a liquid (subphase). The area of the liquid surface is varied by a barrier (or two) that can be laterally moved at a controlled speed. The surface tension is measured at all times by a Wilhelmy plate connected in turn to the controller. The computer registers the values of the surface pressure and the trough area to produce the Langmuir isotherms. To avoid mechanical vibrations the trough is placed on a passive vibration-isolating table and usually enclosed in a protective hood to avoid dirt being deposited on the surface.

5 Adding Label-Free Chemical Spectroscopy

Who Is Who?

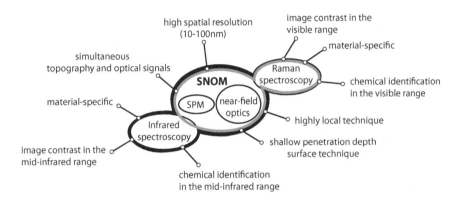

CHEMICAL SPECTROSCOPY

In this chapter we revisit the interaction of matter and electromagnetic radiation with implications other than those connected to optical microscopy. *Spectroscopy* studies the interaction of matter with practically all forms of radiation. The main implication of such study is the chemical identification and/or characterization of any kind of substance. The term *spectroscopy* is in fact quite generic, since it comprises different types of spectroscopies, each one linked to a specific type of radiation. Indeed, the interaction of matter with visible and ultraviolet light is the focus of study in electronic and Raman spectroscopy, whereas vibrational and rotational spectroscopy account for the interactions of matter with infrared (IR) and microwave radiation, respectively. Even radio waves are used to excite matter in nuclear magnetic resonance spectroscopy.

In particular, infrared and Raman spectroscopies have been combined with scanning probe microscopy (SPM) to obtain chemical and topographical maps in a simultaneous manner. Indeed, as in any type of spectroscopy, each substance shows a unique infrared and Raman spectrum that can be considered as a *fingerprint* of its chemical structure as well as of its physicochemical environment. Therefore chemical spectroscopy bestows on SPM a precious gift: the unambiguous identification of sample components.

Raman spectroscopy

1928 -- On studying the light scattered by pure vapours and liquids, C.V. Raman and K.S. Krishnan discovered a small quantity of scattered light of lower frequency than that used to irradiate the samples. This light depends on the vibrational properties of the samples.

Nobel Prize (1930)

 Raman, C.V., Krishnan, K.S. 1928. A new type of secondary radiation. *Nature* 121:501-502

1974 -- J. Fleischmann and coworkers report on the inusual high Raman-scattering of single molecules adsorbed onto roughened metal surfaces. Discovery of the surface-enhanced Raman spectroscopy and first applications on single-molecule detection.

 Fleischmann J., Hendra, P.J., McQuillan, A.J. 1974. Raman spectra of pyridine adsorbed at a silver electrode. *Chem. Phys. Lett.* 26:163-166

1995 -- C.L. Jahncke and coworkers report on the application of near-field optical microscopy to map the Raman intensity of rubidium-doped potassium titanyl phosphate. At the same year D.A. Smith and coworkers present a near-field Raman microscope, which they employed in locally collecting Raman spectra of polymer crystals and diamond films at subwavelength resolution.

 Jahncke, C.L., Paesler, M.A., Hallen H.D. 1995. Raman imaging with near-field optical microscopy. *Applied Physics Letters* 67:2483-2485

Smith, D.A., Webster, S., Ayad, M., Evans, S.D., Fogherty, D., Batchelder, D. 1995. Development of a scanning near-field optical probe for localised Raman spectroscopy. *Ultramicroscopy* 61:247-252

2000 -- C. Stöckle and coworkers and N. Hayazawa and coworkers report on tip-enhanced Raman spectroscopy of organic molecules.

 Stöckle, C., Suh, Y.S., Deckert, V., Zenobi, R. 2000. Nanoscale chemical analysis by tip-enhanced Raman spectroscopy. *Chem. Phys. Lett.* 318:131-136

Hayazawa, N., Inouye, Y., Sekkat, Z., Kawata, S. 2000. Metallized tip amplification of near-field Raman scattering. *Opt. Commun.* 183:333-336
Hayazawa, N., Inouye, Y., Sekkat, Z., Kawata, S. 2001. Near-field scattering enhanced by a metallized tip. *Chem. Phys. Lett.* 335:369-374

FIGURE 5.1 Historical perspective of the Raman effect and Raman spectroscopy in a timeline.

The Raman effect was discovered in 1928, quite recently in comparison with the discovery of infrared radiation, as the timelines of Figures 5.1 and 5.2 show. Named after Chandrasekhara Venkata Raman, the scientist who discovered it, Raman radiation is within the range of visible light, but it also carries information about the vibrational properties of samples. Conversely, the discovery of infrared radiation as well as the groundwork for infrared spectroscopy was mainly done before the end of the eighteenth century. However, the main technical development occurred throughout the twentieth century for both types of spectroscopy, which has greatly contributed to making these techniques among the most widespread analytical tools.

The discussion in the following background section will help the reader to understand the basics of chemical spectroscopy.

IR spectroscopy

1800 -- Frederick William Herschel reports on the discovery of infrared radiation. He detects it through the increase of temperature when a thermometer was irradiated with light beyond the red part of the electromagnetic spectrum. He coined the radiation "the calorific rays"

> Herschel, W. 1800. Investigation of the powers of prismatic colours to heat and illuminate objects (...). *Philos. Trans. Roy. Soc. London* 90:255-283

> Herschel, W. 1800. Experiments of the refrangibility of the invisible rays of the sun (...). *Philos. Trans. Roy. Soc. London* 90:284-392

> Herschel, W. 1800. Expeirments on the solar and on the terrestrial rays that occasion heat *Philos. Trans. Roy. Soc. London* 90:293-326 (part I), 437-538 (part II)

1881 -- Albert Abraham Michelson invents the interferometer and settles the grounds for interferometric spectroscopy

Nobel Prize (1907)

1882-1900 -- William de Wiveleslie Abney and Edward Robert Festing photograph absorption for 52 compounds. They correlate absorption bands with the presence of certain organic groups in the molecules

1903 -- William W. Coblentz continues the work of Abney and Festing and lays the groundwork for infrared spectroscopy. He investigates the spectra of hundreds of substances, both organic and inorganic. He concludes:
- each compound has a unique IR spectrum
- certain groups give absorption bands at approximately the same wavlengths even if they are in different molecules

1949 -- R Barer´s group develop the first combination of microscopy and IR spectroscopy

1953 -- Development of the first commercial microscope for IR spectroscopy

FIGURE 5.2 Historical perspective of infrared radiation and infrared spectroscopy in a timeline.

Background Information

REVISITING THE ELECTROMAGNETIC SPECTRUM

In Chapter 3 we introduced visible light as an electromagnetic wave of varying wavelengths in the range of 400 to 700 nm. However, visible light covers only a tiny portion of the whole electromagnetic spectrum. The latter comprises a much wider range of wavelengths and types of radiation: from the highly energetic and harmful γ-rays of wavelengths below 0.01 nm to the commonplace radio waves of long, kilometric wavelengths (10^{11} nm) (Günzler and Gremlich 2002). Figure 5.3 shows a scheme that represents the electromagnetic spectrum. We

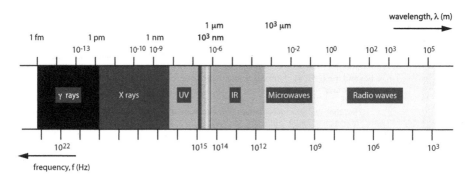

FIGURE 5.3 *(See color insert.)* The electromagnetic spectrum beyond the visible. The electromagnetic spectrum encompasses most diverse types of radiation; from the highly energetic γ-rays of very short wavelengths (from femtometers to picometers) to the widespread radiowaves of long wavelengths (from few centimeters to hundreds of kilometres).

would like to draw the reader's attention to the visible light range as well as to the adjacent regions at longer wavelengths. The large *infrared* region extends over three orders of magnitude in wavelength, from 10^3 to 10^6 nm. This region is in turn divided into three subregions, namely the near-, mid- and far-infrared in order of increasing wavelength. Beyond the far-infrared radiation, we encounter the region of the *microwaves*, encompassing radiation of wavelengths between 10^6 and 10^7 nm.

At this point, it may be useful to introduce parameters other than the wavelength to characterize the electromagnetic radiation that is especially common in spectroscopy. Bearing in mind the representation of an electromagnetic wave in Figure 3.2, the frequency ν is the number of oscillations of the electric or magnetic radiation per time unit, and it is related to the wavelength through the following relation:

$$\nu = \frac{c}{\lambda} \tag{5.1}$$

where c is the speed of the electromagnetic wave in the medium, that is, the speed of light in that particular medium. Frequency has thus units of hertz and is inversely proportional to the wavelength. In vacuum, c is a fundamental constant, c_0, and equal to $2.99793 \cdot 10^{10}$ cm·s^{-1}; in a medium of refractive index n, $c = c_0/n$, is smaller than c_0. As a matter of fact, air has a refractive index very close to 1 (1.00027), and the difference between c and c_0 is negligible.

The *wave number* $\tilde{\nu}$ is the inverse of the wavelength, another parameter that is gaining acceptance in IR and Raman spectroscopy. This magnitude is often expressed in units of cm^{-1}, that is to say, the number of cycles along a 1-cm-long wave train.

$$\tilde{\nu} = \frac{1}{\lambda(cm)} = \frac{10^7}{\lambda(nm)} \tag{5.2}$$

CHEMICAL SPECTROSCOPY: AN INTUITIVE, SIMPLIFIED VIEW

Visible, infrared, and microwave radiation may be absorbed or emitted by matter as a result of an interaction. The absorption and emission of radiation occur only at specific wavelengths or frequencies, though; the particular values are intimately connected to the chemical nature of the material. The extent of absorption or emission of radiation as a function of wavelength or wave number is known as *absorption* or *emission spectrum* and can be considered as a kind of chemical fingerprint of the material, thus constituting a valuable identifying tool. The acquisition, analysis, and interpretation of spectra to infer the chemical structure of materials, characterize their dynamics, or sense their physicochemical environment are considered aims in *chemical spectroscopy.*

As in the case of fluorescence (see Chapter 3), the absorption and emission of electromagnetic radiation involves transitions between discrete energy levels, which are inherent to the material and define a particular dynamic condition of the constituting molecules or atomic network. This condition is described in electronic, vibrational, and rotational states or energy levels. The transition from a lower to a higher energy level—*excitation*—may induce the absorption or radiation, whereas the transition from a higher to a lower energy level—*deexcitation* or *decay*—may result in the emission of radiation. The radiation energy that is emitted or absorbed is equal to the energy difference between the levels involved in the transition.

Figure 5.4 shows a rather oversimplified view of the energy diagram of a molecule, depicting the electronic, vibrational, and rotational levels. Each electronic level comprises a series of vibrational levels, and each vibrational level contains in turn a set of rotational levels. Transitions between electronic states involve radiation of higher energy than those between vibrational levels, and transitions between the latter are in turn more energetic than those between rotational levels. Not every transition is possible, which means that not all the transitions are equally probable. The probability of a transition is dictated by the so-called *selection rules* and by the number of molecules occupying the energy level from which the transition starts.

Visible light can induce transitions between the electronic and, through Raman scattering, the vibrational levels of the molecule, whereas infrared and microwave radiation alter the molecule's vibrational and rotational states, respectively.

Vibrational spectroscopy accounts for energetic transitions between vibrational energy levels brought about by infrared radiation in conventional infrared spectroscopy or by visible light, in the case of Raman spectroscopy.

In practical terms, the spectra of complicated molecules are not interpreted in terms of the whole vibrational activity of the molecule, but of fragments of it. The vibrational spectra of those fragments are well known, and in most cases the assignment consists of the identification of particular peaks as symptomatic of the presence of the corresponding fragment in the molecule. Table 5.1 shows a list of characteristic absorption peaks of the most common chemical functional groups. Each functional group may have a series of vibration modes that

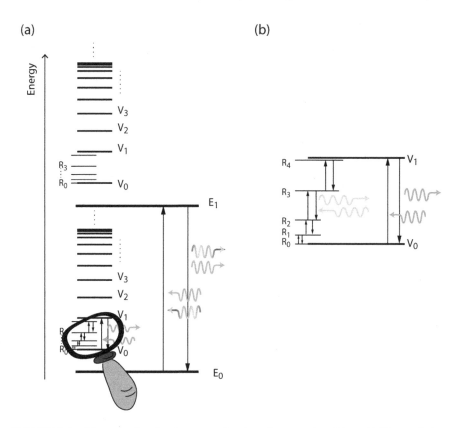

FIGURE 5.4 Diagram of molecular energy levels and energy transitions. (a) Electronic levels are labelled *"E_i"* ($i = 0, 1, 2, ...$); transitions between these levels involve either visible or UV light. Each electronic level comprises a set of non-equally spaced vibrational energy levels, labelled *"V_j"* ($j = 0, 1, 2$). Transitions between these levels involve infrared radiation. The vibrational levels comprise in turn a set of rotational energy levels denoted *"R_k"* ($k = 0, 1, 2, ...$). Transitions between these levels involve microwave radiation. (b) Magnified view of the rotational energy levels between the first vibrational levels V_0 and V_1. The diagram is an oversimplified view of the actual energy landscape of molecules which can be very complex (e.g., due to the presence of degenerate states or the overlap of vibrational levels from different electronic states).

are either IR- or Raman-active. These vibration modes are oscillations of bond lengths (stretching modes) and/or of bond angles (bending modes). Every mode can be expressed as a combination of one or various of the so-called *fundamental vibrations*. The number of fundamental vibrations depends on the *degrees of freedom* of the molecule that are independent modes of motion, including translations and rotations. The number of degrees of freedom is *$3N$*, where N is the number of atoms in the submolecular fragment. After subtracting the translational and rotational modes, linear molecules or fragments possess $3N - 5$ fundamental vibrations, whereas nonlinear molecules or fragments have $3N - 6$.

TABLE 5.1

Infrared Absorption Bands of Some Typical Organic and Inorganic Groups

	Group	Structure	Absorption region (in cm⁻¹)
ORGANIC	C-C triple bond	$-\equiv-$	2000 - 2400
	C-C double bond	$>=<$	1500 - 1800
	C-C single bond	$-\overset{\mid}{\underset{\mid}{}}-\overset{\mid}{\underset{\mid}{}}-$	1300 - 900
	C-H methyl, methylene groups	$-\overset{H}{\underset{H}{\overset{\mid}{C}}}H \quad \overset{H}{\underset{}{>}}\overset{}{\underset{H}{<}}$	2800-3000 (stretching), 1300-1500 (bending, twisting)
	alkyl-halogen compounds	$C-X \quad X= F, Cl, Br, I$	1365-1120 (F), 830-560 (Cl), 680-515 (Br), 610-485 (I)
	ethers and acetals	$R'-O-R'' \quad -\overset{H}{\underset{OR''}{\overset{\mid}{C}}}-OR'$	1100 (ethers) ; 1175-1065 (acetals)
	alcohols	$-\overset{\mid}{\underset{\mid}{C}}-O\text{-}H$	1250-1000 (C-O) ; 3330 (O-H)
	carbonyl compounds	$>=O$	1800-1650
	amides	$R-N>=O$	1670-1620 (C=O); 1400-1600 (C-N)
INORGANIC	phosphate	$PO_4{}^{3-}$	1100-950
	sulphate	$SO_4{}^{2-}$	1130-1080, 680-610
	nitrate	$NO_3{}^-$	1410-1340, 860-800
	perchlorate	$ClO_4{}^-$	1140-1060
	permanganate	$MnO_4{}^-$	920-890, 850-840
	dichromate	$Cr_2O_4{}^{2-}$	950-900
	silicate	$SiO_4{}^{2-}$	1100-900
	ammonium	$NH_4{}^+$	335-3030,1485-1390

SPECTROMETERS

A (light*) spectrometer measures the properties of the electromagnetic radiation, such as intensity or polarization, over a specific region of the electromagnetic spectrum. Most of the instruments available operate in the ultraviolet, visible, infrared, and x-ray regions, and they are accordingly referred to as UV-Vis, IR, and x-ray spectrometers. They all share a basic setup, consisting of an *illumination source*, a series of collimating lenses and/or beam mirrors, a device to select a particular wavelength out of the outgoing light (usually containing a diffraction grating), and a *detector*. Illumination and detection can be either opposed to each other in the 180° configuration or at right angle in the 90° configuration.

Figure 5.5 depicts a general setup that can be applied to a UV-Vis or to an IR spectrometer. The core of the spectrometer is the wavelength selector, a device that can be placed either between the source of radiation and the sample

---

* As opposed to a mass spectrometer.

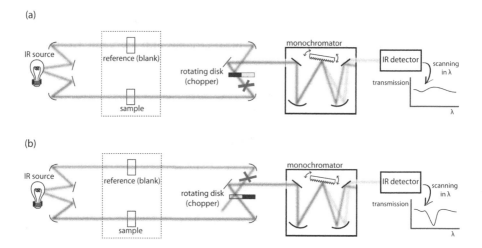

FIGURE 5.5 An IR dispersive spectrometer. Infrared radiation is delivered in two parallel beams through a two-cell compartment containing the sample and a reference. The respective beams reach the monochromator one at a time by means of a rotating disk or chopper. The monochromator selects a particular wavelength and delivers the light towards the infrared detector. The light intensity is acquired as a function of the wavelength. The latter is varied by changing the orientation of the rotating diffraction grid in the monochromator. (a) Acquiring the spectrum of the reference. (b) Acquiring the spectrum of the sample.

or between the sample and the detector, as the figure shows. The wavelength selector usually consists of a diffraction grating and a slit that only lets light of a specific wavelength reach the detector; in this case the device is a *monochromator*. However, the wavelength of the outgoing light can be tuned by slight rotation of the diffraction grating, and therefore the spectrum can be obtained by sequential acquisition of the light intensity at each wavelength. Instead of a monochromator, it is possible to use a *polychromator*. Unlike the monochromator, the device has several exit slots that simultaneously acquire light intensity at multiple wavelengths. The instruments that employ either monochromators or polychromators are called dispersive spectrometers. In the case of Fourier-transform infrared spectroscopy (FTIR), the diffraction grating is replaced by an interferometer (discussed in the next section) (Günzler and Gremlich 2002).

The spectrometer may have a two-cell compartment where the spectrum of the sample can be measured relative to that of a reference in a quasi-simultaneous manner. The reference is usually a substance that does not exhibit any absorption or emission in the range of experimental wavelengths. A rotating disk or chopper allows the alternate acquisition of the sample and reference spectra. As it turns, the disk may block the light beam coming from the sample and direct the light beam coming from the reference to the monochromator and the detector (see Figure 5.5a). In this way the spectrum of the reference is obtained. As it further rotates, the disk will eventually transmit the light beam coming from the sample, which will reach in turn the monochromator and hence the detector. In this case, it is the spectrum of the sample that is collected (see Figure 5.5b).

Both the reference and sample spectra are required to obtain meaningful results. Spectrometers are composed of optical elements that cannot be perfectly manufactured or aligned, and thus create instrumental artifacts. Stray light is also an important source of interference. All these factors may produce spurious signals, which can be easily cancelled out if the spectra are corrected with a suitable reference.

MICHELSON INTERFEROMETER

An interferometer is a device that uses the superposition of electromagnetic waves to extract information about them. The Michelson interferometer is the first of its kind, developed by Albert Abraham Michelson in 1881. It is composed of a coherent light source and four optical elements arranged in a cross, as seen in Figure 5.6. The coherent light source (i.e., a laser) hits the tilted surface of a beam splitter, a half-silvered mirror. Part of the light is reflected, reaching a movable mirror, and part of the light is transmitted, reaching a fixed mirror. Both movable and fixed mirrors are placed at known distances from the beam splitter. The beams are reflected at each mirror and thus return to the beam splitter. At this point, both beams are again partially reflected and transmitted, reaching the detector and creating an interference pattern of concentric dark and bright fringes. Both beams have traveled different paths, but the path length may be the same or different, depending on the position of the movable mirror. If the path length is the same or the path length difference is an even integer of half the wavelength of light, the interference is constructive and the beams form a

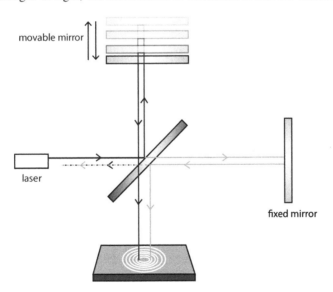

FIGURE 5.6 The Michelson interferometer. A light beam reaches a semi-reflecting mirror that partly reflects the light to a movable mirror and partly transmits the light to a fixed mirror. The first beam travels along a path whose distance varies with the position of the movable mirror. The respective beams are reflected, reaching the detector below and building up an interference pattern of concentric dark and bright fringes.

bright halo. However, if the path-length difference is an odd integer of half the wavelength of light, the interference is destructive and the beams form a dark halo. The interferometer can thus be used to precisely measure distances or light wavelengths.

RAMAN SCATTERING AND SPECTROSCOPY

When an electromagnetic radiation interacts with a molecular sample, the former will be reflected, transmitted, absorbed, or scattered by the latter. Most of the scattered radiation has the same wavelength as that of the incoming radiation; this elastic scattering is known as *Rayleigh scattering*. There is, however, a small amount of the scattered radiation (approximately 1 in 10^7 photons) that would either have longer or shorter wavelengths than that of the incoming radiation. This is the *Raman effect*, and the resulting radiation with longer and shorter wavelength is referred to as *Stokes* and *anti-Stokes Raman* scattering, respectively.

The Raman effect may involve the modification of electronic, rotational, or vibrational states of the molecules, and it can be described in either classical or quantum mechanical terms.

We choose here the classical description, since we consider it to be straightforward and accurate enough to explain the phenomenon (Gucciardi et al. 2007). We will then introduce quantum mechanical concepts to illustrate the previously mentioned modifications of the molecule dynamics as transitions between quantized energy levels.

When an electromagnetic field **E** acts on a molecule, it induces an electric dipole on the latter. The dipole moment **P** can be expressed in a first approximation by

$$\mathbf{P} = \alpha \cdot \mathbf{E} \tag{5.3}$$

where α is the *polarizability tensor*. The polarizability is the sample's property that determines the degree of scattering of an incident radiation. It is a measure of how far the electrons in the molecule would be displaced relative to the nuclei (Hollas 2004). The polarizability is an anisotropic property, which means that it may take different values in different directions starting from the center of the molecule or from each atom's nucleus and can be expressed in the form of a matrix. To express the matrix elements, α_{ij}, as functions of nuclei positions, they can be expanded in a Taylor series near the nuclei equilibrium positions:

$$\alpha_{ij} = a_{ij}^0 + \sum_k \left(\frac{\partial a_{ij}}{\partial q_k} \right)_0 q_k + \frac{1}{2} \sum_{k,l} \left(\frac{\partial^2 a_{ij}}{\partial q_k \partial q_l} \right)_0 q_k q_l + \dots \tag{5.4}$$

where a_{ij}^0 are the polarizability tensor elements at the equilibrium positions, and q_k the dynamic coordinates (vibration, rotation, etc.). The derivatives are summed over all possible dynamic modes. Considering only one dynamic mode, e.g., the kth normal mode of vibration and neglecting the higher-order terms for small vibrations, the element α_{ij} can be expressed as

$$\left(\alpha_{ij}\right)_k = a_{ij}^0 + \left(\frac{\partial a_{ij}}{\partial q_k}\right)_0 q_k \qquad (5.5)$$

and thus the polarizability tensor has the form

$$\alpha_k = \alpha_0 + q_k \alpha_k' \qquad (5.6)$$

Assuming that the vibration coordinate behaves with time as a harmonic oscillator with an amplitude A_0 and a frequency ω_k

$$q_k = A_0 \cos \omega_k t \qquad (5.7)$$

the electric component of the electromagnetic field has the form

$$\mathbf{E} = \mathbf{E}_0 \cos \omega_L t \qquad (5.8)$$

Substituting Equations (5.6, 5.7, and 5.8) into Equation (5.3) gives the expression of the electric dipole moment for the kth vibration mode, which can then be expressed as a sum of three components

$$\mathbf{P} = \alpha_0 \mathbf{E}_0 \cos \omega_L t + \frac{1}{2}\alpha_k' \mathbf{E}_0 q_{k0}\left[\cos(\omega_L + \omega_k)\cdot t + \cos(\omega_L - \omega_k)\cdot t\right] \qquad (5.9)$$

The first term of Equation (5.9) has the same frequency of the electromagnetic field and thus gives rise to the Rayleigh scattering. The second and third terms describe oscillating dipoles at higher ($\omega_L + \omega_k$) and lower ($\omega_L - \omega_k$) frequencies than the electromagnetic field, respectively. These terms give rise to the anti-Stokes and Stokes Raman scattering, respectively, and they both depend on the derivative of the polarizability tensor with respect to the dynamic coordinates. Thus for a molecule to exhibit Raman activity, it must undergo a change of its polarizability in the course of the dynamic mode, e.g., in the course of a vibration mode or rotation

$$\alpha_k' \neq 0 \text{ or } \left(\frac{\partial \alpha_k}{\partial q_k}\right) \neq 0 \qquad (5.10)$$

Equation (5.10) thus constitutes the *selection rule* in Raman spectroscopy.

The quantum mechanical treatment replaces the electric dipole of the classical theory by a transition electric dipole that is generated through the transition of the molecule from an initial quantum state $|i\rangle$ to a final quantum state $|f\rangle$. Each quantum state is associated with an energy level, and the presence

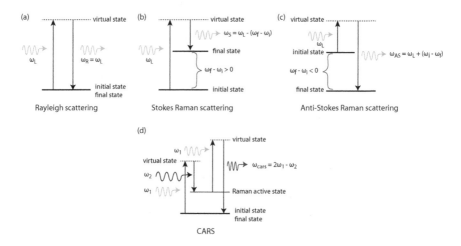

FIGURE 5.7 Transitions between energy levels giving rise to Rayleight and Raman scattering. (a) Rayleigh scattering, (b) Stokes Raman scattering, (c) Anti-stokes Raman scattering, (d) CARS.

of the electromagnetic field induces transitions between different energy levels. Figure 5.7 illustrates those transitions for the case of Rayleigh and Raman scattering. In both processes, two transitions occur. The interaction of electromagnetic radiation is associated with the absorption of a photon of energy $\hbar\omega_L$ that brings the system from the initial state, $|i\rangle$, to an intermediate state, either virtual or real, $|k\rangle$, and to the emission of a photon that brings the system from the intermediate state $|k\rangle$ to the final state, $|f\rangle$. The energy of the initial state is $E_i = \hbar\omega_i$, and the energy of the final state is $E_f = \hbar\omega_f$.

If $\omega_f - \omega_i = 0$, the initial and final states coincide, and the scattered radiation has the same frequency as that of the incident one, as corresponds to Rayleigh scattering (Figure 5.7a). If $\omega_f - \omega_i > 0$, the final state has higher energy than the initial state, and hence a Stokes photon of energy $\hbar[\omega_L - (\omega_f - \omega_i)]$ is emitted, which has lower energy than the incident one (Figure 5.7b). Conversely, if $\omega_f - \omega_i < 0$, the final state has lower energy than the initial state, and an anti-Stokes Raman photon is emitted, having higher energy than the incident one, $\hbar[\omega_L + (\omega_i - \omega_f)]$ (Figure 5.7c).

The quantum mechanical description also predicts that the intensity of the Stokes emission is higher than that of the anti-Stokes emission. In particular, the intensity ratio between Stokes (I_S) and anti-Stokes emission (I_{AS}) can be obtained as

$$\frac{I_S}{I_{AS}} = \frac{(\omega_L - \omega_k)^4}{(\omega_L + \omega_k)} \cdot e^{-\left(\frac{\hbar\omega_k}{k_B T}\right)} \tag{5.11}$$

which also accounts as well for the temperature dependence of the intensity ratio.

SURFACE-ENHANCED RAMAN SCATTERING

If a particular molecule that exhibits Raman active vibration or rotation modes is adsorbed upon a metal surface, the intensity of the corresponding Raman bands is considerably increased. The phenomenon, called *surface-enhanced Raman scattering*, was discovered in 1974 by Fleishmann when studying the Raman signal of pyridine molecules adsorbed on silver electrodes.

The reason for such enhancement has been much debated and is attributed to be a combination of two mechanisms. The first is referred to as the *electromagnetic effect*, mainly caused by the excitation of surface plasmons in the metal structure that in turn enhances the electromagnetic field in the immediate surroundings above the metal surface. This effect is more prominent on rough or particulate surfaces, where the molecules are adsorbed on corrugations of smaller size than the light wavelength. The enhancement of the Raman line intensity can be expressed to a first approximation as

$$M_{SERS} = \left| \frac{E(r,\omega_0)}{E_0(\omega_0)} \right|^4 \tag{5.12}$$

where M_{SERS} is the signal-enhanced Raman scattering (SERS) factor; $E(r,\omega_0)$ is the total electric field at the molecule's position r; and $E_0(\omega_0)$ is the electric field of the incoming radiation, free from the substrate's influence. The ratio

$$\left| \frac{E(r,\omega_0)}{E_0(\omega_0)} \right|$$

is the so-called field-enhancement factor. The second mechanism is referred to as a chemical effect that may cause a shift in the Raman bands as well as signal amplification (Schmid 2008). The chemical effect arises from the interaction between the adsorbed molecules and the surface that alters or generates their electronic states, inducing resonant Raman scattering. The electromagnetic effect can give signal enhancement factors of 104 to 108, whereas the chemical effect accounts for signal enhancements of 10 to 102.

A NONLINEAR EFFECT: COHERENT ANTI-STOKES RAMAN SCATTERING (CARS)

The former classical derivation of the Raman scattering was based on Equation (5.3), which states that an electric field interacting with a molecule induces a dipole moment proportional to the strength of the electric field. However, the expression is a simplification of the following:

$$\mathbf{P} = \alpha \cdot \mathbf{E} + \frac{1}{2}\beta : \mathbf{EE} + \frac{1}{6}\gamma \vdots \mathbf{EEE} + \cdots \tag{5.13}$$

The polarization **P** can thus be expressed as a sum of terms of ever-increasing order. The higher-order terms are progressively smaller, so that the overall polarization remains finite. The first term is responsible for the linear effects of Rayleigh scattering as well as Stokes and anti-Stokes Raman scattering. The second and third terms are nonlinear in **E**; the third term describes the so-called *four-wave mixing processes*, where the coherent anti-Stokes Raman scattering (CARS) is the most common.

The four-wave mixing processes pivot around the concept of three incoming electromagnetic fields that interact to produce a fourth field (Thiel n.d.). In the case of CARS, the process involves the incoming of two distinct electromagnetic fields of frequencies ω_1 (the *pump* field) and ω_2 (the *Stokes* field) that interact with the target molecule and with the incoming beam at ω_1 (the *probe* field) to produce an output field with frequency $\omega_{AS} = 2\omega_1 - \omega_2$ (Figure 5.7d). If the two distinct incoming fields are chosen so that the difference of their respective frequencies $\omega_1 - \omega_2$ coincides with a Raman active transition of the molecule, a CARS signal is generated at the anti-Stokes frequency $\omega_{CARS} = 2\omega_1 - \omega_2$. Furthermore, coupling and hence energy transfer between the fields will only occur if momentum and energy are conserved. The conservation of momentum is expressed as

$$\Delta \mathbf{k} = \mathbf{k}_{CARS} - 2\mathbf{k}_1 + \mathbf{k}_2 = 0 \tag{5.14}$$

where $\mathbf{k}_1$ and $\mathbf{k}_2$ are the wave vectors of the respective electric fields involved. The CARS signal will then attain its maximum value if the so-called phase-matching condition is satisfied.

RAMAN (AND IR) MICROSCOPY

The combination of a Raman spectrometer with an optical, usually confocal, microscope produces Raman spectra of small sample volumes, down to less than one micrometer laterally and from one to a few micrometers in depth (McCreery 2000). Indeed, both techniques constitute a good match, since Raman spectroscopy often involves the use of visible light. Figure 5.8 depicts the combination of an upright microscope with a Raman spectrometer. Light coming from a coherent light source such as a laser is fed into the excitation axis of the optical microscope by means of a beam splitter, a dichroic mirror, or a holographic notch filter (McCreery 2000). The beam is reflected at the sample and collected by the same objective in a 180° backscattered geometry. The reflected light is separated from the excitation path and directed into the entrance slit of the spectrometer. This is done by the same optical device that previously inserted the laser beam into the optical axis of the microscope.

The type of spectrometer defines the different imaging modes in Raman microscopy. The laser spot illuminates a region of the sample within which spectra can be acquired. If the spectrum is acquired at a single point on the sample, the mode is called *single-point Raman microscopy* or *microspectroscopy*; point-to-point mapping results from the *sequential acquisition* of spectra from a series of sample points. If the points are aligned, they can produce a point-to-point profile, where the intensity at a constant Raman shift is plotted as a function of the point's position.

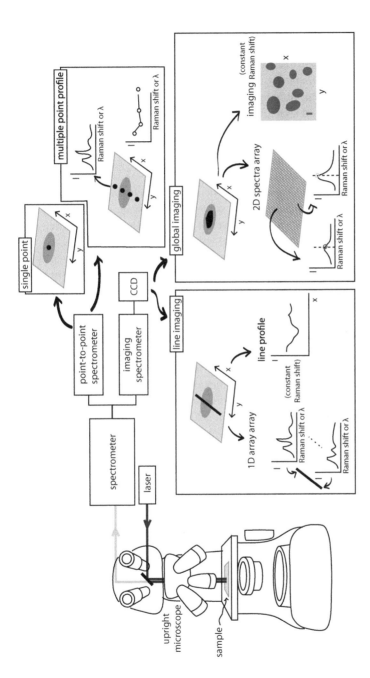

FIGURE 5.8 A Raman (or IR) microscope. The setup combines an upright reflection optical microscope and a spectrometer. The light source coming from a laser—that may be integrated in the spectrometer—is directed to the sample along the optical axis of the microscope. The light coming from the sample is first filtered out to remove any excitation light before it reaches to the spectrometer. Imaging modes depend on the type of the spectrometer. A point-to-point spectrometer enables microspectroscopy at selected positions within the sample. An imaging spectrometer and a CCD allow line or global imaging. In all cases a spectrum (light intensity as a function of Raman shift or of IR wavelength) is acquired at each sample position, that is, characterized by a set of x, y coordinates (x, y and z coordinates in the case of a confocal optical microscope).

In *Raman imaging*, the collection of Raman spectra is done simultaneously at different sample positions with a charge-coupled device (CCD) attached to an imaging spectrometer. The results can be displayed in different ways: in the shape of images, showing the spatial distribution of the intensity of the Raman signal at a particular Raman shift in a certain area of the sample (*global* or *direct imaging*); in the shape of profiles, showing the intensity distribution along a line on the sample (*line imaging*); or in the shape of whole spectra plots (intensity versus Raman shift) at each single point.

The end result in Raman imaging is often termed *hyperspectral imaging*, since light intensity of the Raman signal as well as the Raman shift are recorded at each point within the sample. This point can be characterized by two coordinates or (x,y) pairs in the focus plane, or by three coordinates or (x,y,z) pairs in 3-D space. The latter case applies when a confocal microscope is used.

The resolution in Raman microscopy depends not only on the size on the laser spot, but also on the collection optics, in particular on the objective's magnification and depth of focus. Indeed, materials like silicon may show quite different Raman spectra if the light is collected with objectives of different magnification (McCreery 2000). The best lateral resolution ever attained with those means is 500 nm. The depth resolution can be considerably improved with a confocal optical microscope, though the magnitude depends in a complex manner on the confocal aperture as well as on the objective's magnification and on the objective's numerical aperture. For small confocal apertures (i.e., 100 μm) and high-magnification objectives (i.e., 100×), the depth resolution can be as small as 2 μm (McCreery 2000).

The counterpart to Raman microscopy in the nonvisible infrared range is infrared (IR) microscopy. The combination of optical microscopy and infrared spectroscopy shares most of the instrumental and imaging modes of Raman microscopy mentioned previously. However, the particularities of infrared radiation demand other types of radiation sources, collecting and focusing objectives, spectrometers, and detectors. The most convenient configuration is a reflection optical microscope with a Fourier-transform infrared (FTIR) spectrometer. The infrared beam of the FTIR spectrometer is made collinear with the visible optical path of the microscope, so that samples can be easily positioned and localized with visible light and subsequently analyzed with infrared radiation. Beam focusing onto the sample plane and collection of the reflected infrared radiation are made by Cassegrain objectives (Günzler and Gremlich 2002). Photodetectors such as an MCT detector (a semiconductor detector made of an alloy of mercury, cadmium, and tellurium: HgCdTe) are employed in FTIR microscopy. The main drawback of IR microscopy is its relatively poor lateral resolution. Lateral resolution is mainly determined by the diffraction limit, which in the case of infrared radiation can be as large as several micrometers ($\approx\lambda/2$, ≈ 0.5–400 μm).

Raman Microscopy beyond the Diffraction Limit: Near-Field Raman Spectroscopy

The integration of Raman spectroscopy into the scanning near-field optical microscopy (SNOM) brought the capabilities of Raman microscopy to nanoscale resolution.

Raman SNOM thus makes it possible to map point by point the structural, chemical, and electronic properties along with the sample topography.

Aperture Near-Field Raman Spectroscopy

This configuration was employed in the early years of the technique (Gucciardi et al. 2007), using an aperture SNOM connected to a high-resolution Raman spectrometer in both illumination and collection modes, as illustrated in Figure 5.9 (see Chapter 4 for details about these modes). Both setups make use of aperture SNOM probes consisting of tapered optical fibers. Light coming from an Ar^+ laser (488 or 514.5 nm) in most cases is delivered to the sample either by a subwavelength aperture located at the tip of an optical fiber, which is positioned very close to the specimen, or by an optical objective in the far field. In the first case or illumination mode, reflected light is collected by a long-distance optical objective oriented at 45° from the surface normal (Smith et al. 1995); alternatively, transmitted light can be collected with an objective situated below the sample plane (Jahncke, Paesler, and Hallen 1995). In the second case or collection mode, the reflected light is collected by the aperture SNOM tip. In the illumination mode, the long-distance objective is coupled to the entrance of a monochromator before the light reaches a photomultiplier tube (PMT) in the photon-counting mode. In the collection mode, the SNOM probe delivers the light into a monochromator, which in turn directs it into a charge-coupled device (CCD). Both configurations make use of notch filters at the entrance of the respective collection systems to filter out the elastically scattered light and thus reduce the background signal. The tip–sample distance is kept constant through a shear-force-based feedback mechanism (see details in Chapter 4): A tuning fork attached to the fiber sets the latter to oscillate laterally. Interaction with the surface results in variations in either the amplitude, the phase, or the frequency that the feedback electronics compensate in order to reestablish a constant tip–sample separation.

During raster scanning, an array of Raman spectra is acquired, where a spectrum is taken at each different position within the sample. Whole spectra can thus be obtained as a function of the sample position, and these can be compared with the topography data. This allows one to chemically identify sample components. However, these conditions did not promote fast acquisition times. Sample scans took a few hours to complete, achieving ultimate lateral resolutions of 100 nm (Smith et al. 1995; Deckert 1998). An important drawback of this setup is the limited light transmission of the aperture SNOM probes, which in turn impairs sensitivity and resolution. Another problem is the high reflectivity of the glass of the optical fibers used as SNOM probes. Glass gives an intense, broad Raman signal that extends in the range of 150–550 cm^{-1}, which makes this spectral window impractical for sample identification (Zeisel et al. 1997) and distinct Raman bands at 800 and 1100 cm^{-1} (Deckert 1998).

Tip-Enhanced Near-Field Raman Spectroscopy (TERS)

The application of tip-enhanced SNOM to the Raman radiation is especially convenient in obtaining Raman spectra of materials, especially of poor Raman scatterers that otherwise exhibit very low or undetectable Raman signals.

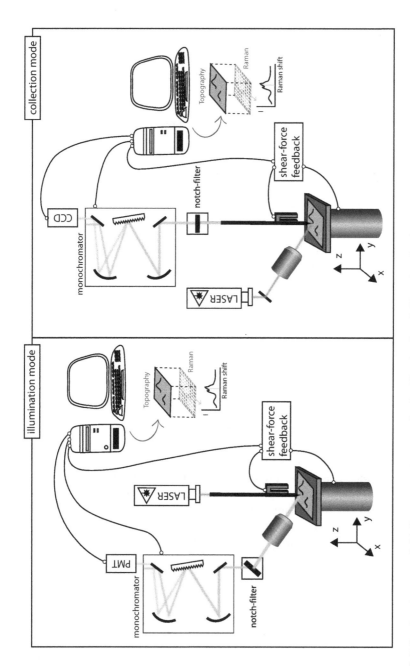

FIGURE 5.9 Aperture near-field Raman spectroscopy in the illumination (left) and collection (right) modes. The reported setups make use of a sample-moving configuration. Probe–sample distance is regulated by a shear-force feedback. The depicted spectrometer is of a dispersive type (a monochromator) that directs the near-field Raman signal to a photomultiplier (PMT). A computer registers the Raman spectra and the topography as a function of the sample position.

Setups working in both transmission and reflection modes have been developed and sketched in Figure 5.10. The setup that operates in transmission mode is based on that of an inverted microscope. Radiation coming from an Ar^+ (488 nm) or a HeNe laser (633 nm) is enhanced by a sharp, apertureless SNOM probe—either an STM-like (scanning tunneling microscope) probe or an AFM-like (atomic force microscope) probe—situated above the sample plane and very close to the specimen. The distance between the SNOM probe and the sample is regulated by shear-force feedback as well. The probe can be illuminated by a Gaussian-mode laser beam (Hartschuh, Anderson, and Novotny 2003), by a p-polarized high-order laser beam, by a p-polarized evanescent field (Hayazawa et al. 2000, 2001), or by a radially polarized beam focused through a high numerical aperture (NA) (e.g., 1.4) optical objective (Quabis et al. 2000; Dorn, Quabis, and Leuchs 2003). Light is collected by the objective situated below the sample plane and detected by a spectrometer or by an avalanche photodiode. In the first case, notch filters are used to cut the Rayleigh scattering; in the second case, band-pass filters centered at the Raman peaks are employed to selectively collect the weak Raman-scattered light.

In reflection mode, the tip is illuminated from the same side of the tip and the light collected likewise by an optical objective, situated at a certain angle from the surface normal. Simulations performed by Downes and coworkers actually found that the highest enhancement is achieved when the angle of illumination is 70° from the tip axis (Downes, Salter, and Elfick 2006). Although this configuration especially eases the illumination of the tip with radiation polarized along the tip axes, it has been scarcely employed, mainly due to the AFM scanner, which impairs the positioning of a low-working-distance objective too close to the surface plane. Instead, long-working-distance objectives with rather low NA (0.5) are used (Elfick, Downes, and Mouras 2010), which impair high lateral resolution. Background signals cannot be removed by lock-in techniques, as in the case of other types of tip-enhanced SNOMs (see Chapter 4; Elfick, Downes, and Mouras 2010), and therefore excitation spots should be as small as possible to illuminate just the tip. Such small excitation spots are difficult to achieve when the incoming laser beam is orientated at 70° from the surface normal and focused with low NA objectives.

The mechanism of the tip-induced enhancement of the Raman effect is still a pending issue, though it is agreed to depend intimately on the tip's material and geometry. The mechanism is believed to be of a dual nature, as in the case of the surface-enhanced counterpart: electromagnetic and chemical (Gucciardi et al. 2007; Stöckle et al. 2000).

The electromagnetic mechanism is related to the increase of the electromagnetic field in the region immediately close to the apex of a metallic tip, due to the lightning-rod effect (see Chapter 4). The effect can be even more pronounced by excitation of plasmon polaritons in tips made of materials that exhibit plasmon resonances, such as gold or silver. The chemical effect is believed to be related to the increased polarizability of the sample molecules due to charge transfer or bond formation with the metal tip. These result in the alteration of the energy levels of the former, leading to shifts or enlargement of the main peaks in the Raman spectra.

Although tip enhancement is produced by p-polarized light along the tip axis, excitation of some Raman modes in certain types of molecules can only occur when

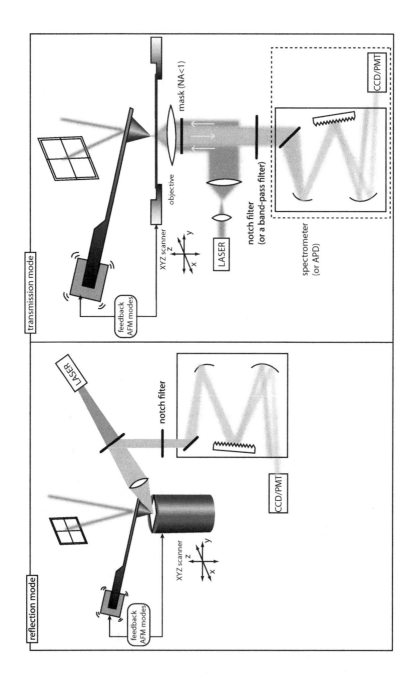

FIGURE 5.10 Tip-enhanced Raman spectroscopy (TERS) in reflection (left) and transmission (right) modes. The light beam coming from the laser illuminates the tip either at an angle from the tip side or from below through a high NA optical objective. A notch filter or a band pass filter blocks the far-field signal and the Rayleigh scattering. The Raman signal enters the spectrometer and is detected by either a CCD or a photomultiplier tube.

irradiated with *s*-polarized light, parallel to the sample plane (Hayazawa and Saito 2007). This is especially true in the case of the G-bands in single-walled carbon nanotubes (Hayazawa and Saito 2007), where polarization measurements reveal the orientation of carbon nanotubes (Hartschuh et al. 2003). This coupling intensifies the strength of the electromagnetic field in the vicinity of the tip and, hence, further increases the tip enhancement.

Tip-Enhanced Coherent Anti-Stokes Raman Scattering (CARS)

Reports on the use of CARS in combination with near-field techniques revolve about two distinct setups: the earlier CARS-SNOM using optical fiber probes (Schaller et al. 2002) and the tip-enhanced CARS (TE-CARS) (Ichimura et al. 2004).

The first uses a collection-mode SNOM, where a noncoated fiber probe collects the reflected CARS emission coming from the sample. The latter is obliquely illuminated by laser pulses coming from a single titanium:sapphire laser that generates a Stokes pulse of fixed frequency (ω_2) and a tunable pump pulse of variable frequency (ω_1, $\omega_1 > \omega_2$). The pulses are focused onto a spot of a few micrometers on the sample surface and centered at the position of the optical fiber's tip—an electrochemically etched end of 50-nm diameter. The collected light is directed through the optical fiber toward a photomultiplier tube through band-pass filters. Complete acquisition of Raman images takes approximately 30 min (Schaller et al. 2002).

The reported tip-enhanced CARS setup is based on a very similar configuration as TERS in transmission mode. In this case, however, the excitation light source is different; it consists of two mode-locked titanium:sapphire pulse lasers that provide the beams at frequencies ω_1 and ω_2 (Hayazawa and Saito 2007; Downes, Mouras, and Elfick 2009). The beams are collinearly overlapped in space and time, and focused by a 1.4-NA optical objective onto the tip of the SNOM probe, usually metallized and attached to the free end of a micromachined cantilever, i.e., an AFM-type probe. The backscattered CARS emission enhanced by the tip is collected by the same objective and detected by an avalanche photodiode (APD) and a photon counter through a cut-off filter that removes the residual excitation light and a monochromator that selects a particular spectral window. Raman images can be collected in 3–12 min (Ichimura et al. 2004).

Sources of Enhancement of the Raman Signal in Near-Field Raman Spectroscopy

Cross sections for bulk Raman scattering (10^{-30} cm^{-2}, Gucciardi et al. 2007; Hayazawa and Saito 2007) are extremely low compared to those of IR adsorption (10^{-19} cm^{-2}) or fluorescence (10^{-16} cm^{-2}). This means that the intensity of Raman scattered light is much weaker than that of fluorescence, for example. To detect a measurable signal it is thus necessary to irradiate with high-power illuminating sources or to use very long exposure times.

The first works on near-field Raman microscopy dealt with very weak signals, mainly due to the small measurement volumes produced by local illumination of the samples or by collecting the scattered light with subwavelength aperture tips (Jahncke, Paesler, and Hallen 1995; Smith et al. 1995). The acquisition times required to obtain the near-field Raman spectra were at least 60 times longer than those required to obtain far-field Raman spectra (Smith et al. 1995; Webster, Smith, and Batchelder 1998), and Raman images were only completed after 10 hours (Jahncke, Paesler, and Hallen 1995).

However, it is possible to increase the intensity of near-field Raman signals. There are several methods available. On the one hand, the presence of a rough metallic surface in the vicinity of a Raman scatterer intensifies the signal of the latter. The surface can either be a planar substrate upon which the specimen is adsorbed (*surface-enhanced*) or a small, sharp-pointed probe scanning over the specimen (*tip-enhanced*). On the other hand, the Raman signal can be amplified by coherent anti-Stokes Raman scattering (CARS). CARS was applied in combination with SNOM 10 years ago (Schaller et al. 2002) and, more recently, with tip-enhanced SNOM (Ichimura et al. 2004; Downes, Mouras, and Elfick 2009).

Surface-Enhanced

As mentioned previously, the Raman signal of a molecule can be intensified if the latter is adsorbed on a rough metal surface. Surfaces that promote such an enhancement are often called SERS substrates (signal-enhanced Raman scattering). Signal enhancement factors induced by such surfaces can be at least 10^6, and some authors claim to have found evidence for single-molecule detection at such amplifications (Kneipp et al. 1997; Nie and Emory 1997). Likewise, the lateral resolution of surface-enhanced Raman has been reported to be one order of magnitude higher than that attained in micro Raman imaging (Deckert et al. 1998).

The application of SNOM on molecules adsorbed on SERS was first proposed and later developed by Zenobi's group (Zeisel et al. 1997; Deckert et al. 1998). Using aperture probes, the authors obtained Raman scattering maps of cresol fast violet (CFV) and p-aminobenzoic acid (PABA) adsorbed on silver-coated substrates (Zeisel et al. 1997) and of DNA strands labeled with brilliant cresyl blue (BCB) on silver-coated Teflon nanospheres (Deckert et al. 1998). Under these conditions, Raman spectra for CFV and BCB-DNA strands were acquired with an exposure time of 60 s, giving signal-to-noise ratios of >10 and >25, respectively. In the case of PABA, the exposure times were 600 s. Lateral resolutions were reported to be in the range of 80–100 nm. Although the authors claimed to have found no significant differences between the near-field and far-field spectra, they did detect a few bright spots in the Raman images that exhibit unusually intense Raman activity. The existence of these spots, denominated *hot sites* by other authors (Nie and Emory 1997), has been hinted at in works of SERS on single molecules adsorbed on granular metal substrates. The hot sites can either be particles, gaps, or cavities in the metal substrate; however none of the studies has been able to find experimental evidence beyond doubt.

Tip-Enhanced

This enhancing effect was reported for the first time in 2000 with the works of Stöckle et al. (2000) and Anderson (2000), although it had been hinted at in the earlier work of Jahncke, Hallen, and Paesler (1996), who tried to explain the high intensities and new bands observed in the near-field spectra of rubidium-doped potassium titanyl phosphate (Rb-KTP). These results had been reported one year earlier by

Jahncke, Paesler, and Hallen (1995), who suggested that the metal coating that surrounds the aperture of the SNOM tip might have been responsible for the higher Raman signals, acting as a sort of surface-enhanced effect due to its proximity to the sample surface.

The tip-enhanced effect is, in fact, a surface-enhanced effect. In this case, the sharply pointed tip of the SNOM probe—rather than a flat, extended metal substrate—acts as a field enhancer. Phenomena such as the lightning-rod and the nanoantenna effects or excitation of surface plasmon polaritons may contribute to intensify the electromagnetic field in a very small region around the SNOM tip or a round surface (Figure 5.11), resulting in signal enhancement factors as high as 10^7 or even higher, up to $5 \cdot 10^9$ (Hartschuh, Anderson, and Novotny 2008) and lateral resolutions of 10 nm (Elfick, Downes, and Mouras 2010). Overall enhancement of the scattered signal is proportional to E^4, as in SERS, since the tip acts as an enhancer of both the excitation field and the Raman scattering emitted by the sample (Elfick, Downes, and Mouras 2010).

However, not all the tips and tip materials are suitable for tip-enhanced Raman scattering. A requirement for large enhancement is that the tip should be a metal with high concentration of free electrons and surface plasmon resonances in the near-UV or visible range. Silver- or gold-coated AFM-type probes, gold nanoparticles, nanorods, or nano bow ties attached to the free ends of cantilevers have proved to be highly efficient. However, STM-type probes made of gold wires and sharpened through electrochemical etching have been found to outperform metal-coated AFM-type probes (Picardi et al. 2007; Pettinger et al. 2004).

Tip-induced near-field effects are often evidenced through comparison of the Raman spectra obtained in the far-field, where the tip is far away from the sample or in bulk samples, and in the near-field, where the tip is in contact or very close to the sample (1–10 nm) (Schmid et al. 2008; Picardi et al. 2007). The field enhancement can be demonstrated in either one of two ways: The first detects the intensity of a certain Raman band of the specimen (i.e., *constant Stokes shift*) as a function of the tip–sample separation; the second acquires the entire Raman spectrum of the specimen as a function of the tip–sample separation. There are several magnitudes that have been employed to quantify the tip-induced enhancement.

Contrast, C, or *observed enhancement*, OE, is the ratio of the Raman intensity of a particular band in the near field to the intensity of the same band obtained in the far field.

$$C = OE = \frac{I_{\text{near-field}}}{I_{\text{far-field}}} = \frac{I_{\text{total,near-field}} - I_{\text{far-field}}}{I_{\text{far-field}}} = \frac{I_{\text{total,near-field}}}{I_{\text{far-field}}} - 1 \qquad (5.15)$$

However, this ratio underestimates the true tip-induced enhancement. The band intensity is proportional to the number of Raman scatterers illuminated by the incoming radiation and hence depends on the sample volume. The sample volume under near-field conditions is much smaller than the sample volume in the far-field. The respective intensities cannot thus be directly compared, and they must then be referred to the same sample volume.

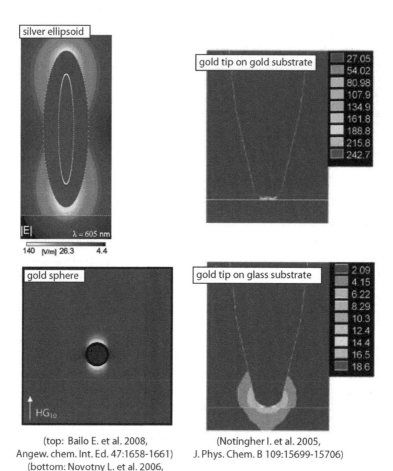

(top: Bailo E. et al. 2008,
Angew. chem. Int. Ed. 47:1658-1661)
(bottom: Novotny L. et al. 2006,
Annu. Rev. Phys. Chem. 57:303-331)

(Notingher I. et al. 2005,
J. Phys. Chem. B 109:15699-15706)

FIGURE 5.11 *(See color insert.)* Tip-enhancement on curved surfaces. Calculated field distribution around a silver ellipsoid, a gold sphere and a gold tip on a gold and on a glass substrate. Colour scale of the left-hand graphics: the electric field increases from dark purple to yellow. (*Top*: reprinted from E. Bailo, V. Deckert, *Chem. Soc. Rev.* 37:921-930, 2008, with permission *of the Royal Society of Chemistry. Bottom*: reprinted with permission from L. Novotny, S.J. Stranik, *Annu. Rev. Phys. Chem.* 57:303-331, 2006. Copyright 2006 Annual Reviews). Colour scale of the right-hand graphics: the electric field increases from purple to red (Reprinted with permission from I. Notingher, A. Elfick, *J. Phys. Chem. B* 109:15699–15706, 2005. Copyright 2005 American Chemical Society).

The *net enhancement*, NE, takes into account the measurement volumes for a fairer comparison

$$NE = \frac{I_{\text{near-field}}}{I_{\text{far-field}}} \cdot \frac{V_{\text{far-field}}}{V_{\text{near-field}}} = \left(\frac{I_{\text{total,near-field}}}{I_{\text{far-field}}} - 1 \right) \cdot \frac{V_{\text{far-field}}}{V_{\text{near-field}}} \quad (5.16)$$

where $V_{\text{far-field}}$ and $V_{\text{near-field}}$ are the sample volumes when the sample is illuminated from the far-field and from the near-field, respectively. The sample volume in the far-field, $V_{\text{far-field}}$, is considered to be that of a disk whose radius R coincides with that of the exciting laser spot, R_{spot}, and whose thickness h is the smallest of the following three magnitudes: the sample thickness, the light penetration depth (usually 0.5 μm in silicon at a wavelength of 514 nm), and the focal depth if confocal optics are used (usually 1 μm)

$$V_{\text{far-field}} = \pi \cdot R_{\text{spot}}^2 \cdot h \quad (5.17)$$

Likewise, the sample volume in the near-field, $V_{\text{near-field}}$, is assumed to be that of a disk whose area coincides with the projected area of the tip over the surface, A_{tip}, and thickness t, which is the smaller of the following magnitudes: the sample thickness or the penetration depth of the near-field radiation (approximately equal to 1–20 nm) (Mehtani et al. 2005; Schmid et al. 2008)

$$V_{\text{near-field}} = \pi \cdot R_{\text{tip}}^2 \cdot t \quad (5.18)$$

Substituting Equations (5.17) and (5.18) into Equation (5.16), the intensity enhancement can be expressed as follows:

$$NE = \left(\frac{I_{\text{total,near-field}}}{I_{\text{far-field}}} - 1 \right) \cdot \frac{R_{\text{spot}}^2 \cdot h}{R_{\text{tip}}^2 \cdot t} \quad (5.19)$$

Literature in TERS is especially prolific, reporting quite diverse intensity ratios ranging from 1.5 to 150, which corresponds to enhancements of 10^3 to 10^7. However, these values are far below those predicted by theoretical calculations, where a 1000-fold field enhancement corresponds to enhancement factors of the Raman signal of about 10^{12} (Gucciardi et al. 2007).

Surface-Enhanced + Tip-Enhanced

Field enhancement can be further amplified by combining surface- and tip-enhanced Raman spectroscopy (Notingher and Elfick 2005; Gucciardi et al. 2007) by the use of either STM-like or AFM-like setups and probes. Indeed, if the specimens are adsorbed on roughened or even flat metal films and if they are probed with metal tips exhibiting plasmon resonances, Raman scattering excitation occurs on those species located in the laser focus. At the same time, surface plasmons of the tip will be excited if the wavelength of the laser approaches that of the resonance, which in turn produces further excitation of only those species located in the area closest to the tip apex. As a result, surface and tip enhancement take place in the cavity created by the metal tip and the metal substrate, producing a combined effect that enables the detection of small molecules and weak Raman scatterers, which otherwise are difficult to detect (Pettinger et al. 2003).

This combined field enhancement depends on the metals that the tip and substrate are made of, apart from the fact that the Raman spectra of the sample may be altered by the metal used as substrate (Pettinger et al. 2004). The tandem Au(tip)/Au(substrate) has been found to outperform other metal heterogeneous pairs such as Ir(tip)/Au(substrate) and Au(tip)/Pt 110 (substrate) at 632 nm excitation (Pettinger et al. 2004). The findings follow the statement that the refractive index of both the tip and the substrate materials may influence the magnitude of field enhancement (Elfick, Downes, and Mouras 2010). However, the latter may be substantially damped by the presence of a dielectric specimen between the metal tip and substrate (Notingher and Elfick 2005). Enhancement factors of the Raman signal encountered in the experimental studies range from 10^4 to 10^7 (Pettinger et al. 2004; Domke, Zhang, and Pettinger 2007).

TE-CARS

Though intensity enhancement factors are much lower than those reported for the two previous mechanisms (10^2–10^6), third-order optical processes are much more localized, since they are only selectively induced where the electric field intensity is highest. Therefore, CARS phenomena are much more localized at the probe apex than linear Raman scattering or second-order optical processes, such as second-harmonic generation (SHG) or sum-frequency generation (SFG) (Hayazawa and Saito 2007). Consequently, CARS images may have higher lateral resolution than topography images. The third-order nonlinearity of CARS easily discriminates between the case where the probe is present and the case where it is not, since it is impossible to detect any CARS signal from the sample without the tip (Ichimura 2004). Consequently, TE-CARS does not need far-field background suppression (Elfick 2010).

TE-CARS was originally employed to image DNA clusters of 20 nm in height and approximately 100 nm in diameter (Ichimura 2004). However, its range of applicability has not been developed since then (Downes et al. 2009).

IR Microscopy beyond the Diffraction Limit: IR Near-Field Spectroscopy

Another means of assigning chemical content to materials at the nanoscale is to combine infrared spectroscopy with scanning near-field optical microscopy. The first attempts in this direction were done with fiber probes in the last years of the twentieth century, but with rather limited performance in the near-infrared region (Piednoir, Licoppe, and Creuzet 1996). However, the development of apertureless SNOM at the turn of the twenty-first century brought about major advances in the technique by expanding the excitation range to the mid-infrared region.

Aperture SNOM + IR Spectroscopy

Though employed in the very first work on the topic (Piednoir, Licoppe, and Creuzet 1996), such a combination found rather limited application and was swiftly replaced by the newcomer apertureless SNOM. The reason lies in the optical fibers used as aperture probes. They cannot transmit excitation beams whose wavelengths are longer than a few micrometers. Silica fibers have a cut-off wavelength of 2 μm, and although they were replaced in this case by fibers made of arsenic sulfide or fluoride glass, the cut-off limit of the latter is not much further extended (6 μm) (Piednoir, Licoppe, and Creuzet 1996; Cricenti et al. 2008). The first illumination sources were globar or cascade arcs[*] (basically an argon plasma; Raghavan 2000) and a pulsed, tunable free-electron laser (FEL). The first two provide continuous, polychromatic radiation, and they may be seen as black-body radiators having temperatures about 1,300K and 13,000K, respectively. The second is a powerful source of monochromatic radiation produced by a wiggling beam of electrons in a periodic magnetic structure called an *undulator*. The radiated wavelength depends on the energy of the electrons, the undulator period, and the applied magnetic field, which can be tuned at any value from ultraviolet radiation to the far-infrared.[†]

The SNOM configuration used in the pioneer study was a photon scanning tunneling microscope in the collection mode (see Chapter 4; Piednoir, Licoppe, and Creuzet 1996). The scheme of such a setup is shown in Figure 5.12a. The infrared radiation produced is totally reflected by the backside of a suitable substrate such as a wafer of silicon that is transparent in the infrared range. The resulting evanescent wave is collected by the tip of a fluoride glass optical fiber (diameter 2–5 μm) that collects the radiation and drives it to the HgCdTe detector if the illumination source is the cascade arc or to a boxcar average in the case of the free electron laser. The movable tip is kept at a close distance to the sample (0.3 μm) to achieve the optimal image contrast and moved in the x- and y-directions for raster scanning at constant height. Far-field background, mainly due to light diffusion and thermal radiation, is suppressed by modulating the tip–sample distance through oscillation of the fiber and subsequently detecting the near-field signal through a lock-in amplifier. In the case of the pulsed source where such a method cannot be applied, background is substantially reduced by employing metal-coated tips.

The setup was able to provide infrared images of silica bands and steps on silicon substrates by collecting the light emitted in the range between 1 and 6 μm at each sample's position or image pixel. The acquisition time for each pixel was 2 s, which resulted in image acquisition times of approximately 1 hour (the images were composed of 32 × 32 or 64 × 64 pixels). Spatial resolutions of $\lambda/4$ were attained when FEL-produced radiation of 4-μm wavelength was employed. The first local IR spectra of a photosensitive resin film coated on glass were acquired with the same

---

[*] Globar (from glowbar) arcs are typical sources of infrared radiation in IR spectrometers. They consist of a silicon carbide rod that is resistively heated through an electric current. Cascade arcs produce plasma out of a gas (typically argon) by applying an electric current through a channel along which this gas flows. The result is a continuous radiation adjustable through the gas flow rate, the current flowing through the arc, and the gas pressure.

[†] See http://clio.lcp.u-psud.fr/clio_eng/FEL.html.

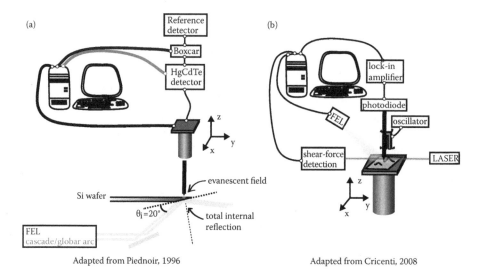

FIGURE 5.12 Setups for aperture infrared SNOM. (a) The first one to be developed makes use of total internal reflection to illuminate the probe by the resulting evanescent field created at the sample substrate (a Si wafer) (Piednoir 2006). (b) A recent setup used to investigate human keratinocytes (Cricenti et al. 2008).

device and FEL as the illumination source. The fiber was kept at 0.2 μm from the film and used to collect the IR emission at particular wavelengths within a wavelength range where the material shows characteristic absorption bands. The near-field IR spectrum, which greatly resembled the conventional, far-field spectrum, was recorded in the range of wavelengths where the film exhibits an absorption doublet (at 4.8–4.5 μm or 2080–2200 cm^{-1}), due to the vibration of the diazoquinone groups ($N=N=N^+$). The sample area probed was defined by the size of the fiber tip, slightly less than 10 μm and hence on the order of 10×10 μm^2 and with a spectral resolution of about 5 cm^{-1}.

Figure 5.12b shows a recent setup operating in the collection mode that was developed 12 years after the study by Piednoir, Licoppe, and Creuzet (1996). The illumination source is a FEL and the probe–sample distance is kept constant by a shear-force feedback. The lock-in amplifier separates the far-field background from the near-field signal. This instrument has been used to investigate human skin keratinocyte cells (Cricenti et al. 2008).

Apertureless SNOM + IR Spectroscopy

The last years of the twenty-first century witnessed the birth of apertureless SNOM, and soon it was employed in mapping the vibrational behavior of materials at nanoscale resolution (Knoll 1998, 1999). As discussed previously, this information is key to chemically identifying specimen components, but the usefulness of the technique is not restricted to that. In the last decade, the technique has been successfully applied to detect local variations in material crystallinity, crystal orientation, or stress at the

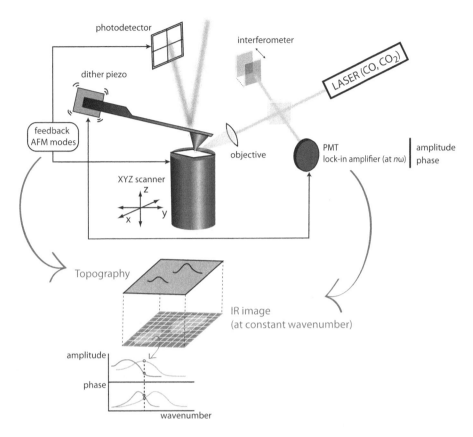

FIGURE 5.13 Setup for apertureless infrared SNOM. The tip of an AFM cantilever is illuminated from the side. Infrared radiation coming from a CO or a CO_2 laser is focussed on the tip by a Cassegrain objective that also collects the near-field signal. An interferometer and a lock-in amplifier extract the amplitude and phase of the near-field signal at each wavelength. The topography is acquired by an AFM-based setup, where the distance between the tip and the sample is controlled by a feedback AFM mode (usually intermittent contact mode, Brehm et al., 2005). The resulting images show the sample's topography together with the spatial distribution of the near-field amplitude and phase at a constant wavelength (or wavenumber).

surface and to identify sample components 50 nm beneath the surface (Ocelic and Hillenbrand 2004; Taubner, Keilmann, and Hillenbrand 2005).

The most popular setup reported to date is based on an apertureless SNOM: a combination of an atomic force microscope operating in intermittent contact mode with an optical system that focuses infrared radiation on the AFM-tip and collects the scattered light by an infrared detector after proper subtraction of the background signal (see Figure 5.13 and Chapter 4 for details). The illumination source consists of either a tunable CO_2 or a CO laser that provides monochromatic radiation in the range of 9.2 to 10.7 μm and of 5.5 to 7 μm, respectively (Brehm et al. 2005). The power of these lasers is often attenuated to a few milliwatts (10–30 mW) (Knoll and Keilmann 1998, 1999) to minimize laser-induced sample damage or to avoid excessive sample heating.

Side illumination of the tip is often preferred where the laser beam is focused on the tip at grazing incidence by a focusing lens of 0.5 NA (numerical aperture). The probes are usually cantilevered tips made of silicon and coated with either platinum, silver, gold, or a Pt/Ir alloy (Knoll and Keilmann 1998, 1999; Akhremitchev et al. 2002; Hillenbrand, Taubner, and Keilmann 2002; Taubner, Keilmann, and Hillenbrand 2005), although tungsten (W) STM-type wires have also been found to be convenient (Formanek et al. 2004). Collection of either the backscattered or near-forward radiation can be done either by the same lens of by an additional Cassegrain objective that drives the signal to an infrared detector (usually a HgCdTe type). Before reaching the detector, the modern setups may include a Michelson interferometer to obtain both the amplitude and the phase from the near-field signal. The detector forwards the signal to a device that filters out the background signal. Apertureless SNOM devices are affected by large background signals, as discussed in Chapter 4. In the context of infrared microscopy the most usual way of background subtraction is the demodulation method, briefly mentioned in this chapter and in Chapter 4. The AFM tip is set to vertically oscillate at its resonance frequency, f_0, which accordingly modulates the tip–sample distance. The near-field signal, an evanescent wave, is highly sensitive to this parameter and will correspondingly increase and decrease in intensity at the same frequency or at the higher harmonics, nf_0. The background signal is not so much affected by the distance modulation and hence remains as a steady signal that can be filtered out by a lock-in amplifier. The latter extracts both the amplitude and the phase of the near-field signal at a particular frequency, either at f_0 or at nf_0, usually at the second ($n = 2$) or the third ($n = 3$) harmonic (Akhemitchev et al. 2002; Hillenbrand, Taubner, and Keilmann 2002; Formanek et al. 2004; Taubner, Hillenbrand, and Keilmann 2004; Ocelic and Hillenbrand 2004).

Likewise for the aperture counterpart, the usual outcome of such setups is in the shape of infrared images in parallel to the sample topography. The infrared images are obtained by mapping the intensity of the near-field signal at the wavelengths where sample components exhibit characteristic absorption bands; in this case, the vibrational contrast is generated by differences in the infrared activity of the different sample constituents. Infrared spectroscopy can be obtained by the same device; however, the range of operational wave numbers (1580–1800 cm^{-1}, 880–950 cm^{-1}) is somewhat restricted due to the limited tunability of infrared lasers (Ocelic and Hillenbrand 2004).

Another Source of Enhancement in Near-Field IR Microscopy: Phonon Excitation

Just as excitation of surface plasmon polaritons (*plasmons*) is a cause of field enhancement in visible-light SNOM, excitation of surface phonon polaritons (*phonons*) has the same effect when operating in infrared frequencies. Phonon resonances can be sharper and stronger than plasmon resonances when small particles of the same size are compared, which may result in more intense enhancement or near-field contrast (Hillenbrand, Taubner, and Keilmann 2002).

The near-field enhancement through phonon resonance depends on the value of the dielectric constant of the material at a particular infrared frequency. Indeed, the phenomenon can be induced either by a polar sphere, the tip, or a polar sample. The near-field interaction between both dipoles can be modeled by assuming the tip to be a polar sphere of radius a and polarizability α at a distance z from a plane sample. The sphere induces an image dipole in the sample with polarizability $\alpha\beta$ where

$$\alpha = 4\pi a^3 \frac{\varepsilon_t - 1}{\varepsilon_t + 2} \quad \text{and} \quad \beta = \frac{\varepsilon_s - 1}{\varepsilon_s + 1} \tag{5.20}$$

where ε_t and ε_s are the dielectric function of the tip and of the sample, respectively. The resonance can thus occur either when the dielectric value of the sample approaches -1 or when that of the tip approaches -2. The near-field interaction between these two dipoles can then be described by an effective polarizability, α_{eff}, that in the electrostatic approximation (when the sphere has a small size compared to the radiation wavelength) has the form

$$\alpha_{\text{eff}} = \frac{\alpha(1+\beta)}{1 - \dfrac{\alpha\beta}{16\pi(a+z)^3}} \tag{5.21}$$

The scattered intensity S of the coupled system and detected as near-field signal is proportional to the square of the effective polarizability, $S \propto |\alpha_{\text{eff}}|^2$.

Tips and surfaces made of or patterned with phonon resonators can thus profit from such field enhancements. Materials that exhibit phonon resonances are crystalline silicon carbide, silicon oxide, II-V or II-VI semiconductors, ceramics, or ferroelectrics, which are of great interest for the electrooptic technologies or for high-temperature and high-power operative devices (Ocelic and Hillenbrand 2004). In particular, the near-field signal produced by crystalline silicon carbide in the near-IR region (around 940 cm^{-1}) is about 20 to 30 times higher than that produced by gold, a plasmon-active material and near-field enhancer in the visible range.

SUMMARY

The application of Raman and infrared spectroscopy in the near field takes chemical microscopy beyond their respective diffraction limits. Obviously, both techniques share most of the characteristics and limitations of SNOM, but they do have some particularities, which are referred to in v 5.2. Samples must be either Raman- or infrared-active and exhibit a distinctive spectroscopic behavior relative to their supports or to other sample constituents. A new contribution to image contrast adds to the already known near-field contrast. It is chemical and brought about by the width

TABLE 5.2

Summary of the Near-Field Raman and Infrared Microscopies

CHARACTERISTICS	Near-field Raman microscopy	Near-field Infrared microscopy
Lateral resolution	100 nm (aperture), 10 nm (apertureless)	1 µm -100 nm (aperture), 20-50 nm (apertureless)
Vertical resolution	10-20 nm (tip-enhanced)	
Sample requirements	Raman active in the visible range - substrate must not be Raman active -	infrared active in the mid-infrared range phonon excitable in the mid-infrared range -substrate must be transparent to IR radiation -
Local/global	local	
Contrast	vertical: f(tip enhancement factor) lateral: size of the near-field probe	
	lateral: width and position of the Raman spectroscopic bands	lateral: width and position of the infrared spectroscopic bands
Other features	non-destructive support imaging and spectroscopy	
		subsurface imaging (approx. 50 nm below the surface)
Limitations	sensitive to tip contaminations spectra of complex systems (biopolymers, cells) difficult to interpret - > impairs correct assignment	
	reflectivity can disguise or interfere with the true Raman activity	restricted to the mid-infrared region and few wavelengths
	tip pressure, chemical effect, gradient field effect or sample heating can alter the position of the Raman bands	
	near-field effect may change the selection rules	
	sample may be damaged due to heating	

and position of the spectroscopic bands. The main drawback of both techniques lies on the high sensitivity of the spectra to the presence of the near-field probe, which may bring contamination or excessive pressure to the sample. Other factors such as chemical and gradient field effects, sample heating, or new selection rules can alter the position of the Raman bands, which impairs the correct interpretation and assignment of the spectra.

APPLICATIONS OF SPM + RAMAN SPECTROSCOPY

Of all the modalities based on combining SPM with Raman spectroscopy, TERS is by far the most prolific. Since its invention, the applicability of TERS has been continuously expanding, as it has been employed in the topological and spectral characterization of very diverse materials, from the very simple to the most complex. Proof comes from the numerous reviews that have been published in recent years on this topic (Bailo and Deckert 2008b; Yeo et al. 2009; Domke and Pettinger 2010; Pettinger 2010). According to most of them, the range of applications of TERS can be classified in research devoted to a particular group of materials.

(Sub-) Monolayers of Dyes

Together with fullerene (C_{60}) films (Anderson 2000), dyes such as brilliant cresyl blue (BCB), cresol fast violet (CFV), rhodamine 6G, and malachite green isothiocyanate (MGITC) were chosen as test samples in the early studies to demonstrate the proof-of-concept of TERS. Later, in an attempt to follow the path led by Kneipp, Nie, and Emory (Kneipp et al. 1997; Nie and Emory 1997), who found evidence of single-molecule detection using surface-enhanced Raman spectroscopy, the very same systems were used to prove the limits of detection of TERS. Indeed, Zhang claimed reaching the single-molecule level after observing intensity fluctuations of a particular band in the TER spectra of brilliant cresyl blue adsorbed on flat gold (Pettinger 2010; Zhang 2007). The number of Raman scatterers that contribute to the observed signal was estimated to be 320. However, the ultimate proof of single-molecule detection has been done under ultra-high vacuum conditions, as shown in Figure 5.14 (Pettinger 2010; Domke and Pettinger 2010).

Single-Wall Carbon Nanotubes

Single-wall carbon nanotubes are strong Raman scatterers and recurrent in the TERS literature (Hartschuh, Anderson, and Novotny 2003; Hartschuh et al. 2003; Roy, Wang, and Welland 2006; Chan and Kayarian 2011). Their well-defined geometry and simple chemistry have made them suitable test samples to evaluate enhancement factors, polarization effects, background interference, or lateral resolution in TERS setups. Single-wall carbon nanotubes (SWCNTs) are scarce materials from which high-resolution Raman images (approx. 15 nm) have been obtained under environmental conditions (see Example A). In addition, TERS has provided valuable information about the distribution of possible local defects along the nanotubes (Hartschuh et al. 2003).

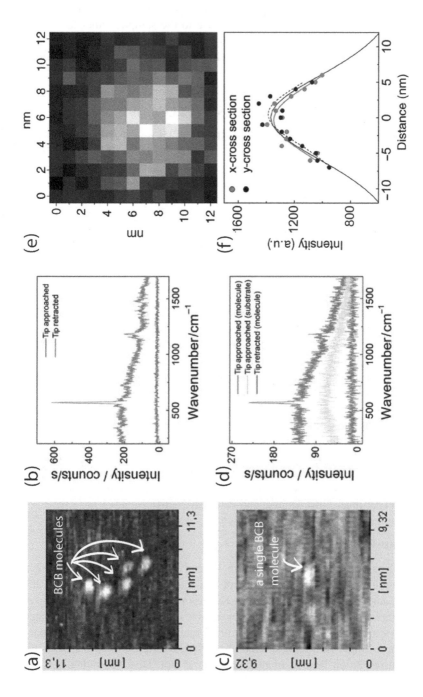

FIGURE 5.14 *(See color insert.)* Single molecule detection with TERS. (a) and (c) show STM images of five and one brilliant cresyl blue (BCB) molecules, respectively, and their corresponding Raman spectra in (b) and (d). Tip enhancement is clearly shown if one compares the Raman *(continued)*

INORGANIC MATERIALS: SILICON AND SEMICONDUCTORS

Combined SNOM and Raman spectroscopy has been employed on either scratched or milled silicon—the latter in grooves of defined depth and width—to localize the strained regions produced by scratching (Webster, Smith, and Batchelder 1998) or by focused-ion-beam milling (FIB) (Zhu et al. 2007), respectively. Using an aperture SNOM and in spite of the strong background of the silica optical fiber, Webster and coworkers (1998) observed that scratching resulted in the extinction of the crystalline silicon (Si I) band at 520 cm^{-1}, presumably due to the formation of a new phase, namely metallic Si II that could revert into amorphous Si. The presence of the latter was evident in the region of the scratch. Shifts in the center of the Si I band were attributed to the presence of residual stresses in the silicon around the scratch. More recently, lines of strained silicon (sSi) 100 nm wide were properly resolved with TERS, illustrating the capability of TERS to provide structural as well as chemical information at the appropriate spatial resolution (Bailo and Beckert 2008b). The Raman spectra showed an additional band at 516 cm^{-1}, indicative of the presence of sSi (Zhu et al. 2007). Ascertaining the stress distribution of micromachined silicon is key to optimize the electron mobility in transistors and hence to improve the quality of integrated circuits (Bailo and Beckert 2008b). In addition, TERS has also been used to characterize semiconductor surfaces, such as SiC, GaN nanowires, and graphene to determine the ferroelectric order of barium titanate crystals and to evaluate the pyrite content in fossil stones (Domke and Pettinger 2010; Kaupp 2006).

BIOMOLECULES

Challenging as they may be due to their inherent spatial and compositional complexity (Elfick, Downes, and Mouras 2010), biomolecules have been repetitively chosen in TERS studies due to their importance in biosystems and physiological processes. At the single-molecule level it has been possible to identify vibrational bands of submolecular fragments, such as amino acids and the heme group of the protein cytochrome C (cytC, Yeo et al. 2009). The authors additionally observed alterations in the TER spectrum of the protein and attributed this to the different orientations of cytC at the surface (Yeo et al. 2008). Raman fingerprints of the different DNA bases adsorbed on flat gold have been identified even at picomolar quantities (Domke, Zhang, and Pettinger 2007). From all the nucleobases, adenine gives a prominent Raman signal at 734 cm^{-1}, which constituted experimental evidence of the chemical effect in the tip-enhanced Raman emission.

One of the most problematic applications for TERS is the sequencing of nucleic acids (Hartschuh 2008; Bailo and Deckert 2008a). The reason is twofold: on the

FIGURE 5.14 *(continued from facing page.)* spectra in (b) and (d). Tip enhancement is clearly shown if one compares the Raman spectra taken when the tip is close to the molecule(s) (tip approached, red curve) and when the tip is far away (tip retracted, grey curve). (e) and (f) show the TERS image of a single BCB molecule on flat gold and the respective intensity profile along the vertical and the horizontal (x and y) axes. (Reprinted with permission of J. Steidtner, B. Pettinger, *Phys. Rev. Lett.* 100:236101, 2008. Copyright 2008 by the American Physical Society, URL http://link.aps.org/abstract/PRL/v100/e236101.)

one hand, sequencing requires an extremely high lateral resolution and sensitivity to resolve the succession of nucleotides at both submolecular and single-nucleotide levels; on the other hand, the presence of sugars and phosphate groups in these molecules blocks the Raman signal of the nucleobases, namely the sequence vehicles. A recent report proposes the use of Raman-active dyes that selectively bind the nucleobases; the approach was tested on an RNA consisting of a homopolymer of cytosine (Bailo and Deckert 2008a).

Other reports on biomolecules deal with the compositional distribution of phase-separated lipid monolayers (Opilik et al. 2011) or the characterization of calcium-alginate fibers (Schmid et al. 2008). These works pave the way to chemically identify constituents of complex biosystems, since they can be considered as main components of cell membranes and biofilms, respectively. Native membranes of burst erythrocytes under physiological conditions have been successfully used for studies in TERS (Yeo et al. 2009). Working in liquids appears to be advantageous in two ways: first, it preserves the conditions under which biosystems remain functional and viable; second, the aqueous environment considerably reduces carbon contamination of tips, radical-mediated sample damage, as well as sample heating due to the excitation light (Yeo et al. 2009).

VIRUS, BACTERIA, AND HUMAN CELLS

The potential of TERS to chemically identify and spatially localize constituents of cell surfaces and virus strains has already been demonstrated in recent reports. The main difficulty of these studies resides in the complexity of such systems, which hinders the accurate assignment of bands in the Raman spectra. Yet the latter are specific enough to allow the identification of distinct virus varieties (Hermann et al. 2011), to detect the contribution of coat proteins and RNA of single tobacco mosaic virus particles (Cialla et al. 2009), and to resolve the local distribution and label-free chemical characterization of protein domains and other components on bacterial cell surfaces (Biju et al. 2007; Neugebauer et al. 2006; Budich et al. 2008). Attempts to employ TERS on fixed eukaryotic cells are still scarce (Böhme et al. 2010), and to our knowledge no reports have appeared where TERS has been employed in the study of living eukaryotic cells.

Example A: TERS on Single-Wall Carbon Nanotubes

Single-wall carbon nanotubes show a well-known and relatively simple Raman spectrum characterized by the following bands: the radial breathing-mode band (RBM) at 100–300 cm^{-1}, a sharp band at 1596 cm^{-1} corresponding to the tangential stretching mode (G band), and a broader band around 2615 cm^{-1} (the G′ band) that is an overtone of the disorder-induced mode around 1310 cm^{-1}. In a work published in 2003, Hartschuh and coworkers showed for the first time images of single isolated SWNTs with a spatial resolution of 30 nm (Hartschuh et al. 2003). Simultaneously with topography data, Raman maps were acquired

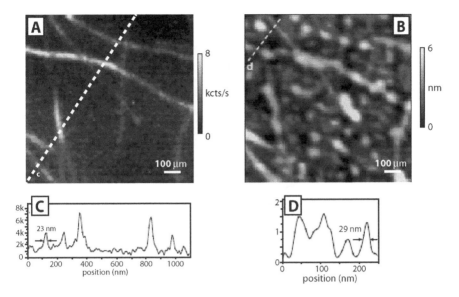

FIGURE 5.A1 Simultaneous near-field Raman image (A) and topographic image (B) of SWNTs grown by chemical vapour deposition on glass. The Raman image has been acquired by registering the intensity of the G′ band as a function of the sample position. (C) and (D) show line profiles depicted in the respective images. (Reprinted with permission from A. Hartschuh, E. J. Sánchez, X. Sunney Xie, L. Novotny. *Phys. Rev. Lett.* 90:095503, 2003. Copyright 2003 by the American Physical Society, URL http://link.aps.o. rg/abstract/ PRL/v90/e095503)

by detecting the intensity of the band G′ (2615 cm⁻¹) upon laser excitation at 633 nm at a slightly better resolution than that of the topography images (25 nm) (see Figure 5.A1). Variations of the relative intensity of the G and G′ bands along a single nanotube were detected only in the near field and attributed to several origins (see Figure 5.A2). Since the nanotube was produced by arc discharge using Ni/Y catalyst particles, the authors considered external stress due to the presence of those catalysts to be one of the possible reasons for the spectral fluctuations, the other being local defects in the tube structure.

Example B: Toward Nucleic-Acid Sequencing

Aiming toward nucleic-acid sequencing, Bailo and coworkers applied TERS on single RNA strands adsorbed on mica (Bailo et al. 2008a). The RNA strands consisted of linear homopolymers of cytosine, a nucleobase of well-characterized Raman response. These molecules were used as test samples to evaluate the instrument's sensitivity, lateral resolution and local enhancement, since all these factors contribute to determine the feasibility of TERS as a sequencing tool. More specifically, the requisites are: high enhancement and high sensitivity down to the single-base level; and high lateral resolution to detect a reduced number of bases simultaneously.

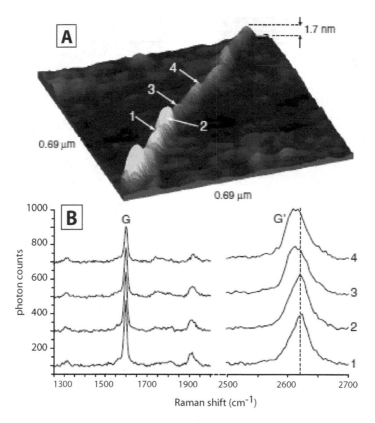

FIGURE 5.A2 (A) Topographical image of a single SWNT on glass. The bumps are attributed to the presence of Ni/Y catalyst particles; (B) Raman spectra at four different locations on the SWNT and marked from 1 to 4 in (A). At positions 1 and 2, the G band is significantly stronger that the G′ band. Moving along the tube at positions 3 and 4, the amplitude of the G band decreases with respect to the G′ band. Whereas the position of the G band remains fixed, the G′ band broadens and shifts 5 cm⁻¹ to the left. (Reprinted with permission from A. Hartschuh, E. J. Sánchez, X. Sunney Xie, L. Novotny. *Phys. Rev. Lett.* 90:095503, 2003. Copyright 2003 by the American Physical Society, URL http://link.aps.org/abstract/PRL/v90/e095503)

The authors claim to achieve single-base sensitivity with overall enhancements of at least 10^4. The signal-to-noise ratio of the enhanced Raman signal was estimated to be about 200 with respect to the reference spectrum, the unenhanced Raman signal of one 1-μm-long RNA strand (ca. 3000 bases). Since the fragment of RNA responsible for the TER spectrum is 20 nm long (ca. 30–60 bases), each base contributes with a signal-to-noise ratio of 3–7 to the spectrum, which makes every base distinguishable.

Figure 5.B1 shows successive TER spectra along a single RNA strand, acquired at adjacent positions. The positions were chosen to be approximately equidistant in 100 nm. The TER spectra show the characteristic fingerprint of cytosine together with slight fluctuations in both peak intensity and peak position. The authors claim

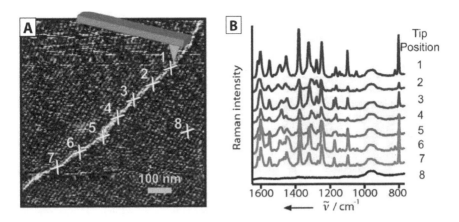

FIGURE 5.B1 (A) Topography image of a single RNA strand of poly-cytosine on mica. The positions upon which TER spectra are acquired are marked with crosses and consecutively numbered from 1 to 7. Position number 8 corresponds to the tip position of the reference spectrum. (B) TER spectra corresponding to the different positions depicted in (A). All Raman spectra show the fingerprint of cytosine as well as slight variations in both intensity and position of the peaks. (Reprinted with permission from E. Bailo, V. Deckert. *Angew. Chem. Int. Ed.* 47:1658–1661, 2008. Copyright 2008 John Wiley & Sons).

these minute alterations in the TER spectra to be caused by slight variations in either the relative orientation or the location of the probing tip with respect to the probed fragment of the molecule at each position.

Example C: Identifying Virus Strains

Hermann and coworkers have recently evaluated the capacity of TERS to differentiate virus particles of different varieties (Hermann et al. 2011). The authors acquired simultaneously topographic and TER spectra of various virus particles coming from two different strains: that of the avipoxvirus (APV) and that of the adeno-associated virus (AAV). The APV particles are bigger, with a nominal height of 100 nm and a diameter of about 450 nm when adsorbed on mica. Contrarily, the AAV particles are noticeable smaller, with heights of about 12 nm and diameters of about 60 nm. The TER spectra of both virus strains were complex and subjected to certain variations when different particles of the same virus strain were compared. Tentative assignment of the numerous peaks (shown in Figures 5.C1 and 5.C2) reveal the presence of the characteristic amide I (1630–1680 cm^{-1}), amide III (1290, 1245 cm^{-1}), and CH_2-stretching (2850–3050 cm^{-1}) bands, among others, revealing the presence of proteins and lipids. The spectral variations encountered on the large APV particles were more pronounced than those found on the smaller AAV, revealing a certain correlation between the spectra variability and the size of the virus particle. The work illustrates the local variability in the near-field Raman response when the sizes of the probed specimen are large compared to the tip size.

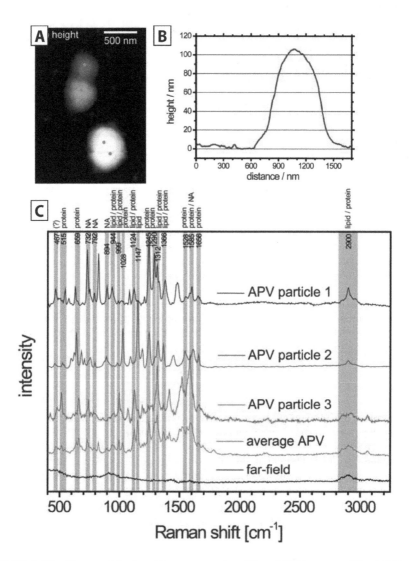

FIGURE 5.C1 Topographic data (A), cross section (B) and TER spectra (C) of different APV particles. Raman spectra of three APV particles are in dark blue, blue and red, whereas the average spectrum of five different spectra is in green. The far-field spectrum is shown for comparison. (Reprinted with permission from P. Hermann, A. Hermelink, V. Lausch et al. *Analyst* 136:1148–1152, 2011. Copyright 2011 Royal Society of Chemistry.)

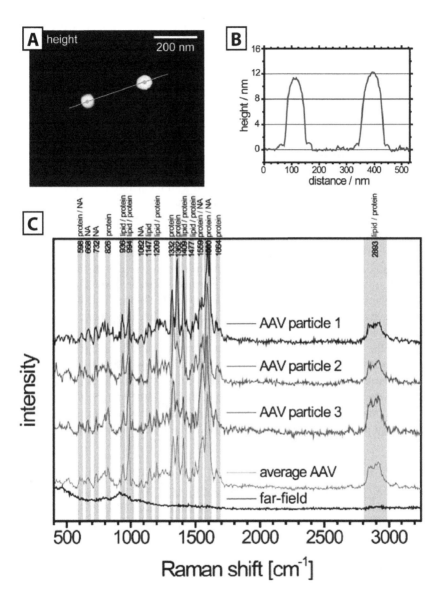

FIGURE 5.C2   Topographic data (A), cross section (B) and TER spectra (C) of different AAV particles. Raman spectra of three AAV particles are in dark blue, blue and red, whereas the average spectrum of seven different spectra is in green. The far-field spectrum is shown for comparison. (Reprinted with permission from P. Hermann, A. Hermelink, V. Lausch et al. *Analyst* 136:1148–1152, 2011. Copyright 2011 Royal Society of Chemistry.)

APPLICATIONS OF SPM + IR SPECTROSCOPY

The bulk of applications are encountered in the mid-infrared region and focus in the characterization of structured materials that exhibit distinct vibrational signatures. These can be particular infrared absorption bands or phonon resonances at certain wavelengths.

INORGANIC MATERIALS

Mostly metals and silicon have been repetitively used as test samples to demonstrate the instrumental capabilities of the technique, such as chemical specificity, lateral resolution, or enhancement. The first works focused on structured samples made of silica bands on silicon (Piednoir, Licoppe, and Creuzet 1996), aluminum stripes in IR polarizers, or plain (100) GaAs surfaces (Knoll and Keilmann 1998). Samples composed of gold deposited on silicon have revealed vibrational contrast when imaged at wavelengths of 9.7 and 10.7 µm, distinguishing both materials from their distinct infrared reflectivities (S(Au):S(Si) 1.5:1; Taubner, Keilmann, and Hillenbrand 2005). On the other hand, silicon carbide (SiC) has been the standard material to test field enhancements caused by phonon resonance and is frequently present in samples together with gold, a typical material that exhibits plasmon resonance, which eased comparison between both resonant phenomena (Hillenbrand, Taubner, and Keilmann 2002; Formanek et al. 2004). Phonon resonators such as SiC have been found to exhibit more intense infrared activities than plasmon resonators like gold (Au) in the range of wave numbers 800–950 cm^{-1} (10.5–12.5 µm in wavelength). Indeed, the signal coming from SiC can be 30 times higher than the infrared response of gold at 940 cm^{-1} (Ocelic and Hillenbrand 2004). The tandem SiC-Au has also been exploited as a test sample to check the effect of the vibration amplitude of the tip in the image contrast (Taubner, Keilmann, and Hillenbrand 2005): Small amplitudes favor background suppression and contrast in the amplitude and phase images of the near-field signal.

POLYMERS

Polystyrene (PS) and poly(methylmethacrylate) (PMMA) have been ubiquitous in studies dealing with macromolecules. Composite films of PS and Au on Si wafers have been used to illustrate the capability of scattering-type SNOM to identify each component when operating at mid-infrared wave numbers. Indeed, infrared images acquired at 9.7 and 10.7 µm (1030 cm^{-1} and 934 cm^{-1}, respectively) revealed the regions of PS, Au, and Si as areas of distinct infrared reflectivity in the order S(Au) > S(Si) > S(PS) (Taubner, Hillenbrand, and Keilmann 2003). Similar contrast was found on PMMA films partially coating a silicon substrate. The images taken at 5.8 and 5.7 µm (1721 and 1744 cm^{-1}, respectively) revealed the different infrared activity of both materials (Taubner, Hillenbrand, and Keilmann 2004).

Nanoparticles of polystyrene and gold were successfully resolved by scattering-type SNOM in the infrared region, illustrating the capacity of the technique to identify materials at the level of single particles and high spatial resolutions (the particles were smaller than 100 nm in diameter) (Cvitkovic, Ocelic, and Hillenbrand 2007). Likewise, the technique proved to be sensitive enough to resolve metallic nanostructures buried by a layer of PMMA film of 50-nm thickness, showing that subsurface imaging was possible under those conditions (Taubner, Keilmann, and Hillenbrand 2005).

VIRUSES AND CELLS

Works on biosystems are still scarce, probably due to their complex chemistry and the need to work in liquids, which remains a challenge. Nevertheless, there are a few qualitative studies on single tobacco mosaic virus (TMV) tubular particles of 18 nm diameter and approximately 200 nm long (Brehm et al. 2006) as well as fixed human cells, such as keratinocytes (HaCaT) (Cricenti et al. 2008), and rat pancreatic cells (INS-1) (Vobornik et al. 2004). The works rather demonstrate that infrared imaging of such systems is possible at high resolution (10–90 nm); the absorption band at 6.1 μm (1640 cm^{-1}) corresponding to the stretching of the C=O bond, usually referred to as amide I, has been repetitively used as a fingerprint of the presence of biomaterial. The very same band was monitored by Cricenti and coworkers to detect chemical modifications on cell membranes when exposed to electromagnetic fields (Cricenti et al. 2008) and by Brehm and coworkers to localize the virus particles from a sample containing a mixture of virus and PMMA particles (Brehm et al. 2006). In the field of biomolecules, the work of Akhremitchev and coauthors identified the striped regions of DNA grafted chains in a sample of DNA and hexadenethiol monolayers on gold (Akhremitchev et al. 2002). The DNA-grafted regions were clearly visible as dark stripes in the IR images when radiation of 10.2 μm (980 cm^{-1}) was employed, which corresponds to the absorption band of the phosphate groups of the DNA strand.

Example D: Naming Particles with Scattering-Type SNOM and IR Imaging

The near-field interaction at infrared frequencies between metal tips and nanoparticles was modeled and probed as an identifying tool of particle materials of diameters above 7 nm (Cvitkovic, Ocelic, and Hillenbrand 2007). Particles of gold and polystyrene (PS) immobilized on a layer of bovine serum albumin (BSA) were successfully recognized from their distinct near-field signal amplitudes recorded at the mid-infrared wavelength of 10.6 μm. The infrared images show the nanoparticles as dark round spots in a bright background (see the BSA-coated silicon substrate in Figure 5.D1), with the PS particles appearing noticeably darker than the gold counterparts. The negative contrast encountered in such images, especially between gold and silicon ($S(Au) < S(Si)$) is opposite to the positive contrast typical of planar films of gold on silicon, where gold appears brighter ($S(Au) > S(Si)$). The authors attribute this result to the fact that the nanoparticles are smaller in size than the probing tip, and the overall signal is thus a contribution of two near-field interactions, the one arising between the tip and the particle and the other between the tip and the substrate. Contrast arises from the differing dielectric values of gold ($\varepsilon_s = -5000 + 1000i$), polystyrene ($\varepsilon_s = 2.4 + 0.1i$), and silicon ($\varepsilon_s = 11.7$), but also from the particle size, as Figure 5.D2 shows. The authors applied the extended dipole model to predict the amplitude signals of both gold and polystyrene as a function of the particle size.

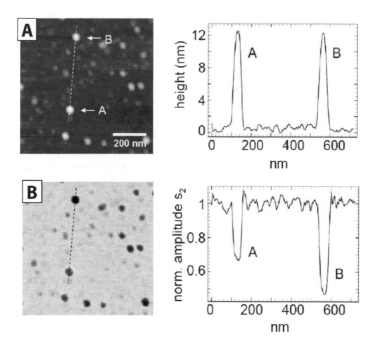

FIGURE 5.D1 Topography (A) and IR image (B) of Au and PS nanoparticles immobilized on a Si substrate. Two equally-sized particles are depicted as A and B. The height profile along the depicted dotted line shows that both are 12.8 nm high. IR images show that the particles appear as dark spots in a bright background. The IR profile along he dotted line shows however that the optical signals of particles A and B are different: the signal coming from B is less intense than signal coming from A. B is thus assigned to polystyrene and A to gold. (Reprinted with permission from A. Cvitkovic, N. Ocelic, R. Hillenbrand, Nano Lett. 7:3177–3181, 2007. Copyright 2007 American Chemical Society).

Example E: Revealing the State of Crystallinity in Phonon-Resonant Materials

Hillenbrand's group applied scattering-type SNOM in the infrared region to detect surface modifications in the crystallinity of silicon carbide (Ocelic and Hillenbrand 2004). Figure 5.E1 shows the behavior of two structural forms of silicon carbide. In its crystalline form (c-SiC), silicon carbide exhibits phonon resonance in a wavelength range where its dielectric value is smaller than –1 (788–965 cm^{-1}). Conversely, amorphous silicon carbide (a-SiC) is nonresonant, since its dielectric value is positive at all infrared wavelengths. Thus changes in the structure of silicon carbide can readily be detected by infrared near-field microscopy. In this case, the image contrast is due to the occurrence (or not) of phonon resonances in crystalline (and amorphous) areas patterned on the material's surface by focused ion beam (FIB) milling. The so-called *implantation* modifies the material's

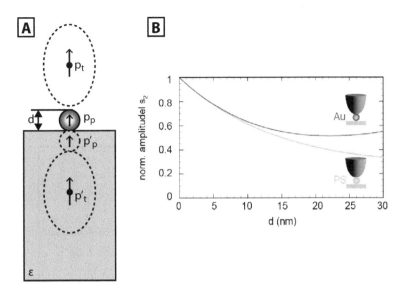

FIGURE 5.D2 Extended dipole model (A) and plot of the near-field signal amplitude as a function of particle size at $\lambda = 10.6$ µm. (B). Particles of gold and PS can be distinguished from one another as long as they are not smaller than 7 nm in diameter. (Reprinted with permission from A. Cvitkovic, N. Ocelic, R. Hillenbrand, Nano Lett. 7:3177–3181, 2007. Copyright 2007 American Chemical Society).

chemical ordering without altering the surface topography. Figure 5.E2 shows the topography image of a rather flat, randomly scratched surface, whereas the IR image acquired at phonon resonance frequencies reveals the FIB pattern in both amplitude and phase signals: The bright and dark areas are indicative of the presence of c-SiC and a-SiC, respectively.

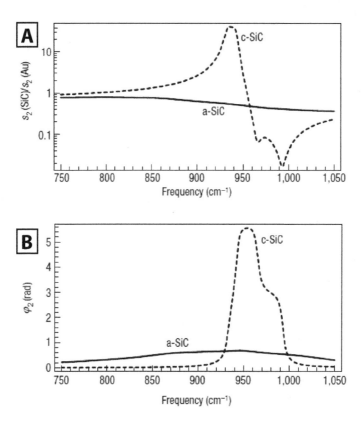

FIGURE 5.E1 Calculated amplitude (A) and phase (B) curves of the near-field signal in amorphous and crystalline silicon carbide (a-SiC and c-SiC, respectively). The amplitude is referred to that of gold. The c-SiC curves show a maximum at approximately 950 cm^{-1} where phonon resonance occurs. (Reprinted by permission from Macmillan Publishers Ltd: Nat. Mat., Ocelic, N. Hillenbrand, 3:606–609, copyright 2004.)

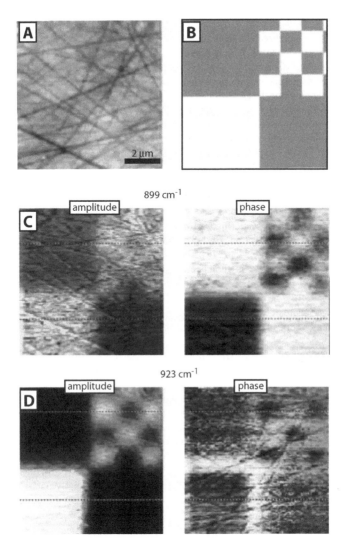

FIGURE 5.E2 Topography (A), scheme of the FIB-pattern (B), and amplitude and phase of the near-field signals recorded at 899 (C) and 925 cm^{-1} (D). Amplitude contrast is enhanced and phase contrast is reversed at the higher wavenumber, according to the calculated values shown in figure b1. (Adapted by permission from Macmillan Publishers Ltd: Nat. Mater., Ocelic, N. Hillenbrand, 3:606–609, copyright 2004.)

REFERENCES

Akhremitchev, B. B., Y. Sun, L. Stebounova, and C. G. Walker. 2002. Monolayer-sensitive infrared imaging of DNA stripes using apertureless near-field microscopy. *Langmuir* 18:5325–28.

Anderson, M. S. 2000. Locally enhanced Raman spectroscopy with an atomic force microscope. *Appl. Phys. Lett.* 76:3130–32.

Bailo, E., and V. Deckert. 2008a. Tip-enhanced Raman spectroscopy of single RNA strands: Towards a novel direct-sequencing method. *Angew. Chem. Int. Ed.* 47:1658–61.

———. 2008b. Tip-enhanced Raman scattering. *Chem. Soc. Rev.* 37:921–30.

Biju, V., D. Pan, Y. A. Gorby, et al. 2007. Combined spectroscopic and topographic characterization of nanoscale domains and their distributions of a redox protein on bacterial cell surfaces. *Langmuir* 23:1333–38.

Böhme, R., M. Richter, D. Cialla, P. Rösch, V. Deckert, and J. Popp. 2010. Towards a specific characterisation of components on a cell surface-combined TERS-investigations of lipids and human cells. *J. Raman Spectrosc.* 40:1452–57.

Brehm, M., H. G. Frey, R. Guckenberger, et al. 2005. Consolidating apertureless SNOM. *J. Korean Phys. Soc.* 47:S80–S85.

Brehm, M., T. Taubner, R. Hillenbrand, and F. Keilmann. 2006. Infrared spectroscopic mapping of single nanoparticles and viruses at nanoscale resolution. *Nano Lett.* 6:1307–10.

Budich, C., U. Neugebauer, J. Popp, and V. Deckert. 2008. Cell wall investigations utilizing tip-enhanced Raman scattering. *J. Microsc.* 229:533–39.

Chan, K. L. A., and S. G. Kayarian. 2011. Tip-enhanced Raman mapping with top-illumination AFM. *Nanotechnology* 22:175701-1/175701-5.

Cialla, D., T. Deckert-Gaudig, C. Budich, et al. 2009. Raman to the limit: Tip-enhanced Raman spectroscopic investigations of a single tobacco mosaic virus. *J. Raman Spectrosc.* 40:240–43.

Cricenti, A., R. Generosi, M. Luce, et al. 2008. Low-frequency electromagnetic field effects on functional groups in human skin keratinocytes cells revealed by IR-SNOM. *J. Microsc.* 229:551–54.

Cvitkovic, A., N. Ocelic, and R. Hillenbrand. 2007. Material-specific infrared recognition of single sub-10 nm particles by substrate-enhanced scattering-type near-field microscopy. *Nano Lett.* 7:3177–81.

Deckert, V., D. Zeisel, R. Zenobi, and T. Vo-Dinh. 1998. Near-field surface-enhanced Raman imaging of dye-labelled DNA with 100-nm resolution. *Anal. Chem.* 70:2646–50.

Domke, K. F., and B. Pettinger. 2010. Studying surface chemistry beyond the diffraction limit: 10 years of TERS. *ChemPhysChem* 11:1365–73.

Domke, K. F., D. Zhang, and B. Pettinger. 2007. Tip-enhanced Raman spectra of picomole quantities of DNA nucleobases at Au(111). *J. Am. Chem. Soc.* 129:6708–9.

Dorn, R., S. Quabis, and G. Leuchs. 2003. Sharper focus for a radially polarized light beam. *Phys. Rev. Lett.* 91:233901-1/233901-4.

Downes, A., R. Mouras, and A. Elfick. 2009. A versatile CARS microscope for biological imaging. *J. Raman Spectrosc.* 40:757–62.

Downes, A., R. Mouras, M. Mari, and A. Elfick. 2009. Optimising tip-enhanced optical microscopy. *J. Raman Spectrosc.* 40:1355–60.

Downes, A., D. Salter, and A. Elfick. 2006. Finite element simulations of tip-enhanced Raman and fluorescence spectroscopy. *J. Phys. Chem. B* 110:6692–98.

Elfick, A. P. D., A. R. Downes, and R. Mouras. 2010. Development of tip-enhanced optical spectroscopy for biological applications: A review. *Anal. Bioanal. Chem.* 396:45–52.

Formanek, F., Y. De Wilde, L. Aigouy, W. K. Kwok, L. Paulius, and Y. Chen. 2004. Nanometer-scale probing of optical and thermal near-fields with an apertureless NSOM. *Superlattice Microstr.* 35:315–23.

Gucciardi, P. G., S. Trusso, C. Vasi, S. Patanè, and M. Allegrini. 2007. Near-field Raman spectroscopy and imaging. In *Applied scanning probe methods V—Scanning probe microscopy techniques*, ed. B. Bhushan, H. Fuchs, and S. Kawata. New York: Springer Verlag.

Günzler, H., and H.-U. Gremlich. 2002. *IR spectroscopy*. Weinheim, Germany: Wiley-VCH.

Hartschuh, A. 2008. Tip-enhanced near-field optical microscopy. *Angew. Chem. Int. Ed.* 47:8178–8191.

Hartschuh, A., N. Anderson, and L. Novotny. 2003. Near-field Raman spectroscopy using a sharp metal tip. *J. Microsc.* 210:234–40.

Hartschuh, A., E. J. Sánchez, X. S. Xie, and L. Novotny. 2003. High-resolution near-field Raman microscopy of single-walled carbon nanotubes. *Phys. Rev. Lett.* 90:095503-1/095503-4.

Hayazawa, N., Y. Inouye, Z. Sekkat, and S. Kawata. 2000. Metallized tip amplification of near-field Raman scattering. *Opt. Commun.* 183:333–36.

———. 2001. Near-field Raman scattering enhanced by a metallized tip. *Chem. Phys. Lett.* 335:369–74.

Hayazawa, N., and Y. Saito. 2007. Tip-enhanced spectroscopy for nano investigation of molecular vibrations. In *Applied scanning probe methods VI—Characterization*, ed. B. Bhushan and S. Kawata. Berlin, Heidelberg, New York: Springer Verlag.

Hermann, P., A. Hermelink, V. Lausch, et al. 2011. Evaluation of tip-enhanced Raman spectroscopy for characterizing different virus strains. *Analyst* 136:1148–52.

Hillenbrand, R., T. Taubner, and F. Keilmann. 2002. Phonon-enhanced light-matter interaction at the nanometre scale. *Nature* 418:159–62.

Hollas, J. M. 2004. Modern spectroscopy, 4th ed. Chichester, U.K.: John Wiley & Sons.

Ichimura, T., N. Hayazawa, M. Hashimoto, Y. Inouye, and S. Kawata. 2004. Tip-enhanced coherent anti-Stokes Raman scattering for vibrational nanoimaging. *Phys. Rev. Lett.* 92:220801-1/22801-4.

Jahncke, C. L., H. D. Hallen, and M. A. Paesler. 1996. Nano-Raman spectroscopy and imaging with the near-field scanning optical microscope. *J. Raman Spectrosc.* 27:579–86.

Jahncke, C. L., M. A. Paesler, and H. D. Hallen. 1995. Raman imaging with near-field scanning optical microscopy. *Appl. Phys. Lett.* 67:2483–85.

Kaupp, G. 2006. Scanning near-field optical microscopy on rough surfaces. Applications in chemistry, biology and medicine. *Int. J. Photoenergy* 2006:1–22.

Kneipp, K., Y. Wang, H. Kneipp, et al. 1997. Single molecule detection using surface-enhanced Raman scattering (SERS). *Phys. Rev. Lett.* 78:1667–70.

Knoll, B., and F. Keilmann. 1998. Scanning microscopy by mid-infrared near-field scattering. *Appl. Phys. A* 66:477–81.

———. 1999. Near-field probing of vibrational absorption for chemical microscopy. *Nature* 399:134–37.

McCreery, R. L. 2000. Raman spectroscopy for chemical analysis. New York: John Wiley & Sons.

Mehtani, D., N. Lee, R. D. Hartschuh, et al. 2005. Nano-Raman spectroscopy with side-illumination optics. *J. Raman Spectrosc.* 36:1068–75.

Neugebauer, U., U. Schmid, K. Baumann, et al. 2006. On the way to nanometer-sized information of the bacterial surface by tip-enhanced Raman spectroscopy. *ChemPhysChem.* 7:1428–30.

Nie, S., and S. R. Emory. 1997. Probing single molecules and single nanoparticles by surface-enhanced Raman scattering. *Nature* 275:1102–6.

Notingher, I., and A. Elfick. 2005. Effect of sample and substrate electric properties on the electric field enhancement at the apex of SPM nanotips. *J. Phys. Chem. B* 109:15699–706.

Novotny, L., and S. J. Stranick. 2006. Near-field optical microscopy and spectroscopy with pointed probes. *Annu. Rev. Phys. Chem.* 57:303–31.

Ocelic, N., and R. Hillenbrand. 2004. Subwavelength-scale tailoring of surface phonon polaritons by focused ion-beam implantation. *Nat. Mat.* 3:606–9.

Opilik, L., T. Bauer, T. Schmid, J. Stadler, and R. Zenobi. 2011. Nanoscale chemical imaging of segregated lipid domains using tip-enhanced Raman spectroscopy. *Phys. Chem. Chem. Phys.* 13:9978–81.

Pettinger, B. 2010. Single-molecule surface- and tip-enhanced Raman spectroscopy. *Mol. Phys.* 108:2039–59.

Pettinger, B., G. Picardi, R. Schuster, and G. Ertl. 2003. Surface-enhanced and STM tip-enhanced Raman spectroscopy of CN⁻ ions at gold surfaces. *J. Electroanal. Chem.* 554:293–99.

Pettinger, B., B. Ren, G. Picardi, R. Schuster, and G. Ertl. 2004. Nanoscale probing of adsorbed species by tip-enhanced Raman spectroscopy. *Phys. Rev. Lett.* 92:096101-1/096101-4.

Picardi, G., Q. Nguyen, J. Schreiber, and R. Ossikovski. 2007. Comparative study of atomic force mode and tunneling mode tip-enhanced Raman spectroscopy. *Eur. Phys. J.–Appl. Phys.* 40:197–201.

Piednoir, A., C. Licoppe, and F. Creuzet. 1996. Imaging and local infrared spectroscopy with a near-field optical microscope. *Opt. Commun.* 129:414–22.

Quabis, S., R. Dorn, M. Eberler, O. Glöckl, and G. Leuchs. 2000. Focusing light to a tighter spot. *Opt. Commun.* 179:1–4.

Raghavan, R. 2000. Infrared characterization of a cascaded arc light source. Master thesis, Case Western Reserve University, Cleveland, OH.

Roy, D., J. Wang, and M. E. Welland. 2006. Nanoscale imaging of carbon nanotubes using tip-enhanced Raman spectroscopy in reflection mode. *Faraday Discuss.* 132:215–25.

Schaller, R. D., J. Zielgelbauer, L. F. Lee, L. H. Haber, and R. J. Saykally. 2002. Chemically selective imaging of subcellular structure in human hepatocytes with coherent anti-Stokes Raman scattering (CARS) near-field scanning optical microscopy (SNOM). *J. Phys. Chem. B* 106:8489–92.

Schmid, T., A. Messner, B.-S. Yeo, W. Zhang, and R. Zenobi. 2008. Towards chemical analysis of nanostructures in biofilms II: Tip-enhanced Raman spectroscopy of alginates. *Anal. Bioanal. Chem.* 391:1907–16.

Smith, D. A., S. Webster, M. Ayad, S. D. Evans, D. Fogherty, and D. Batchelder. 1995. Development of a scanning near-field optical probe for localised Raman spectroscopy. *Ultramicroscopy* 61:247–52.

Stöckle, R. M., Y. D. Suh, V. Deckert, and R. Zenobi. 2000. Nanoscale chemical analysis by tip-enhanced Raman spectroscopy. *Chem. Phys. Lett.* 318:131–36.

Taubner, T., R. Hillenbrand, and F. Keilmann. 2003. Performance of visible and mid-infrared scattering-type near-field optical microscopes. *J. Microsc.* 210:311–14.

———. 2004. Nanoscale polymer recognition by spectral signature in scattering infrared near-field microscopy. *Appl. Phys. Lett.* 85:5064–66.

Taubner, T., F. Keilmann, and R. Hillenbrand. 2005. Nanoscale-resolved subsurface imaging by scattering-type near-field optical microscopy. *Opt. Express* 13:8893–99.

Thiel, C. W. n.d. Four-wave mixing and its applications. http://www.physics.montana.edu/students/thiel/docs/FWMixing.pdf.

Vobornik, D., G. Margaritondo, S. Vobornik, et al. 2004. Very high resolution chemical imaging with infrared scanning near-field optical microscopy (IR-SNOM). *Bosn. J. Basic Med. Sci.* 4:17–21.

Webster, S., D. A. Smith, and D. N. Batchelder. 1998. Raman microscopy using a scanning near-field optical probe. *Vib. Spectrosc.* 18:51–59.

Yeo, B. S., S. Mädler, T. Schmid, W. H. Zhang, and R. Zenobi. 2008. Tip-enhanced Raman spectroscopy can see more: The case of Cytochrome C. *J. Phys. Chem. C* 112:4867–73.

Yeo, B. S., J. Stadler, T. Schmid, R. Zenobi, and W. Zhang. 2009. Tip-enhanced Raman spectroscopy: Its status, challenges and future directions. *Chem. Phys. Lett.* 472:1–13

Zeisel, D., B. Dutoit, V. Deckert, T. Roth, and R. Zenobi. 1997. Optical spectroscopy and laser desorption on a nanometer scale. *Anal. Chem.* 69:749–754.

Zhang, W. H., B. S. Zeo, T. Schmid., and R. Zenobi. 2007. Single molecule tip-enhanced Raman spectroscopy with silver tips. *J. Phys. Chem. C* 111:1733–1738.

Zhu, L., C. Georgi, M. Hecker, et al. 2007. Nano-Raman spectroscopy with metallized atomic force microscopy tips on strained silicon structures. *J. Appl. Phys.* 101:104305.

6 Combining the Nanoscopic with the Macroscopic

SPM and Surface-Sensitive Techniques

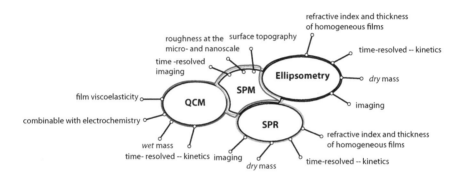

MODEL-BASED SURFACE TECHNIQUES

Surface plasmon resonance, ellipsometry, and the quartz crystal microbalance are surface-sensitive techniques that provide valuable information of optical and mechanical properties of interfaces. They all are considered to be macroscopic techniques in the sense that they supply global parameters—such as refractive index, thickness, or surface stiffness—as average magnitudes over relatively large sample areas. Another feature shared by all these techniques is that they all rely on appropriate mathematical models to get meaningful values of surface parameters. However, in many cases, choosing the right model is not straightforward.

The combination of scanning probe microscopy (SPM) with such surface techniques is a profitable match. Coupling a macroscopic technique with a nanoscopic technique such as SPM can be especially advantageous in unraveling the relation between the nanostructure and global properties of matter. In addition, SPM can act as a validating tool to test the correctness of the mathematical models employed in data analysis.

The history of surface plasmon resonance (SPR) is briefly accounted for in Figure 6.1. The discovery of light effects associated with surface plasmons occurred

1902 -- Wood reports the anomalous behaviour of light when diffracted by metallic gratings. The phenomenon is now known to be caused by surface plasmons.

Wood, R.W. 1902. On a remarkable case of uneven distribution of light in a diffraction grating spectrum. *Phil. Mag.* 4:396-408

1957 -- Ritchie predicts the existence of coherent electron oscillations on the surface of a conductor.

Ritchie, R.H. 1957. Plasma losses by fast electrons in thin films. *Phys. Rev.* 106:874-881

1959 -- Powell and Swann demonstrate the predictions of Ritchie.

Powell, J., Swan, J.B. 1959. Origin of the characteristic electron energy losses in aluminium. *Phys. Rev.* 115: 869-875

Powell, J., Swan, J.B. 1959. Origin of the characteristic electron energy losses in magnesium. *Phys. Rev.* 116: 81-83

1968 -- Kretschmann and Otto develop excitation and observation techniques that greatly facilitate SPR studies.

Kretschmann, E. , Raether, H. 1968. Radiative decay of non radiative surface plasmons excited by light. *Z. Naturforschung* 23a: 2135-2136

Otto, A. 1968. Excitation of nonradiative surface plasma waves in silver by the method of frustrated total reflection. *Z. Phys.* 216: 398-410

FIGURE 6.1   History of the discovery of surface plasmons and the development of the SPR technique.

at the beginning of the twentieth century, but the effects were not properly interpreted until the second half of the century. In the late 1960s, Kretschmann and Raether (1968) and Otto (1968) developed the instrumental basis that made SPR possible. Their instrumental configurations are still employed today.

Ellipsometry is a technique as old as SPR (see Figure 6.2). Both instrumental setup and theoretical background were conceived by Drude in the last years of the nineteenth century, but the major technical breakthroughs did not occur until 70 years later.

The quartz crystal microbalance (QCM) is the youngest technique of all mentioned in this chapter (see Figure 6.2). Since the original discovery of piezoelectric resonators, it has steadily developed in the last half of the twentieth century.

The discussion in the following background section will help the reader to understand the concepts introduced in this chapter.

Ellipsometry

1887-1888 -- Drude develops the ellipsometric technique and derived the ellipsometry equations.

Drude, P. 1887. Über die Gesetze der Reflexion und Brechung des Lichtes and der grenze absorbierende Kristalle. *Ann. Phys.* 32: 584-625

Drude, P. 1888. Beobachtungen über die reflexion des lichtes am Antimonglanz. *Ann. Phys.* 34: 489-531

1960 -- Ellipsometry develops to provide the sensitivity required to measure nanoscale layers, i.e., those used in microelectronics.

1975 -- Aspnes and Studna develop a fully automated spectroscopic ellipsometer

Aspnes, D.E., Studna, A.A. 1975. High precision scanning ellipsometer. *Appl. Optics* 14: 220-228

1984 -- Muller and Farmer develop a spectroscopic ellipsometer for real-time monitoring.

Muller, R.H. and Farmer, J.C. 1984. Fast, self-compensating spectral-scanning ellipsometer. *Rev. Sci. Instr.* 55: 371-374

1990 -- Kim and coauthors develop an instrument for simultaneous measurements at multiwavelengths.

Kin, Y.-T., Collins, R.W., Vedam, K. 1990. Fast scanning spectroelectrochemical ellipsometry: in-situ characterization of gold oxide. *Surf. Sci.* 233:341-350

QCM

1948 -- Mason describes the use of piezoelectric resonators for torsional sensors in the kHz range.

Mason, W.P. 1948. *Piezoelectric crystals and their applications in ultrasonics.* Princeton: Van Nostrand.

1959 -- Sauerbrey shows the utility of high-frequency thickness shear mode (TSM) resonators. The sensitivity of mass deposition reaches molecular monolayers.

Sauerbrey, G. 1959. Verwendung von Schwingquarzen zur wägung dünner schichten und zur mikrowägung. *Z. Phys.* 155: 206-222

1980s -- First applications of QCM in liquids.

1996 -- Rodahl and Kasemo develop QCM with dissipation monitoring (QCM-D).

FIGURE 6.2 History of ellipsometry (top) and the QCM (bottom).

NOMENCLATURE IN OPTICAL REFLECTION

As mentioned in Chapter 3, light approaching a flat interface between two distinct media may be reflected. The typical configuration in reflection optics is reviewed in Figure 6.3. The incoming and reflected light beams lie on a plane that is normal to the interface. This plane is called the plane of incidence. According to Snell's law of reflection, the angle at which the light approaches the interface is equal to the angle at which the light is reflected. The angle of incidence, θ, is defined in optics as the angle between the light beam and the surface normal. An angle of incidence of $0°$ means that the beam approaches normally to the surface.

FRAME OF REFERENCE: s- AND p-COMPONENTS OF LIGHT IN REFLECTION OPTICS

We already know from previous chapters that light is a transversal electromagnetic wave where the electric and magnetic field vectors are perpendicular to one another. Those vectors have a magnitude and a direction that depend on time and space. At a particular time and position, any of these vectors can be expressed as a sum of two mutually perpendicular vector components. These components are in turn vectors, whose directions define the frame of reference in reflection optics.

$$\mathbf{E}(r,t) = \mathbf{E}_s + \mathbf{E}_p = E_s(r,t)\mathbf{u}_s + E_p(r,t)\mathbf{u}_p$$

$$= E_{0s}e^{i(\omega t + kr + \delta_s)}\mathbf{u}_s + E_{0p} \cdot e^{i(\omega t + kx + \delta_p)}\mathbf{u}_p$$

$$(6.1)$$

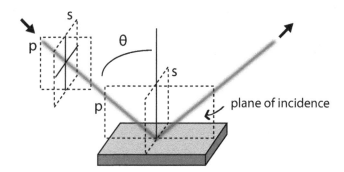

FIGURE 6.3 Notation and frame of reference in reflection optics. Light is reflected at the sample surface. The plane of incidence is defined by the incident and the reflected beam, and it is normal to the interface. The electromagnetic wave associated to the light beam may be decoupled into two vector components, one parallel to the plane of incidence ($\mathbf{E}_p$) and one perpendicular ($\mathbf{E}_s$). The p and s directions are frequently referred to in reflection optics.

Figure 6.3 depicts the frame of reference in reflection optics. One direction or axis is parallel to the plane of incidence. The projection of the electric-field vector onto this axis is the p-component (p stands for parallel). The other direction or axis is perpendicular to the plane of incidence and referred to as s (s stands for *senkrecht*, the German word for perpendicular). The projection of the electric field vector onto this axis is the s-component. The decomposition of light into mutually perpendicular components is key to understanding the different states of light polarization.

STATES OF LIGHT POLARIZATION

A transversal wave means that the electric field oscillates perpendicularly to the direction of the propagation. Once the direction of propagation is defined, however, there are infinite perpendicular planes along which the electric field may oscillate.

WHICH PLANE OSCILLATES THE ELECTRIC FIELD?

If a light beam is composed of waves where the electric field oscillates asynchronously at random planes, the light is said to be *unpolarized*. Conversely, if the light beam is composed of waves whose electric field synchronously oscillates in the same manner, the light is said to be *polarized*. In the latter case, the plane or planes at which the electric field oscillates defines the state of polarization of the light. Light can also be *partially polarized* if it is composed of a mixture of polarized and unpolarized waves. The degree of polarization is denoted as p, which can have any value between 0 (for unpolarized light) and 1 (for polarized light).

Figure 6.4 lists the most common states of polarization of light, namely linear, circular, and elliptical (Fujiwara 2007). In the following subsections, we will briefly describe the characteristics of each one.

Linear Polarization

Light is said to be linearly polarized if the electric field oscillates in the same plane. If one could track the end point of the electric field vector during an oscillation cycle, one would find that the latter follows a straight line perpendicular to the line of propagation (top of Figure 6.4). In terms of light components, E_x and E_y—where x and y can be s and p, respectively—light is linearly polarized when

$$\varphi = \delta_y - \delta_x = 0, \pi, 2\pi \qquad (6.2)$$

i.e., when both components are in phase.

Circular Polarization

Light is said to be circularly polarized if the end point of the electric field vector follows a circle during one oscillation cycle (middle of Figure 6.4). In terms of light components, this occurs when

$$E_x = E_y \quad \text{and} \quad \varphi = \delta_y - \delta_x = \pm \frac{\pi}{2} \qquad (6.3)$$

i.e., when both components are equal and are shifted $\pi/2$ to one another.

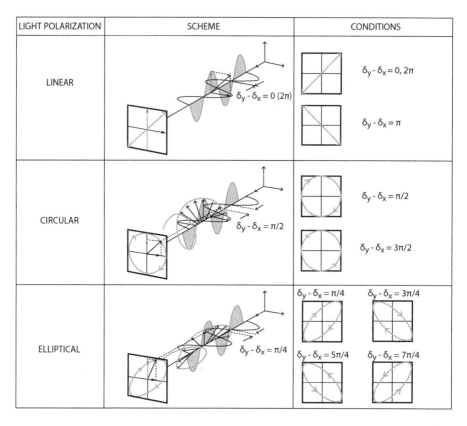

FIGURE 6.4 The most typical states of light polarization. The field components (x and y) of linearly polarized light propagate in phase. The electromagnetic field describes a line as it propagates along the z-direction. In circularly polarized light the field components are out of phase. The electromagnetic wave describes a circle as it propagates. The circle rotates clockwise if the phase shift is $\pi/2$ and anticlockwise if the phase shift is $3\pi/2$. Elliptical polarized light is produced when the components are neither in phase nor out of phase. The electromagnetic wave thus describes an ellipse whose orientation and sense of rotation are determined by the phase shift as the figure shows. (Adapted from Fujiwara 2007. With permission.)

Elliptical Polarization

Light is said to be elliptically polarized if the end point of the electric field vector follows an ellipse during one oscillation cycle (bottom of Figure 6.4). In terms of light components, this occurs when

$$E_x \neq E_y \quad and \quad \varphi = \delta_y - \delta_x \neq 0, \pi, 2\pi \qquad (6.4)$$

i.e., when both components are different in magnitude and they are not in phase. In this regard, circular polarization can be regarded as a particular type of elliptical polarization.

OPTICAL ELEMENTS

In reflection optics, light passes through diverse optical elements that may change its state of polarization. These optical elements operate in the regime of linear optics and operate according to a basic principle: *p-polarized light remains p-polarized and s-polarized light remains s-polarized.*

POLARIZER

A *polarizer* (Figure 6.5) is an optical element that converts unpolarized light into polarized light. The polarizer only transmits light that oscillates in the same plane of its *transmission axis* and blocks all the other polarization planes. For visible light, this optical element is made of two prisms of a birefringent material—a Glan-Taylor (Figure 6.5a) or a Glan-Thompson (Figure 6.5b) prism. For infrared light, the polarizer consists of a metal wire grid on top of a nonabsorbing substrate for infrared light (Fujiwara 2007).

COMPENSATOR

A *compensator* or *retarder* (Figure 6.6) converts linearly polarized light into circular polarized light or vice versa. The compensator is made of a birefringent material, an anisotropic crystal characterized by two transmission axes, the fast and the slow axes. Light propagates at different speeds if its polarization plane is parallel or normal to the fast axis. The fast and slow axes of the compensator are perpendicular to each other and determined by the inner structure of the crystal. If we decompose light into components parallel to the fast and slow

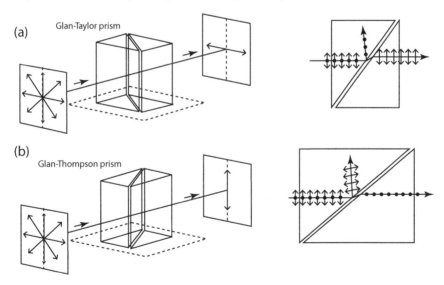

FIGURE 6.5 Polarizers for visible light. A polarizer converts depolarized light into linearly polarized light. (a) The Glan-Taylor prism transmits the light that oscillates along the horizontal axis and reflects light oscillating in the vertical axis. (b) The Glan-Thompson prism transmits the light that oscillates along the vertical axis and reflects light oscillating in the horizontal axis. (Adapted from Fujiwara 2007. With permission.)

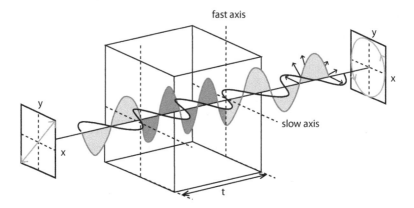

FIGURE 6.6 Compensator or retarder. A compensator converts linear into circularly polarized light and vice versa. The material is birefringent and thus has a different refractive index along its fast and slow axis. The slow axis delays the propagation of the light component along that direction, thus producing a phase shift. The phase shift is proportional to the refractive index difference and to the compensator thickness, *t*. (Adapted from Fujiwara 2007. With permission.)

axes, respectively, the former will propagate faster than the latter. The difference in speed results in a phase difference between the two components that can be expressed as follows:

$$\delta = \frac{2\pi}{\lambda} |n_f - n_s| \cdot t \qquad (6.5)$$

where n_f and n_s are the refractive indices of the fast and the slow axes, respectively, and t is the compensator thickness.

When the phase difference δ is equal to $\pi/2$, the compensator converts linear to circular polarized light and vice versa. In this case, the product $|n_f - n_s| \cdot t$ should be equal to $\lambda/4$, and the compensator is referred to as a *quarter-wave plate*.

PHOTOELASTIC MODULATOR

A *photoelastic modulator* (Figure 6.7) is a type of compensator that utilizes photoelasticity to generate a phase difference between light components. Optical anisotropy and hence birefringence can be produced by the application of a mechanical stress in some materials, such as fused quartz. The advantage is that any phase difference can be created if one can control the extent of the applied stress. This control is achieved by a piezo transducer or piezoelectric material. A modulator thus consists of a piezo transducer (quartz) and a photoelastic material (fused quartz) cemented together. When an ac[*] potential is applied across

---

[*] ac = alternating current.

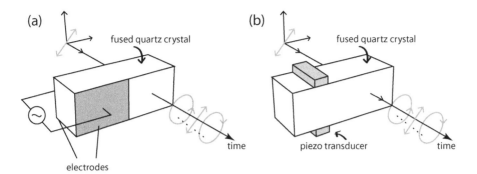

FIGURE 6.7 Photoelastic modulator. The piezoelectric device changes the phase shift of the incoming light as the former expands and compresses with time. The mechanical oscillation is brought about by an oscillating electrical potential. (a) The piezo transducer is a quartz crystal half-sandwiched between two electrodes across which the oscillating potential is applied. The other half is optically transparent and thus transmits light. (Adapted from Jasperson and Schnatterly 1969. With permission.) (b) The piezo element is made of a different material and placed on both sides of a fused quartz crystal. (Adapted from Canit and Badoz 1983. With permission.)

the piezo transducer, a periodic stress is generated in the photoelastic material. If light reaches the modulator at a polarization angle of $\pi/4$, a phase difference between the light components can be created, which varies continuously with time according to

$$\delta = F \cdot \sin \omega t \propto \frac{V}{\lambda} \cdot \sin 2\pi f t \qquad (6.6)$$

where V is the voltage amplitude applied across the piezo transducer, λ is the wavelength of light and, f is the frequency of the ac potential (usually 50 kHz).

FRESNEL'S FORMULAE

The Fresnel equations relate the light reflectivity and transmittance of an interface separating two distinct media into their respective optical constants. In this case, the optical constant is the refractive index of the material, N, which may be either a real or a complex quantity for a nonabsorbing or for an absorbing material, respectively. In the first case $N = n$, and in the second case $N = n - ik$ where i is the imaginary number and k the extinction coefficient.

Reflection and transmission can be characterized by the amplitudes of the s- and p-components of the electric field in the incident (i), reflected (r), and transmitted (t) light beams. The Fresnel equations thus relate the so-called *amplitude reflection* or *transmission coefficients* for the s- and the p-components of light with the refractive indices of the respective media

$$r_s = \frac{E_{rs}}{E_{is}} = \frac{N_i \cos\theta_i - N_t \cos\theta_t}{N_i \cos\theta_i + N_t \cos\theta_t}$$

$$r_p = \frac{E_{rp}}{E_{ip}} = \frac{N_t \cos\theta_i - N_i \cos\theta_t}{N_t \cos\theta_i + N_i \cos\theta_t}$$

$$t_s = \frac{E_{ts}}{E_{is}} = \frac{2N_i \cos\theta_i}{N_i \cos\theta_i + N_t \cos\theta_t}$$

$$t_p = \frac{E_{tp}}{E_{ip}} = \frac{2N_i \cos\theta_i}{N_t \cos\theta_i + N_i \cos\theta_t}$$

(6.7)

where r_s and r_p are the reflection coefficients for the s- and p-polarized light, respectively, whereas t_s and t_p are the respective transmission coefficients. N_i and θ_i represent the refractive indices of the medium of the incident beam and the incidence angle, respectively. N_t and θ_t are the refractive indices of the second medium and the transmission angle, respectively.

IMPEDANCE

The concept of impedance appears in different areas of physics, and it can be qualitatively defined as the extent of a cause to produce an effect. In other words, the impedance is defined as the ratio of two magnitudes: a perturbation (*stress*) and a response (*motion*).

In electrical circuits, the *electrical impedance* is defined as the opposition to the passage of an electric current. The stress in this case is an applied voltage, and the response is the motion of electrons, or electric current. In acoustics, the *acoustic impedance* is the opposition to the sound transmission, and it is defined as the ratio of stress to the speed of sound. Likewise, the mechanical counterpart or *mechanical impedance* measures the opposition to motion upon the application of a force, and it is defined as the ratio of the applied stress to the motional velocity.

SOME CONCEPTS OF ELECTROCHEMISTRY

An electrical current can induce a chemical reaction or vice versa, and that is what electrochemistry is about (Atkins and de Paula 2006). The chemical reactions thus involve the interchange of electrons from the species that loses electrons (or *oxidizes*) to the species that gains electrons (or *reduces*). These reactions are called *redox*.

An electrochemical cell is a container in which electrochemical reactions take place. It consists of two electrodes in contact with an electrolyte. One of them is the *anode*, where oxidation occurs; the other electrode is the *cathode*, where reduction occurs; consequently, *the electrons flow from the anode to the cathode*. The electrodes may be immersed in different compartments or share a common one. In the case of a two-compartment cell, a salt bridge completes the circuit.

There are two types of electrochemical cells. A *galvanic cell* generates an electrical current as a consequence of a spontaneous chemical reaction; consequently, the latter generates an electrical potential between the electrodes. In an *electrolytic cell*, however, a nonspontaneous chemical reaction takes place when an electrical potential is applied between the electrodes.

Chemical reactions in electrochemistry are heterogeneous, since they occur at the interface between an electrode of interest, a solid support, and the electrolyte solution, a liquid. The processes occurring at the electrode surfaces can be studied by *voltammetry*, in which the current is monitored as the potential of the electrode is changed.

The experimental cell arrangement in voltammetry is similar to that used to measure electrochemical rates (Atkins and de Paula 2006). It consists of a compartment and three electrodes: the *working electrode*, the *counter electrode*, and the *reference electrode*.

The electrodes that carry the current are the working electrode, which is the electrode of interest, and the counter electrode, which completes the circuit. The potential at the working electrode is measured relative to the reference electrode. In voltammetry, a potentiostat applies a defined potential between the working and the reference electrodes and measures the current between the working and the counter electrodes. The output is a *voltammogram*, where the current is plotted against the potential.

In linear-sweep voltammetry the potential is increased linearly with time, as seen in Figure 6.8a. If reduction occurs at the working electrode, a cathodic current is detected due to the reduction (and likely deposition) of reducible ions (Ox) onto the electrode surface. This cathodic current grows as the potential increases and approaches the reduction potential of the species. Soon after the value is reached, the cathodic current reaches a maximum value, which is proportional to the molar concentration of the species in solution.

In cyclic voltammetry, the potential is linearly increased and decreased in a sawtooth manner, as Figure 6.8b shows. The current is monitored as a function of the potential. The first half of the cycle resembles the process described in linear-sweep voltammetry, where the current is cathodic as a consequence of the reduction of the reducible species (Ox for oxidized). In the second half, the current changes sign and turns anodic due to the oxidation of the oxidizable species (Red for reduced), which is now at high concentrations in the proximity of the electrode. When the potential reaches that required to oxidize the species, the current drops, reaching a minimum and eventually going to zero when oxidation is completed.

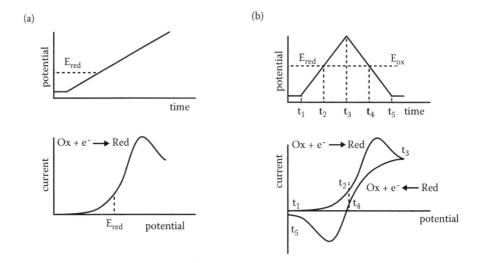

FIGURE 6.8 Experiments in voltammetry. (a) Linear-sweep voltammetry, where the potential is linearly increased with time as the current is measured. The latter reaches a maximum shortly after the potential overcomes the value required for the reducible species (Ox) to gain electrons (E_{red}). The current is said to be *cathodic* as the result of the reduction reaction Ox + e^- → Red. (b) In cyclic voltammetry, the potential is linearly increased and then decreased to its initial value in a sawtooth manner. In the first half cycle, the curve is similar to case (a); in the second half cycle, the high concentration of oxidizable species (Red) close to the working electrode reverses the reaction (Red → Ox + e^-). The current is then turned *anodic*, reaching a maximum when the potential crosses the value required for the oxidizable species (Red) to give away electrons (E_{Ox}).

FUNDAMENTALS OF SURFACE PLASMON RESONANCE

Surface Plasmons on Metal Surfaces

Surface plasmons or surface plasmon polaritons (SPPs) are electromagnetic waves that propagate along the surface of a conductor, usually a metal (Barnes, Dereux, and Ebbesen 2003). They are sort of trapped at the interface between the conductor and a dielectric medium, and they are a consequence of the resonant interaction between light photons and the free electrons of the conductor. In this interaction, the free electrons collectively oscillate at the same frequency and in phase with the light wave alongside the conductor's surface, as seen in Figure 6.9a.

The electromagnetic field associated with surface plasmons (SPs) has the shape of a longitudinal wave that is highly confined at the interface between both media. Indeed, the field strength decays exponentially with distance from the interface along the normal direction into *both* media, and therefore it is said to be *evanescent*. The wave is thus composed of a propagating and an evanescent term along the parallel (x) and normal (z) directions, respectively, according to the following expression:

$$E_{SP} = E_0 \cdot \exp^{i(k_x x \pm k_z z - \omega t)} \tag{6.8}$$

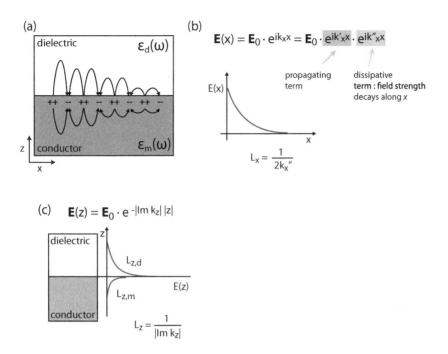

FIGURE 6.9 Surface plasmon polaritons (SPPs). (a) Coherent electron oscillations along the interface between a dielectric and a conductor (i.e., a metal). The resulting electromagnetic wave is longitudinal. It decays along the propagation direction (x) and along the normal to the interface (z). (b) Decay of the electromagnetic wave along the propagation direction. The wave vector along x, k_x has a complex magnitude. The propagative term depends on the real part of k_x, k_x', whereas the dissipative term depends on the imaginary part of k_x, k_x''. (c) Decay of the electromagnetic wave along the normal to the interface. The electric field exponentially decreases with the distance to the interface, z. The characteristic decay length is the skin depth, L_z. There is a skin depth on the side of the conductor $L_{z,m}$ and on the side of the dielectric, $L_{z,d}$. $L_{z,m}$ is slightly lower than $L_{z,d}$, since metals have a higher permittivity, according to Equation (6.11). SPPs are confined at interfaces, since the skin depths are on the order of nanometers.

where k_x and k_z are the wave vectors along the x- and z-directions, respectively, i is the imaginary number. The wave vector along the propagating direction, k_x, satisfies the so-called *dispersion relation* that is derived from the application of Maxwell's equations subjected to the boundary conditions of the interface

$$k_x = k_0\sqrt{\frac{\varepsilon_d\varepsilon_m}{\varepsilon_d + \varepsilon_m}} = k_x' + ik_x'' \tag{6.9}$$

where ε_d and ε_m are the dielectric permittivities of the dielectric and the metal conductor, respectively. Both should have opposite signs in order to produce surface plasmons (Barnes, Dereux, and Ebbesen 2003), and therefore k_x is of complex magnitude (see Figure 6.9b). For metals, ε_m is negative and complex ($\varepsilon_m = \varepsilon'_m + i\varepsilon''_m$), and since dielectrics usually have positive permittivities, the condition is fulfilled. In this case, k_0 is the wave vector of the traveling photons of the same frequency, and it is equal to the ratio of the oscillation frequency to the speed of light in vacuum, ω/c. Due to the complex nature of k_x, the electric field along the x-direction also decays in an exponential manner. The propagation length along the x-direction, L_x, is thus finite and depends on the imaginary part of k_x. L_x has the following form:

$$L_x = \frac{1}{2k''_x} = \frac{c}{\omega} \cdot \left(\frac{\varepsilon'_m + \varepsilon_d}{\varepsilon'_m \varepsilon_d} \right)^{3/2} \cdot \frac{(\varepsilon'_m)^2}{\varepsilon''_m} \tag{6.10}$$

Some Numbers: Propagation Lengths along the x-Direction

Silver is the metal with the lowest losses when irradiated with visible light. When the dielectric medium is air, the propagation length at $\lambda = 514.5$ nm is $L_x = 22$ μm, but it can be even larger at near-infrared wavelengths ($L_x = 500$ μm at $\lambda = 1060$ nm) and beyond ($L_x \approx 1$ mm at $\lambda = 1550$ nm) (Barnes, Dereux, and Ebbeson 2003).

As mentioned previously, the electromagnetic field associated with SPPs is evanescent and decays exponentially with the separation from the interface in both media. The decay length, also called *skin-depth*, L_z, characterizes the extent of penetration of the electromagnetic field in each medium. L_z depends on the imaginary part of the wave vector k_z (Im k_z) and on the respective dielectric permittivities of the metal and the dielectric (Pitarke et al. 2007)

$$L_{z,i} = \frac{1}{|\mathrm{Im}\, k_z|} = \frac{\omega}{c} \cdot \sqrt{\frac{\varepsilon_m + \varepsilon_d}{-\varepsilon_i^2}} \tag{6.11}$$

where $_i$ denotes the medium into which the wave penetrates, and thus i can be either m or d: In the former case, $L_{z,m}$ is the skin depth at the metal side; in the latter case, $L_{z,d}$ is the skin depth at the side of the dielectric (see Figure 6.9c).

Some Numbers: Skin Depths

The evanescent nature of the SPPs implies that the skin depths $L_{z,i}$ are far smaller than the propagation lengths L_x. If light of $\lambda = 600$ nm were irradiated at the interface between air and silver, the penetration of the SP field into the air would be $L_{z,\mathrm{air}} = 390$ nm and $L_{z,\mathrm{silver}} = 24$ nm into the metal. If gold were the metal chosen, $L_{z,\mathrm{air}} = 280$ nm and $L_{z,\mathrm{gold}} = 31$ nm.

Photon Creation/Excitation of SPPs

To create or excite surface plasmon polaritons through light, it is essential to use *p*-polarized light and achieve a resonant coupling between the oscillating surface charges and the oscillating electromagnetic wave. The condition is that the wave momentum of both oscillations should coincide. Since the wave momentum is proportional to the wave vectors through the reduced Max Planck constant, $\hbar = h/2\pi$, the resonant condition is achieved when

$$k_{photon} = k_{SP} \Rightarrow \frac{\omega}{c_d} = \frac{\omega}{c}\sqrt{\frac{\varepsilon_m \varepsilon_d}{\varepsilon_m + \varepsilon_d}} \tag{6.12}$$

If the dielectric medium is vacuum or air, $k_{photon} = \omega/c$, and the condition described in Equation (6.12) is never satisfied, i.e., the resonant coupling cannot be attained as long as the dielectric medium is vacuum or air and the interface is perfectly flat. The dispersion diagram shown in Figure 6.10 illustrates the issue. The dispersion curve of photons traveling in free space (in black) and that of the SPPs (in dark gray) never cross one another. The wave vector of the SPPs, k_{SP}, is higher than k_{photon} at all frequencies. However, there are ways to enhance the wave vector of light and match both momenta under these conditions.

One common way is to use a dielectric prism using the so-called *Kretschmann configuration* (Kretschmann 1968). Such a geometry is depicted in Figure 6.11a; the metal film is illuminated through the prism at an angle of incidence greater than the critical angle for total internal reflection. At a particular angle of incidence θ, the in-plane component of the wave vector of light in the prism coincides with the SPP wave vector at the air–metal interface. The resonant condition is thus fulfilled and takes this form

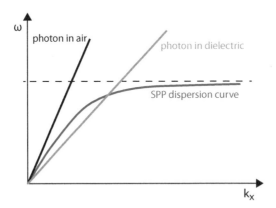

FIGURE 6.10 Dispersion curves of a photon traveling through air (black curve), and a dielectric (light gray curve), compared to that of a surface plasmon polariton (SPP, dark gray curve). In air, the dispersion curve of the photon and that of the SPP never cross and thus SPP excitation does not occur. If the photon travels through a dielectric, it crosses the dispersion SPP curve, which makes excitation possible.

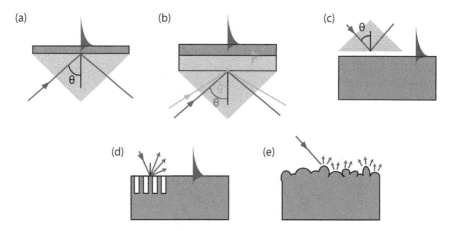

FIGURE 6.11 Excitation of SPPs. Images (a), (b), and (c) illustrate the Kretschmann configuration for flat films of increasing thickness. A prism made of a dielectric is mounted at the back side of the film in (a), at the back side of a dielectric spacer in (b), and at a small gap on top of thick films in (c). Images (d) and (e) illustrate excitation on rough surfaces, where no prism is required, with diffracting gratings in (d) and randomly rough surfaces in (e). (Adapted from Zayats, Smolyaninov, and Maradudin 2005. With permission.)

$$k_{\text{photon}} = k_{\text{SP}} = \frac{\omega}{c}\sqrt{\varepsilon_{\text{prism}}} \sin\theta \qquad (6.13)$$

where $\varepsilon_{\text{prism}}$ is the dielectric constant of the prism. Returning to the dispersion diagram of Figure 6.10, the presence of the dielectric prism reduces the slope of the photon dispersion line and thus forces it to cross the dispersion curve of the SPPs. Under these conditions, total reflection no longer occurs, and instead the surface reflectivity is dramatically reduced to almost zero. The Kretschmann geometry is valid for flat and thin metal films (i.e., approximately 60 nm thick).

For slightly thicker films, a modified Kretschmann geometry can be employed (Figure 6.11b), where an additional dielectric layer with a refractive index smaller than that of the prism is deposited between the prism and the metal film. Excitation of SPPs at both inner and outer interfaces of the metal occurs thanks to the presence of the intermediate layer (Zayats, Smolyaninov, and Maradudin 2005).

Another popular configuration is that developed by Otto for even thicker films (Otto 1968). In this configuration, the prism is situated close to the metal surface so that photon tunneling occurs through the air gap between the prism and the surface, producing the excitation of SPPs at the metal–air interface (Figure 6.11c).

On structured surfaces like diffraction gratings or randomly rough surfaces (Figures 6.11d and 6.11e), the conditions for SPP excitement are met without any special arrangements (Zayats, Smolyaninov, and Maradudin 2005). In the case of surfaces exhibiting regular patterns such as diffraction gratings, the resonant condition can still be defined as

$$\mathbf{k}_{SP} = \frac{\omega}{c} \cdot n_s \cdot \sin\theta \cdot \delta_p \cdot \mathbf{u}_{12} \pm p \cdot \frac{2\pi}{D} \cdot \mathbf{u}_1 \pm q \cdot \frac{2\pi}{D} \cdot \mathbf{u}_2 \qquad (6.14)$$

where $\mathbf{u}_1$ and $\mathbf{u}_2$ are the unit lattice vectors of a periodic structure, D is the period, p and q are integer numbers denoting the different propagation directions of the excited SPPs, and n_s is the refractive index of the medium through which the film is illuminated. However, randomly rough surfaces exhibit low-efficiency SPP excitation, due mainly to ill-defined resonant conditions. The field distribution on rough interfaces is far more complex than in the previous cases; the interface is geometrically heterogeneous at a local scale, and the boundary conditions may differ from one point to another in the metal surface, which results in nonresonant excitations. Consequently, there is a strong presence of reflected exciting light that can in turn interfere with the electromagnetic field of the SPPs. In these cases, surface reflectivity is not so strongly reduced, hence the low efficiency of SPP–light coupling.

The main potential of the technique based on surface plasmons lies on the enhancement of the electromagnetic field at the boundary between the two media, which results in an extremely high sensitivity of the SPPs to any changes in the surface conditions. One of the possible changes is the adsorption or deposition of matter over the metal surface, which has been extensively exploited to study adsorption processes.

EXPERIMENTAL SPR

A common experimental setup for surface plasmon resonance is based on the Kretschmann (1968) configuration, onto which we will focus our discussion. A thin metal layer, usually 50 nm thick gold or silver, is situated at the base of a prism and illuminated with a coherent light source, either in the visible or in the infrared. The inner side of the metal film faces the base of the prism, whereas the outer side of the film is exposed to a fluid, which in turn may be delivered through a perfusion cell, as seen in Figure 6.12a. The sample is situated at the center of a goniometer. Attached to the goniometer are the *laser arm* and the *detector arms*. They, together with the sample, define the plane of incidence. The intensity of the laser beam reflected by the outer side of the metal is measured as a function of the angle of incidence, θ, which can be varied with the goniometer. The result of an angle scan is a reflectivity curve, as shown in Figure 6.12b. Two features are to be seen in such curves: the first one is a steplike change in intensity at an angle θ_c, the critical angle where total internal reflection occurs; the second one is a pronounced minimum at higher angles. This minimum corresponds to the excitation of the surface plasmons at the metal–fluid interface, where the resonant conditions are met. The shape and the position of the minimum depend mainly on the type of metal and fluid, the thickness of the metal film, and the interface roughness.

When the metal–fluid interface is altered, for example by the presence of a thin film or any other adsorbate, the resonance minimum shifts to higher angles, as illustrated in Figure 6.12c. The angle shift for a particular type of adsorbate depends on the optical properties as well as on the amount of adsorbed material. For the particular case of smooth, homogeneous dielectric films deposited on the metal surface,

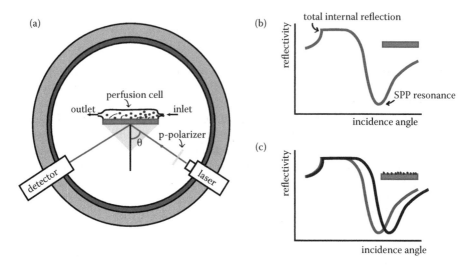

FIGURE 6.12 *(See color insert.)* SPR setup. (a) The film sample rests on a prism according to the Kretschmann configuration. The interface of interest is exposed to a fluid carrying an adsorbate that is delivered through a perfusion cell. The light beam coming from a laser at an incidence angle θ is p-polarized by the polarizer, traverses the prism, reaches the sample and is reflected back towards the detector. Both laser and detector arms can be rotated by means of a computer controlled goniometer and hence θ can be varied. In this way it is possible to monitor the reflectivity as a function of the incidence angle. (b) Typical reflectivity curve on a bare surface, showing the angles at which total internal reflection and the SPP resonance occur. (c) Reflectivity curves of a bare surface (in red) and on the same surface after material has been deposited (in brown). Whereas the angle of total internal reflection does not vary, the resonance angle shifts to a higher value.

the position of the minimum depends on the refractive index and the thickness of the layer. Knowing thus the angle at which resonance occurs and either the refractive index of the material or the layer thickness, it is possible to calculate the layer thickness (in the first case) or the refractive index (in the second case). Data interpretation is commonly based on the Fresnel theory, where the deposited films are treated as infinite and continuous dielectric layers. The treatment can be applied to stratified systems as well, where layers are arranged one on top of another. In most cases, however, neither the refractive index nor the thickness is known, and this may result in multiple possible values for both parameters. Finding the optimal pair of values that reproduce the results often relies on the user's experience with similar materials of known parameters and hence on common sense.

Another typical procedure to monitor changes on surfaces is by tracking the reflected light intensity as a function of time. These curves are obtained at a constant angle of incidence, and they are particularly revealing of the kinetics of adsorption processes. Changes in the reflectivity can be associated with modifications at the interface due to deposition of matter or to a chemical transformation. Such types of measurements are routine in the field of biosensing, especially in the detection of specific binding between biomolecules.

FUNDAMENTALS OF ELLIPSOMETRY

As mentioned in Chapter 3, when light impinges on an interface separating two distinct media, it can be transmitted or reflected. If one of these media is a fluid and the other is a solid film, reflection by the surface changes the state of polarization of light. An ellipsometry experiment measures this change through two characteristic quantities that can be related to the optical properties of the surface.

BASIC EQUATION OF ELLIPSOMETRY

The key quantities provided by an ellipsometry experiment are the *ellipsometric angles* Ψ and Δ, which are related to the reflectivity properties of a surface through the so-called *basic equation of ellipsometry*

$$\tan\Psi \cdot e^{i\Delta} = \frac{r_p}{r_s} \tag{6.15}$$

where r_p and r_s are the reflectivity coefficients for the s- and p-components of light. The reflectivity coefficients are ratios of two magnitudes: the reflected electric field and the incident electric field associated with the light wave. Indeed, if the electric field of the incident (i) and reflected light (r) can be respectively expressed as

$$\mathbf{E}^i = \mathbf{E}_p^i e^{i\delta_p^i} + \mathbf{E}_s^i e^{i\delta_s^i} \tag{6.16}$$

and

$$\mathbf{E}^r = \mathbf{E}_p^r e^{i\delta_p^r} + \mathbf{E}_s^r e^{i\delta_s^r} \tag{6.17}$$

then the reflectivity coefficients r_p and r_s are defined as

$$r_p = \frac{E_p^r}{E_p^i} \cdot e^{i\left(\delta_p^r - \delta_p^i\right)} \quad r_s = \frac{E_s^r}{E_s^i} \cdot e^{i\left(\delta_s^r - \delta_s^i\right)} \tag{6.18}$$

According to these expressions, the ellipsometry angles Ψ and Δ depend on the magnitude and the phase of the incident and reflected electric field as follows:

$$\Delta = \left(\delta_p^r - \delta_s^r\right) - \left(\delta_p^i - \delta_s^i\right) \tag{6.19}$$

$$\tan\Psi = \frac{E_p^r / E_p^i}{E_s^r / E_s^i} \tag{6.20}$$

where Δ depends exclusively on the relative phases, whereas Ψ depends on the amplitudes of the incident and reflected electric fields. The ratio of the reflectivity

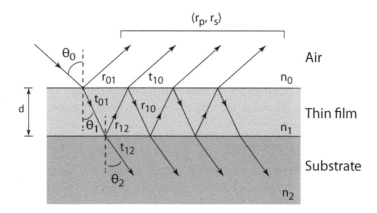

FIGURE 6.13 Thin-layer model. The figure illustrates the case of a single-layer separated by two semi-infinite, continuous media such as air (above) and the substrate (below). The refractive index of the air, the film and the substrate are n_0, n_1, and n_2, respectively. An incoming beam can be reflected as well as diffracted at each interface between two media, A and B. In the film, the beam undergoes multiple reflection and diffraction. Single reflection or refraction is characterized by either a reflectivity or a transmission coefficient, r_{AB} and t_{AB}. The individual phenomena contribute to generate a total reflectivity coefficient, r, which is defined along the p and s directions as r_p and r_s, respectively.

coefficients in turn depends on the optical properties of the film sample, such as thickness and refractive index, as well as the morphological structure and composition of the sample. The relation between the ellipsometric angles and the sample structure is not straightforward, though, and it is always based on model calculations. Most of the models assume that the film is composed of a single or a stack of homogeneous layers of defined thickness and refractive index separated by two infinite, continuous media, as seen in Figure 6.13.

SINGLE-LAYER MODEL

We will consider the case of a single, homogeneous, and nonabsorbing layer of refractive index n_1 and thickness d in between two semi-infinite media, a fluid with refractive index n_0, and a substrate with refractive index n_2. Such a model is depicted in Figure 6.13. Light incident from the fluid at an angle θ_0 is partly reflected and partly refracted at the fluid–layer interface. The refracted beam is deviated from the surface normal an angle θ_1 and is in turn reflected and refracted—in this case, at an angle θ_2—at the layer–substrate interface. The beam in the layer undergoes multiple reflection and refraction as it is bounced back and forth from one interface to the other. The final result is characterized by the overall reflection coefficients of the light beam

$$r_p = \frac{r_{01,p} + r_{12,p} \cdot e^{-i2\beta}}{1 + r_{01,p} \cdot r_{12,p} \cdot e^{-i2\beta}} \tag{6.21}$$

for the *p*-component and

$$r_s = \frac{r_{01,s} + r_{12,s} \cdot e^{-i2\beta}}{1 + r_{01,s} \cdot r_{12,s} \cdot e^{-i2\beta}} \tag{6.22}$$

for the *s*-component of light.

The parameter β has the form

$$\beta = 2\pi \frac{d}{\lambda} \sqrt{n_1^2 - n_0^2 \sin^2 \theta_2}$$

and represents the phase shift at each boundary or interface. The reflectivity coefficients $r_{01,p}$, $r_{12,p}$, $r_{01,s}$, and $r_{12,s}$ describe the reflection of the light beam at the fluid–layer interface (01) and at the layer–substrate interface (12), and they are given by the Fresnel laws (Röseler 1990)

$$r_{01,p} = \frac{n_1 \cos\theta_0 + n_0 \cos\theta_1}{n_1 \cos\theta_1 + n_0 \cos\theta_1} = \frac{n_1 \cos\theta_0 - n_0 \sqrt{1 - \dfrac{n_0^2}{n_1^2} \sin^2\theta_0}}{n_1 \cos\theta_0 - n_0 \sqrt{1 - \dfrac{n_0^2}{n_1^2} \sin^2\theta_0}} \tag{6.23}$$

and

$$r_{12,p} = \frac{n_2 \cos\theta_1 - n_1 \cos\theta_2}{n_2 \cos\theta_1 + n_1 \cos\theta_2} = \frac{n_2 \sqrt{1 - \dfrac{n_0^2}{n_1^2} \sin^2\theta_0} - n_1 \sqrt{1 - \dfrac{n_0^2}{n_2^2} \sin^2\theta_0}}{n_2 \sqrt{1 - \dfrac{n_0^2}{n_1^2} \sin^2\theta_0} + n_1 \sqrt{1 - \dfrac{n_0^2}{n_2^2} \sin^2\theta_0}} \tag{6.24}$$

for the *p*-component of the light beam and

$$r_{01,s} = \frac{n_0 \cos\theta_0 - n_1 \cos\theta_1}{n_0 \cos\theta_0 + n_1 \cos\theta_1} = \frac{n_0 \cos\theta_0 - n_1 \sqrt{1 - \dfrac{n_0^2}{n_1^2} \sin^2\theta_0}}{n_1 \cos\theta_0 - n_1 \sqrt{1 - \dfrac{n_0^2}{n_1^2} \sin^2\theta_0}} \tag{6.25}$$

$$r_{12,s} = \frac{n_1 \cos\theta_1 - n_2 \cos\theta_2}{n_1 \cos\theta_1 + n_2 \cos\theta_2} = \frac{n_1 \sqrt{1 - \dfrac{n_0^2}{n_1^2} \sin^2\theta_0} - n_2 \sqrt{1 - \dfrac{n_0^2}{n_2^2} \sin^2\theta_0}}{n_1 \sqrt{1 - \dfrac{n_0^2}{n_1^2} \sin^2\theta_0} + n_2 \sqrt{1 - \dfrac{n_0^2}{n_2^2} \sin^2\theta_0}} \tag{6.26}$$

for the s-component of light. If the layer is so thin that its thickness is very small compared to the wavelength of light ($d \ll \lambda$), it is possible to expand the reflectivity coefficients in a power series in terms of the ratio d/λ. If all the high-order terms are negligible compared to the first term in this expansion, the expression for Δ results in

$$\Delta \approx \frac{4\sqrt{\varepsilon_0}\,\varepsilon_2\pi\cos\theta_0\sin^2\theta_0}{(\varepsilon_0-\varepsilon_2)\cdot\left[(\varepsilon_0+\varepsilon_2)\cos^2\theta-\varepsilon_0\right]}\cdot\frac{(\varepsilon_1-\varepsilon_0)\cdot(\varepsilon_2-\varepsilon_1)}{\varepsilon_1}\cdot\frac{d}{\lambda} \tag{6.27}$$

where $\varepsilon_0 = n_0^2$, $\varepsilon_1 = n_1^2$, and $\varepsilon_2 = n_2^2$ are the dielectric permittivities of the fluid, the layer, and the substrate, respectively. If the refractive index varies over the thickness of the layer, the term

$$\frac{(\varepsilon_1-\varepsilon_0)\cdot(\varepsilon_2-\varepsilon_1)}{\varepsilon_1}\cdot d$$

has to be replaced by an integral across the layer

$$\eta = \int_0^d \frac{(\varepsilon-\varepsilon_0)\cdot(\varepsilon_2-\varepsilon)}{\varepsilon}\cdot dz \tag{6.28}$$

An ellipsometry experiment on an ultrathin film such as a monolayer yields, therefore, a quantity proportional to η. In the case of an organic layer on a solid support where $\varepsilon_2 > \varepsilon_1$ and ε_1 is very similar to ε_0, η can be further simplified

$$\eta = \frac{\varepsilon_2-\varepsilon_0}{\varepsilon_0}\int_0^d (\varepsilon-\varepsilon_0)dz \tag{6.29}$$

The dielectric permittivity of the layer can be considered to be proportional to the concentration of the organic component, c, through

$$\varepsilon = \varepsilon_0 + c\cdot\frac{d\varepsilon}{dc} \tag{6.30}$$

and therefore the integral of Equation (6.29) is in turn proportional to the amount of organic compound in the layer, Γ

$$\eta = \frac{\varepsilon_2-\varepsilon_0}{\varepsilon_0}\cdot\frac{d\varepsilon}{dc}\cdot\int_0^d c\cdot dz = \frac{\varepsilon_2-\varepsilon_0}{\varepsilon_0}\cdot\frac{d\varepsilon}{dc}\cdot\Gamma \tag{6.31}$$

Therefore the ellipsometric angle Δ is a quantity proportional to the mass of the layer.

EXPERIMENTAL DETERMINATION OF ELLIPSOMETRIC ANGLES

Figure 6.14 shows the typical setup of an ellipsometer where the sample is vertically and horizontally oriented (Figures 6.14a and 6.14b, respectively). The sample rests on a platform at the center of the goniometer. As in the SPR setup, both laser and detector arms are attached to the goniometer and, together with the sample, define the plane of incidence. The laser arm produces the incident beam with a well-defined state of polarization. This is usually achieved by two optical elements: a circular prepolarizer, a linear *polarizer* (*P*), and a *compensator* (*C*). The circular prepolarizer transforms the laser output into circularly polarized light. The linear polarizer is a rotary linear polarizer whose transmission axis can be set at any angle and transforms, in turn, the circularly polarized light into linearly polarized light. The

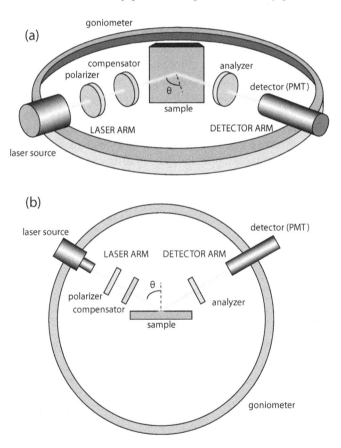

FIGURE 6.14 Setup of an ellipsometer. (a) Vertical sample configuration. (b) Horizontal sample configuration. The sample rests at the centre of a goniometer that holds the laser and detector arms. The laser arm produces circularly polarized light by means of a polarizer and a compensator. Reflection at the sample surface changes the state of polarization of the light. The detection comprises another polarizer (the analyzer) and a piezo multiplier tube at the detector arm.

compensator, which is itself a quarter-wave plate or retarder,[*] transforms the linearly polarized light into circularly polarized light before the latter irradiates the sample surface. The detector arm is composed of the *analyzer* (A) and a detector, a photomultiplier tube. The analyzer is another rotary linear polarizer whose transmission axis can be set at any angle.

The ellipsometry experiments basically consist of obtaining the ellipsometric angles from a certain pair of magnitudes: the transmission angles of the polarizer and of the analyzer. These pairs of values may determine the states of polarization of the incident and reflected light or directly yield Ψ and Δ. The first case refers to a set of measurement modes that modulate the light beam by rotating one of the optical elements (the *element-rotating* methods) or by introducing a (photoelastic) modulator (the *phase-modulation* method). The second case refers to a widespread method based on the extinction of the reflected light and is called *null ellipsometry*.

According to the wavelength of light used in ellipsometry, the technique can also be classified in single-wavelength ellipsometry and multiple-wavelength or *spectroscopic* ellipsometry. The latter is usually carried out throughout the ultraviolet/visible or infrared regions and used in conjunction with element-rotating or phase-modulation methods. Due to the spectral nature of the results, the technique gives a more complete picture of the optical behavior of films, especially in those cases where the refractive index of the sample varies with the wavelength of light or where the sample is composed of multilayers of different materials.

ROTATING-ELEMENT TECHNIQUES

Rotating Polarizer/Analyzer

In this configuration, the reflected light beam is received at the detector as an oscillatory signal due to either a rotating polarizer or a rotating analyzer. If the optical element rotates at a speed $2\pi\omega$, where ω is the angular frequency, the time variation of the intensity at the detector can be expressed as follows:

$$I(t) = I_0 \cdot \left(1 + \alpha \cdot \cos 2\omega t + \beta \sin 2\omega t\right) \tag{6.32}$$

where α and β are the so-called *Fourier coefficients*. These parameters are in turn functions of the ellipsometry angles Ψ and Δ through the following equations:

$$\alpha = \frac{\tan^2 \Psi - \tan^2 X}{\tan^2 \Psi + \tan^2 X} \quad \beta = \frac{2\tan \Psi \cos \Delta \tan X}{\tan^2 \Psi + \tan^2 X} \tag{6.33}$$

where X refers either to the polarizer settings ($X = P$) in the rotating analyzer configuration or to the analyzer settings ($X = A$) in the rotating polarizer configuration. The ellipsometry angles can thus be expressed in terms of α and β

---

[*] In the case of null ellipsometry, the compensator is normally set at 45° or −45° with respect to the plane of incidence to simplify.

$$\tan \Psi = \sqrt{\frac{1+\alpha}{1-\alpha}} |\tan X|; \quad \cos \Delta = \frac{\beta}{\sqrt{1-\alpha^2}} \tag{6.34}$$

In an ellipsometry measurement using either the rotating polarizer or the rotating analyzer, α and β are first determined from the Fourier analysis of the light intensity signal at the detector. The ellipsometry angles can thus be obtained by substituting the respective values into Equations (6.34).

Rotating Compensator

In this case, the light intensity at the detector is modulated by rotating the compensator in the laser arm. If the speed of rotation can be likewise expressed as $2\pi\omega$, the intensity at the detector can be described by the following general formula:

$$I(t) = I_0 \left(1 + \alpha_2 \cos 2\omega t + \beta_2 \sin 2\omega t + \alpha_4 \cos 4\omega t + \beta_4 \sin 4\omega t\right) \tag{6.35}$$

The expression is more complicated than Equation (6.32), and it contains additional terms dependent on higher-order harmonics. In the case of a polarizer angle $P = 45°$, the Fourier coefficients α_2, β_2, α_4, and β_4 can be described in terms of the ellipsometry angles as

$$\tag{6.36}$$

$$\alpha_2 = \frac{p \cdot \sin \delta \cdot \sin 2\varepsilon \cdot \sin 2A}{\alpha_0} \qquad \beta_2 = -\frac{p \cdot \sin \delta \cdot \sin 2\varepsilon \cdot \cos 2A}{\alpha_0}$$

$$\alpha_4 = \frac{p \cdot \sin^2 (\delta/2) \cdot \cos 2\varepsilon \cdot \cos 2(A+\theta)}{\alpha_0} \qquad \beta_4 = \frac{p \cdot \sin^2 (\delta/2) \cdot \cos 2\varepsilon \cdot \sin 2(A+\theta)}{\alpha_0}$$

where $\alpha_0 = 1 + p \cdot \cos 2(\delta/2) \cdot \cos 2\varepsilon \cdot \cos 2(A - \theta)$, p represents the degree of polarization, A is the analyzer angle, δ is the phase retardation introduced by the compensator, and ε and θ are a pair of angles that describe the state of the polarization of the light and are thus related to the ellipsometry angles according to the expressions

$$\sin 2\varepsilon = -\sin 2\Psi - \sin \Delta; \quad \tan 2\theta = -\tan 2\Psi - \cos \Delta \tag{6.37}$$

Through Fourier analysis of the intensity signal, it is thus possible to obtain the Fourier coefficients and calculate the ellipsometry angles by applying Equations (6.36) and (6.37).

Phase Modulation

Inserting a photoelastic modulator between the sample and the analyzer is another way of modulating the reflected light at the detector. If the modulator has a phase shift δ that oscillates with time according to $F - \cos \omega t$, the intensity at the detector can be described as

$$I(t) = I_0 \cdot \left[\alpha_0 + \alpha_1 \sin \delta + \alpha_2 \cos \delta\right] \tag{6.38}$$

The coefficients α_0, α_1, and α_2 depend on the angles of the modulator M, the analyzer A, and the polarizer P as well as the ellipsometry angles according to

$$\alpha_0 = 1 - \cos 2\Psi \cdot \cos 2A + \cos 2(P - M) \cdot \cos 2M(\cos 2A - \cos 2\Psi)$$

$$+ \sin 2A \cdot \cos \Delta \cdot \cos 2(P - M) \cdot \sin 2\Psi \cdot \sin 2M$$

$$\alpha_1 = \sin 2(P - M) \cdot \sin 2A \cdot \sin 2\Psi \cdot \sin \Delta \tag{6.39}$$

$$\alpha_2 = \sin 2(P - M) \cdot \left[(\cos 2\Psi - \cos 2A) \cdot \sin 2M + \sin 2A \cdot \cos 2M \cdot \sin 2\Psi \cdot \cos \Delta \right]$$

Likewise, $\sin\delta$ and $\cos\delta$ can be expressed as a sum of harmonics of ωt time as follows:

$$\sin \delta = \sin(F \cdot \sin \omega t) = 2 \sum_{m=0}^{\infty} J_{2m+1}(F) \sin\left[(2m+1)\omega t\right]$$

$$\tag{6.40}$$

$$\cos \delta = \cos(F \cdot \sin \omega t) = J_0(F) + 2 \sum_{m=1}^{\infty} J_{2m}(F) \cos(2m\omega t)$$

The terms J_i are Bessel functions with respect to F. The expressions of Equation (6.40) are complicated for a Fourier analysis, and normally only the low-frequency components are taken into account. Equation (6.38) can be further simplified if, in addition, F is set to $138°$, $P - M = 45°$, $M = 0°$, and $A = 45°$. In that case, $J_0(F) = 0$, and the expression for the light intensity results in

$$I(t) = I_0 \cdot \left\{ 1 + \sin 2\Psi \cdot \sin \Delta \left[2J_1(F) \sin \omega t \right] + \sin 2\Psi \cdot \cos \Delta \left[2J_2(F) \cos 2\omega t \right] \right\} \tag{6.41}$$

The ellipsometry angles can thus be estimated from the Fourier coefficients of $\sin \omega t$ and $\cos 2\omega t$.

Null Ellipsometry

The method consists of obtaining the pair of analyzer and polarizer angles for which the intensity at the detector is minimal or even zero. Under these conditions, the basic equation of ellipsometry can be expressed as a function of the settings of all the optical elements, that is, the transmission axis of the analyzer, A, and the polarizer, P, as well as the orientation of the retardation axis in the compensator, C.

$$\frac{r_p}{r_s} = -\tan A_0 \cdot \frac{\tan C - i \tan(C - P_0)}{1 + i \tan C \cdot \tan(C - P_0)} \tag{6.42}$$

If C is set to either $-45°$ or $45°$ the expression can be further simplified, giving the corresponding expressions

$$\frac{r_p}{r_s} = \tan \Psi e^{ii\Delta} = \tan A_0 e^{i\left(sP_0 + \frac{\pi}{2}\right)}$$

(6.43)

$$\frac{r_p}{r_s} = \tan \Psi e^{i\Delta} = -\tan A_0 e^{i\left(\frac{\pi}{2} - 2P_0\right)}$$

to directly identify the values for Ψ and Δ for $C = -45°$

$$\Psi = A_0$$

(6.44)

$$\Delta = 2P_0 + \frac{\pi}{2}$$

and for $C = 45°$

$$\Psi = A_0$$

(6.45)

$$\Delta = \frac{\pi}{2} - 2P_0$$

Once the pair of null settings (A_0, P_0) has been determined that completely cancels out the reflected intensity, there is another pair (P_0', A_0') that satisfies the same condition

$$\left(P_0', A_0'\right) = \left(P_0 + 90°, 180° - A_0\right)$$

(6.46)

These pairs of settings are referred as ellipsometry zones. Measurements in various zones are especially convenient to improve accuracy in the determination of the ellipsometric angles as well as to cancel instrumental errors due to misalignment (Motschmann and Teppner 2001).

FUNDAMENTALS OF QUARTZ CRYSTAL MICROBALANCE

Piezoelectric resonators have long been recognized as sensing devices (Mason 1948). Indeed, their resonating properties, such as resonant frequency or the Q-factor, are extremely sensitive to their immediate surroundings. In the same sense, they readily respond to any alteration that may occur in the vicinity. The alteration can be a mass deposition on their surface or the presence of a viscous fluid. As a consequence, they experience either a shift in the resonance frequency or a shift in the Q-factor that can be related to the amount of mass deposited or to the viscoelastic properties of the deposited material and the surrounding fluid.

The core of the quartz crystal microbalance (QCM) is one type of such piezoelectric resonators, the so-called *thickness-shear mode* or *shear-wave resonator* (TSM) (see Figure 6.15). The sensor consists of a plate or a disc made of piezoelectric crystal whose lateral dimensions are much larger than its thickness. The TSM resonators support bulk acoustic modes that have a transverse polarization. This means that

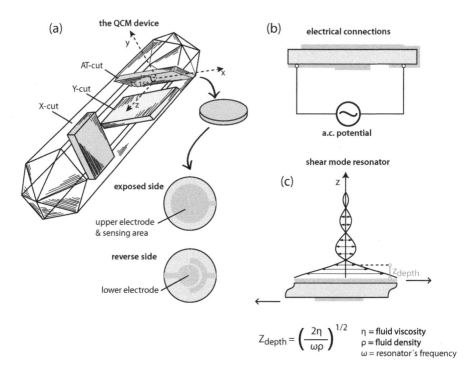

$$Z_{depth} = \left(\frac{2\eta}{\omega\rho} \right)^{1/2}$$

η = fluid viscosity
ρ = fluid density
ω = resonator's frequency

FIGURE 6.15 Basics of QCM. The technique relies on the properties of the QCM sensor, a shear-mode resonator usually a quartz disk of polished sides. (a) The sensor is produced from AT cuts of quartz crystals, which are nowadays artificially grown. Gold is deposited on both sides of the disk to produce the upper and lower electrodes; the upper (exposed) side faces the sample whereas the lower (reverse) side faces the electrical connections of the respective electrodes, as shown in (b). The sensor shear vibrates when an ac potential is applied across the electrodes, the shear amplitude reaches a maximum at the resonance frequency or at its (odd-numbered) overtones. (c) Vertical profile of the resulting mechanical wave. The penetrating depth Z_{depth} depends on the properties of the surrounding fluid (viscosity and density) as well as the resonator's frequency.

the acoustic oscillation is parallel to the sensor surface, a shear wave, where half the crystal moves in opposite directions, as seen in Figure 6.15c. The crystal has been previously cut in a special orientation so that only the acoustic mode or modes of interest can be electrically excited. In other words, the modes should be piezoelectrically coupled. A most convenient cut is the so-called *AT cut* generally encountered in quartz devices (see Figure 6.15a). The AT cut makes only one of the possible acoustic modes piezoelectrically coupled. The piezoelectric coupling is fairly high (8.8%; Schmidt et al. 2001) and the mode affected is almost pure shear, which makes the sensor particularly appropriate for sensing applications in fluids that do not support transverse modes. Another important advantage of the AT cut is its high-temperature stability. The temperature coefficient of the frequency is zero at 25°C, which makes the resonant frequency very stable upon thermal changes, as long as the device operates near room temperature (Schmidt et al. 2001).

To electrically excite the TSM of the sensor, the latter is sandwiched between two electrodes. usually made of gold and deposited on each side of the crystal plate or disc (see Figures 6.15a and 6.15b). The application of an alternating potential between these electrodes causes the sensor to oscillate (Figures 6.15b and 6.15c). If the acoustic wavelength is twice the sensor thickness, then resonance occurs. For a piezoelectric crystal between two electrodes, there are two resonant frequencies instead of one: that of the *series resonance* f_s and that of the *parallel resonance* or *antiresonance* f_p. The first resonance is obtained when the electrodes are connected (short circuit), whereas the second resonance occurs when the electrodes are unconnected (open circuit). In the case of an AT cut crystal, these resonances lie very close to one another and are approximated by (Schmidt et al. 2001)

$$f = M \cdot \frac{v_m}{2h} \tag{6.47}$$

where M is the overtone number, v_m is the velocity of the vibration mode, and h is the plate thickness. The crystal can resonate at its fundamental frequency ($M = 1$) or at higher frequencies or *overtones* ($M > 1$), which can only be odd integers of f_s and f_p.

Some Numbers: Amplitude and Frequency of AT-Cut Crystals

AT cut quartz crystals have a typical fundamental resonance frequency of 5 MHz. The amplitude of vibration at the fundamental frequency decreases radially from the center of the sensor. At this position, the vibration amplitude reaches a maximum, which is (Mecea 2005)

$$A_0 = CQV_d \tag{6.48}$$

where $C = 1.4 \cdot 10^{-12}$ m/V, Q is the quality factor of the quartz resonator, and V_d is the peak voltage applied on the electrodes. For a Q-factor of 1.5×10^5 and V_d of 0.0548 V, A_0 is 11.6 nm.

ELECTROMECHANICAL ANALOGY OF A QCM SENSOR: EQUIVALENT CIRCUIT OF A RESONATOR

A widespread and intuitive approach is to consider the electrical analogy of an acoustic resonator. The analogy is an electrical circuit composed of an arrangement of passive electrical elements. The concerted behavior of those elements is equivalent to the mechanical behavior of the crystal resonator.

The equivalent circuit that represents the QCM sensor is shown in Figure 6.16a and is commonly referred as the Butterworth-van Dyke model (BvD). It is a parallel arrangement of a capacitor—the *electrical* branch—and a set of three elements connected in series—the *acoustic* branch. The elements of the acoustic branch are an

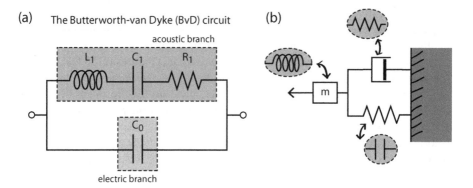

FIGURE 6.16 The Butterworth-van Dyke (BvD) circuit (a) and its mechanical equivalent (b). The BvD circuit represents the electromechanical response of the QCM sensor. It comprises an acoustic and an electric branch. The acoustic branch consists of an inductor of inductance L_1, a capacitor of capacitance C_1, and a resistor of resistance R_1 connected in series. They represent the mass of the sensor, its elasticity and its viscous drag, respectively. The electric branch represents the gold electrode pair that contributes to the electrical response as a capacitor of capacitance C_0.

inductor, a capacitor, and a resistor. In the mechanical representation of the crystal resonator, these elements are in turn equivalent to a mass, a spring, and a dashpot (Figure 6.16b). The first is the crystal mass, the second is the crystal's elasticity, and the third is the viscous loss or dissipation in the crystal. The capacitor in parallel represents the pair of electrodes at both sides of the crystal, completing the picture of the QCM sensor.

The impedance of the crystal, denoted as Z_q, is a function of the inductance L_1, the capacitance C_1, and the resistance R_1 of the acoustic branch, which in turn can be described in terms of the structural and mechanical parameters of the crystal according to (Johannsmann 2007)

$$C_1 = 4\phi^2 \cdot \frac{1}{k} = \frac{8Ae_{26}^2}{d_q(M\pi)^2 G_q} \cdot \left(1 - \frac{8\kappa^2}{(M\pi)^2}\right)^{-1}$$

$$L_1 = \frac{1}{4\phi^2} \cdot m = \frac{\rho_q d_q^3}{8Ae_{26}^2}$$

$$R_1 = \frac{1}{4\phi^2} \cdot \xi = \frac{d_q}{8Ae_{26}^2} \cdot (M\pi)^2 \eta_q = \sqrt{\frac{L_1}{C_1}} \cdot \frac{1}{Q}$$

$$\phi = \frac{Ae_{26}}{d_q}$$

(6.49)

where $\kappa = [e_{26}^2/\varepsilon\varepsilon_0 G_q]^{1/2}$ is a dimensionless coefficient of piezoelectric coupling; e_{26} is the piezoelectric stress coefficient; M is the overtone number; G_q, ρ_q, d_q, and η_q

are the shear modulus, the density, the thickness, and the viscosity of the crystal, respectively; and A is the effective area of the sensor. The variables k, m, and ξ are, respectively, the spring constant, the mass, and the drag coefficient of the crystal.

Some Numbers: Mechanical Parameters of AT-Cut Quartz Resonators

The mass density of the quartz mass, ρ_q, is 2.648 g·cm^{-3}, and the shear modulus depends on the type of cut. For an AT cut quartz sensor, G_q is 2.947·10^{11} dyn·cm^{-2} (2.947·10^{10} Pa).

In the context of equivalent circuits, the sample is usually referred to as *load* and introduced in the BvD circuit as an additional electrical element with impedance Z_L. This element is usually incorporated into the acoustic branch of the circuit, as depicted in Figure 6.17a. If the load is a rigid film, the element will be an inductor, since it only introduces an inert mass to the acoustic system (Figure 6.17b). If the load is a viscoelastic film, the element will additionally have a capacitor and a resistor representing the elasticity and the viscosity of the material, respectively (Figure 6.17c). If the sensor is exposed to a viscous fluid other than air, the circuit will be further modified with the addition of an inductor (L_l) in series with a resistor (R_l), representing the fluid's mass and the fluid's viscosity, respectively (Figure 6.17d).

The load impedance Z_L can be calculated from frequency shifts in a QCM setup, providing that the magnitude is small compared to the impedance of the crystal. This is referred to as the small-load approximation.

SMALL-LOAD APPROXIMATION: CONNECTING FREQUENCY SHIFTS TO LOAD IMPEDANCE

QCM studies rely on the small-load approximation (Johannsmann 2008). A small load implies that the frequency shift is small compared to the fundamental resonance frequency f_f of the couple sensor load ($\Delta f/f_f \ll 1$). In terms of equivalent circuits, the load impedance Z_L is small compared to the acoustic impedance of the quartz sensor Z_q, and the frequency shift can be approximated as

$$\frac{\Delta f^*}{f_f} = \frac{\Delta f + i\Delta\Gamma}{f_f} \approx \frac{i}{\pi} \cdot \frac{Z_L}{Z_q} \tag{6.50}$$

where the frequency shift is depicted here as a complex quantity comprising a real part, the shift of the resonance maxima, Δf, and an imaginary part, the variation of half the bandwidth, $\Delta\Gamma$. Equation 6.50 thus states that the frequency shift is proportional to the load impedance, and since Z_q is a material's constant, it depends linearly on the load.

(a)
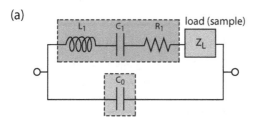

load (sample)

(b) rigid load
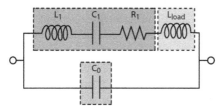

(c) non-rigid load
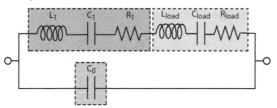

(d) non-rigid load in liquid
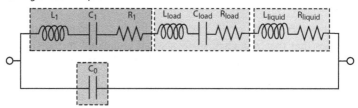

FIGURE 6.17 Modifications to the BvD circuit to account for the presence of the sample. (a) The sample or load is included as an element of impedance Z_{load} in the acoustic branch. (b) A rigid load is represented as an inductor of inductance L_{load} or simply L_L. (c) A viscoelastic load is additionally represented by its elasticity (a capacitor of capacitance C_{load} or C_L) and by its viscosity (a resistor of resistance R_{load} or R_L). (d) If the load is in a fluid characterized by a mass and a viscosity, the circuit is further modified by an inductor of inductance L_{liquid} or L_l and a resistor of resistance R_{liquid} or R_l.

MEASURING FREQUENCY SHIFTS AND MORE: MODES OF OPERATION IN QCM

QCM setups measure frequency shifts or resonant parameters of resonators as a function of time and are mainly based on either one of two instrumental configurations. The *impedance analysis* method applies a standing oscillating voltage to the sensor and measures in detail the impedance spectra around the resonant frequency. The electromechanical parameters are thus obtained from curve fittings to the appropriate equivalent circuit. The *ring-down* method measures the decay of the sensor's oscillation after the application of a transient oscillating voltage. The frequency and the decay constant, which are related to the damping of the oscillation, are obtained from such measurements.

Both procedures are thus able to track any changes occurring at the surface of the sensor that may affect its resonant properties around the fundamental resonance as well as around its higher-order overtones.

Impedance Analysis

The electrodes of the sensor are connected to a network analyzer that measures the complex electrical impedance of the device, $Z(f)$, as a function of frequency. The impedance is the ratio between the applied voltage and the measured current through the sensor. Both voltage and current are sinusoidal signals of the same frequency f, though shifted in phase. The amplitude ratio of both signals is related to the real part of the impedance, whereas the phase shift is related to the imaginary part. The frequency window should be carefully chosen so that it contains the resonance frequency.

Since voltages are more accurately measured than currents, the electrical impedance is usually expressed as a function of voltage ratios. In a network connection such as the one shown in Figure 6.18a, the ratio of voltages V_1 and V_2 depends on the impedance Z through three constants, A_0, A_1, and A_2 (Lucklum and Eichelbaum 2007)

$$\frac{V_1}{V_2} = r_z = \frac{A_0 + Z}{A_1 \cdot Z + A_2} \tag{6.51}$$

The constants can be determined at each frequency by acquiring the voltage ratios for three calibration standards. These standards are usually the open circuit, where the sample is replaced by an air gap; the closed circuit, where the electrodes are connected to one another; and a reference measurement, where the sample is a resistor of known resistance, R_{ref}. In the first case, the impedance is infinite; in the second case, the impedance is zero; and in the third case, the impedance is equal to R_{ref}. The expression for the unknown impedance Z can then be described in terms of the voltage ratios r_{open}, r_{short}, and r_{ref} as follows:

$$Z = R_{ref} \cdot \frac{r_{open} - r_{ref}}{r_{ref} - r_{short}} \cdot \frac{r_{short} - r_z}{r_z - r_{open}}$$

$$r_{open} = \frac{1}{A_1} \quad r_{short} = \frac{A_0}{A_2} \quad r_{ref} = \frac{A_0 + R_{ref}}{A_1 \cdot R_{ref} + A_2} \tag{6.52}$$

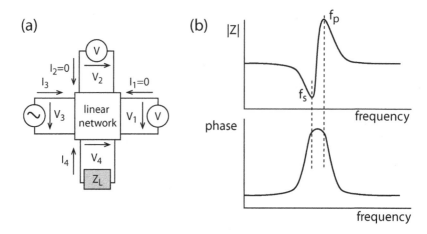

FIGURE 6.18 Measuring the load impedance with a network analyzer. (a) A network or impedance analyzer is the generalized version of a Wheatstone bridge to measure impedance in a branched circuit. (b) Impedance and phase plots of a resonator. A resonator exhibits two resonance frequencies, the series frequency f_s and the parallel frequency f_p. The closeness of these values defines the width of the peak in the phase plot.

All these relations result in experimental graphs of the type shown in Figure 6.18b, where the impedance magnitude and phase of the resonator are plotted as a function of frequency. These resonance curves can be fitted to a BvD-type model so as to obtain the acoustic branch parameters L_1, C_1, and R_1 for a bare sensor and the stray capacitance of the electrode connections, C_0. If material of any sort is deposited on the active area of the sensor, the resonance will shift to lower frequencies and may change shape. The load impedance Z_L will in this case be obtained from that measurement and from the bare sensor results by fitting the new resonance curves to the modified BvD models of Figure 6.17. From these fittings, the resonance frequencies as well as the bandwidth of the resonance—a measure of the energy losses within the sensor or surroundings—can be precisely determined.

Last but not least, it should be mentioned that real-life electrical measurements are not free of parasitic contributions (currents, capacitances), due to nonideal electrical contacts and wiring that can greatly disturb the results. In this sense, the calibration standards as well as the bare sensor measurements serve a double function. On the one hand, they are essential to obtain the load impedance, as we have just seen. On the other hand, they can minimize, or cancel out in the best case, the effect of parasitics.

Ring-Down Technique

Though the term *ring-down* was coined relatively recently (Johannsmann 2007), the technique was developed in the latter half of the 1990s (Rodahl and Kasemo 1996) and readily applied to biomolecular film formation and the substrate anchoring of adherent living cells (Rodahl et al. 1997). Nowadays, setups based on this technology are commercial and denoted *QCM-D*, which stands for *quartz crystal microbalance with dissipation monitoring* (Q-sense, Sweden). The sensor is oscillated by means

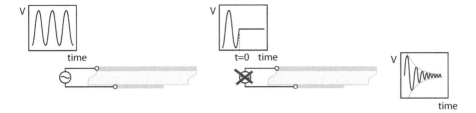

FIGURE 6.19 Ring-down technique. The QCM sensor is set to oscillate by the application of a sinusoidal potential. At time = 0, the excitation stops; however, the sensor continues shear oscillating with less and less amplitude until the oscillation finally fades away. The output signal is thus a time-dependent voltage signal of decaying amplitude from which the frequency and dissipation are extracted.

of a driving circuit—a waveform generator—that supplies a sinusoidal voltage at the series resonant frequency. At time = 0, the sensor is disconnected from the driving circuit, and the decay of the sensor's oscillation is registered on an oscilloscope as a time-dependent voltage signal. As depicted in Figure 6.19, the decay signal is usually an exponentially damped sinusoidal of the form

$$V(t) = V_0 \cdot e^{-t/\tau} \cdot \sin(2\pi ft + \varphi) \quad \text{at } t \geq 0 \tag{6.53}$$

where V_0 is the starting amplitude of the oscillation, f is the oscillation frequency, τ is the decay time constant, and ϕ is the phase. An important magnitude—dissipation factor, D—is defined from the parameters of Equation (6.53) as

$$D = \frac{1}{\pi f \tau} \tag{6.54}$$

Changes occurring at the surface of the sensor can be tracked if frequency and dissipation data are recorded as a function of time. This is possible as long as the sensor can be periodically disconnected and its decay registered and analyzed in real time. For a current ring-down-based setup, an $f - D$ measurement takes approximately less than a second, which defines the time resolution of the technique.

Knowing f and D makes it possible to obtain the electrical parameters of the BvD model by the following relations (Rodahl and Kasemo 1996)

$$f_s = \frac{1}{2\pi\sqrt{C_1 \cdot (L_1 + L_L)}} \quad f_p = \frac{1}{2\pi\sqrt{L_1 \frac{(C_0 + C_L) \cdot C_1}{C_0 + C_L + C_1}}} \tag{6.55}$$

for both the series and the parallel resonant frequencies and

$$D = \frac{R_1 + R_L}{2\pi f_s L_1} \approx \frac{R_1 + R_L}{2\pi f_p L_1} \tag{6.56}$$

for the dissipation. The load is, in this case, represented by a three-component element: a capacitor, a resistor, and an inductor of respective constants C_L, R_L, and L_L connected in series.

FREQUENCY SHIFTS AND VISCOELASTIC PARAMETERS: SOME CASE EXAMPLES

The relation between the QCM experimental parameters and the viscoelastic constants of the deposited films is, in most cases, not a simple one. The difficulty scales with the structural complexity of the film. The experimental outcome is an interplay of a plethora of factors: the film composition, the viscoelastic properties of the film components and the surrounding fluid, film roughness, the nature of the film–electrode contact as well as the fluid–film interaction, film slippage over the sensor, film swelling, etc. For an accurate characterization, each individual factor needs to be properly modeled or known through other means, usually involving independent measurements.

In this section we will briefly account for the simplest systems, that is, films composed of a single layer of a nonconducting material, ideally flat and homogeneous.

The Special Case of Rigid Films in Air: Sauerbrey Equation and Film Thickness

The simplest relation of all relates frequency shifts caused by a rigid film resting on a QCM sensor in air. The relation was derived by Sauerbrey more than 50 years ago (Sauerbrey 1959), and it states that the frequency shift at the fundamental resonant frequency or at any overtone, Δf, is directly proportional to the deposited mass per unit area, Δm

$$\Delta f = -\frac{2 f_f^2 M}{\sqrt{\rho_q G_q}} \cdot \Delta m \qquad (6.57)$$

where f_f is the fundamental resonance frequency, M is the overtone number, and ρ_q and G_q are the density and the shear modulus of the crystal, respectively.

Some Numbers: Calculation of the Mass Adsorbed Using Sauerbrey Equation for an AT-Cut Quartz Sensor

The proportionality factor in Equation (6.57) depends mainly on the crystal, and it can be considered as a material's constant. The values of ρ_q and G_q for an AT cut quartz crystal are known, and therefore Equation (6.55) can be expressed as

$$\Delta f = -K \cdot \Delta m \text{ or } \Delta m = -C \cdot \Delta f \qquad (6.58)$$

with $K = 56.6 \text{ Hz} - \mu g^{-1} - cm^2$ and $C = 17.7 \text{ ng} - cm^{-2} - Hz^{-1}$ at the fundamental resonance frequency ($M = 1$, $f_q = 5 \text{ MHz}$). Therefore, if Equation (6.58) is applicable, a frequency change of 1 Hz corresponds to the adsorption of 17.7 ng/cm². The minimum detectable mass in the center of the quartz resonator depends on the experimental frequency stability, $\Delta f/f$. Best frequency stabilities have been reported to be $(1 - 0.5) \cdot 10^{-11}$, which lead to minimum detectable mass of less than 0.1 pg (Mecea 2005).

If the area of the sensor A and the density of the deposited material ρ_F are known, the thickness of the film t_F can be calculated from Δm as $t_F = \Delta m/\rho_F$. The variable t_F is referred to as *Sauerbrey thickness*.

Though the Sauerbrey equation is strictly valid for rigid, inert films in air, it has been extensively applied in cases where it may not be supposed to apply. In practice, the Sauerbrey equation can give reasonable results in those cases where the energy losses within the sensor or surroundings (experimentally determined by increases in the dissipation factor or resonance bandwidth) are negligible and the frequency shift does not depend on the overtone.

Viscoelastic Films and the Importance of Measuring at Different Overtones

In those cases where the energy losses are noticeable and the frequency shift depends on the overtone number, the films may be viscoelastic, and the Sauerbrey equation cannot be applied any more. Instead, the frequency shift depends on the viscoelastic parameters of the film, usually represented by a film complex shear modulus, G_F^*. The real part of such modulus (G'_F) contains the elastic contribution, the fraction of the energy that is stored during the shear oscillation. The imaginary part (G''_F) contains the viscous contribution, the energy that is lost during the shear oscillation.

If the viscoelastic film oscillates in air, the associated complex frequency shift is

$$\Delta f^* = -\frac{f_f \cdot Z_F}{\pi \cdot Z_q} \cdot \tan(k_F t_F) = -\frac{f_f \cdot \left(\rho_F G_F^*\right)^{1/2}}{\pi \cdot \left(\rho_q G_q\right)^{1/2}} \cdot \tan(k_F t_F) \tag{6.59}$$

where k_F is the wave vector of the oscillation. For thin films where the thickness is much smaller than the wavelength of the oscillation ($k_F t_F \ll 1$), the expression can be approximated to

$$\Delta f^* \approx -\frac{2f_f^2}{\left(\rho_q G_q\right)^{1/2}} \cdot \Delta m \cdot \left[1 + \frac{1}{3} \cdot \left(\frac{\rho_q G_q}{\rho_F G_F^*} - 1\right) \cdot \left(\frac{\Delta m}{m_q} \cdot \pi M\right)^2\right] \tag{6.60}$$

The expression is the sum of two terms: The first term is the Sauerbrey contribution, and the second term is a sort of viscoelastic correction. Viscoelasticity increases the frequency shift and the dissipation (or bandwidth) signals.

The presence of a liquid further modifies the QCM signals, since the shear wave is influenced by the liquid hydrodynamics, represented by its density (ρ_l) and its complex shear modulus (G_l^*). The expression for the complex frequency shift is, in this case

$$\Delta f^* = -\frac{f_f \cdot \left(\rho_F G_F^*\right)^{1/2}}{\pi \cdot \left(\rho_q G_q\right)^{1/2}} \cdot \frac{\left(\rho_F G_F^*\right)^{1/2} \tan\left(k_F t_F\right) - i \cdot \left(\rho_l G_l^*\right)^{1/2}}{\left(\rho_F G_F^*\right)^{1/2} + i \cdot \left(\rho_l G_l^*\right)^{1/2} \tan\left(k_F t_F\right)} \tag{6.61}$$

The complex shear modulus of the liquid is mainly determined by its viscosity η_l and, hence, $G_l^* = i\omega\eta_l = i2\pi f_f\eta_l$. For thin films, the expression of Equation (6.61) can be approximated by

$$\Delta f^* \approx -\frac{2 \cdot f_f^2 \cdot \Delta m}{\left(\rho_q G_q\right)^{1/2}} \left(1 - M \cdot \frac{2\pi i f_F \rho_l \eta_l}{\rho_F G_F}\right) \tag{6.62}$$

At this point, it is important to mention that QCM experiments are not only sensitive to the deposited mass of the material, but also to the amount of liquid that may be incorporated in the film (*trapped*) or that may synchronously shear above the material's surface (*bound*). The results thus may differ from those obtained through the optical techniques of ellipsometry or SPR, where the film optical properties are mainly determined by the material and not so much by its environment, i.e., by the surrounding fluid. In this case, the latter techniques are said to provide with a *dry mass* or *thickness*, in contrast to the higher *wet* or *fluid-dynamic thickness*, the QCM results (Glasmästar et al. 2002).

To estimate structural and mechanical parameters of deposited viscoelastic films in air or any other fluid, Equations (6.59) and (6.61) or their respective approximations, Equations (6.60) and (6.62), should be employed. At a given shear-oscillation frequency, the experimental variables are two at the most: the frequency and dissipation (or bandwidth) shift. However, the unknowns are more numerous. The deposited mass and the complex shear modulus of the film that encompasses two undetermined parameters—namely the film's elasticity and the film's viscosity—are three unknown quantities. What is more, the deposited mass is in turn the product of the film's density and the film thickness, and hence may introduce two additional unknowns.

Measuring at different overtones M will increase the number of experimental variables and hence balance the number of known magnitudes to that of the unknowns. In this case, and providing that the complex shear modulus of the film does not depend on the frequency, it may be possible to obtain the film parameters with two to three overtone measurements. However, this is mostly not the case, and viscoelastic films do have complex shear moduli that depend on the frequency. This leads to the fact that overtone measurements further increase the number of unknowns to that of the experimental variables, and the given set of equations will be insufficient to accurately determine all of the film parameters. In these cases, QCM should be complemented with independent measurements that can provide some of the required values. Alternatively, the parameters can be obtained from the best fit of the QCM results to the given equations or to a particular model, provided that a relatively good set of initial values is at hand. In particular, rheological experiments may assist in estimating the complex shear modulus that can be plugged into

Equations (6.59) through (6.62) to obtain the deposited mass. Conversely, ellipsometry, SPR, or atomic force microscopy (AFM) experiments may provide approximate values of the film thickness that can be used as initial fitting quantities.

MAIN DRAWBACK: THE IMPORTANCE OF QUALITATIVE INTERPRETATION

When films are nonhomogeneous or the adsorbed mass is not enough to cover the whole sensor area, the models described in this section cannot be applied. In these cases, a usual approach is the qualitative interpretation of the experimental data, namely the shifts in frequency and dissipation (bandwidth) as a function of the overtone and time. This approach is useful, especially if the behaviors of different adsorbates are to be compared to one another. A widespread way to look at the data in such studies consists of plotting the dissipation shift ΔD as a function of the frequency shift Δf (Otzen, Oliveberg, and Höök 2003) to track mechanical changes during or after film formation. The changes observed in plots of ΔD versus Δf can be interpreted as alterations in the softness of the deposited material as the film is being formed or film swelling, among other possible reasons.

SUMMARY

Table 6.1 comprises the characteristics as well as the limitations of the techniques described here. Although they all are surface-sensitive, they may have either microscopic lateral resolution (as in the case of SPR and ellipsometry) or not at all (as in QCM). In this sense, they are considered macroscopic and give average magnitudes. SPR and ellipsometry are optical techniques and require reflecting surfaces: Both are able to provide optical constants of thin films and multilayers as long as the right model is chosen. In the case of ellipsometry, only the spectroscopic mode can unambiguously characterize the constituent layers of a stratified sample. The QCM relies on the piezoelectric response of a shear-mode resonator to detect any changes occurring on this sensor's surface. The conversion of electromechanical data into mass or thickness requires modeling and sometimes independent measurements to avoid overparameterization. Whereas the optical techniques are sensitive to dry thickness irrespective of the surrounding fluid, the electromechanical response of a film immersed in a fluid cannot be decoupled from the fluid's influence, and hence a "wet thickness" is measured that comprises that of the film and that of the bound liquid.

THE COMBINATION SPM AND MODEL-BASED SURFACE TECHNIQUES

SPM + SPR

The first studies employing such a combination date from the middle of the 1990s (Chen et al. 1995; Shakesheff et al. 1995), when AFM and SPR were simultaneously used to track the chemical degradation of polymer films. The SPR setup is based on the Kretschmann (1968) configuration, which makes it particularly easy to complement with an AFM instrument. The sample substrate is an SPR sensor

TABLE 6.1

Summary of the Characteristics and Limitations of the Model-Based Surface Techniques Reviewed in This Chapter

CHARACTERISTICS	SPR	Ellipsometry	QCM
Lateral resolution	≈ 1 μm (imaging)		
Vertical resolution	molecular dimensions (≈ nm)		
Sample requirements	optically reflecting supports		
	support exhibiting plasmon resonance in the visible or IR (gold, silver)		supports are shear-mode resonators (i.e., AT-cut quartz crystal)
Local/global	global: along the interface; local: perpendicular to the interface (surface-sensitive)		
Contrast	optical		electromechanical
	width of the SP resonance (in degrees)	width of the ellipsometric minimum (in degrees, null ellipsometry)	width of the shear resonance (Q factor)
Other features	non-destructive, time-resolved (interface kinetics)		
	imaging capabilities optical constants (thickness, refractive index) of single-layer films sensitive to the dry thickness	micrometer-thick films are measurable multilayers through spectroscopy	- high mass sensitivity (< 0.1 pg) - frequency stability $\Delta f/f \approx 10^{-11}$ - sensitive to wet thickness: film + bound liquid - dry thickness if Sauerbrey equation is applicable - viscoelastic parameters (elastic shear moduli and viscosity of films)
	outcome: reflectivity vs θ reflectivity vs time	outcome: ellipsometric angles vs time or λ (spectroscopy)	
Limitations	model is required – accuracy depends on the choice of the right model limited applicability on rough surfaces – difficulty of modeling roughness limited applicability on samples exhibiting submonolayer or partial coverage (non-homogeneous films)		
	require iterative analysis		-overparametrization requires independent measurements to unambiguously determine structrural and viscoelastic parameters - requires to account for slippage, point contacts - viscoelastic parameters of films often unknown or frequency-dependent

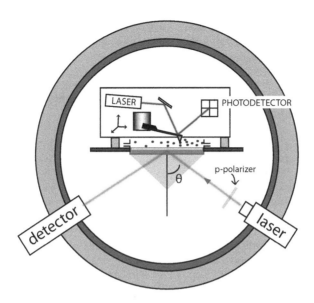

FIGURE 6.20 The combined AFM-SPR technique. An AFM head encompassing the piezo scanner, the cantilever, the aligning laser, and the photodetector can be mounted on the sample platform, whereas SPR is performed from the underside.

coated with a gold or a silver layer on its exposed side (e.g., LaSFN9; Baba, Knoll, and Advincula 2006). SPR experiments are performed from the back side of the SPR sensor, whereas the AFM cantilever accesses the sample from above, as seen in Figure 6.20. The laser beam, usually a helium–neon laser of $\lambda = 632.8$ nm, hits the back side of the sample at an angle above the critical angle to produce total internal reflection and excite the surface plasmons of the gold layer. To produce the excitation, a hemispherical or pyramidal prism made of the same material as the SPR sensor is attached to the back side of the latter. Light reaches the sample without suffering alterations due to the presence of a layer of index-matching fluid between the prism and the SPR sensor. The reflected light is collected by the detector at the detector arm. Both the laser and detector arms are mounted on a goniometer that allows change of the incidence angle and search for the resonance. The AFM instrument is based on a tip-moving configuration and scans the sample from above.

A later report (Baba, Knoll, and Advincula 2006) describes a three-in-one setup where AFM and SPR are combined with cyclic voltammetry to monitor the process of electroformation of a polymer film (*eletropolymerization*). In this case, the techniques are complemented by the presence of the three electrodes—working, counter, and reference—and a potentiostat, which comprise the electrochemical instrumentation. In those experiments, the sample substrate acts simultaneously as an SPR sensor, as the AFM sample, and as a working electrode.

The applications of the SPR + AFM combinations are scarce and focused on the field of polymer science. The first studies applied the techniques to correlate the morphological transformations with the mechanisms of film erosion in biodegradable polymer films subjected to either acidic or basic conditions (Chen et al. 1995;

Shakesheff et al. 1995). To our knowledge, the most recent application consists of tracking the mechanism and morphology of an in situ electrodeposited polymer film from its constituent monomer in solution (Baba, Knoll, and Advincula 2006).

Example A: Tracking the Degradation of a Polymer Blend Film Using a Combined SPR + AFM

The dissolution of biodegradable polymer coatings is a key issue in the process of encapsulated or embedded drug release. In particular, the use of polymer blends as drug vehicles can be useful in tuning the rate of drug release through the rate of degradation of the polymer constituents. Shakesheff et al. (1995) simultane-ously used AFM and SPR to track the morphological changes and the mechanisms of film erosion of a layer composed of two immiscible biodegradable polymers: poly(lactic acid) (PLA) and poly(sebacic anhydride) (PSA). Degradation was brought about by hydrolysis in basic conditions (pH 11). A preliminary combined study on single-component films reveals that PSA degrades much faster than PLA. Films of polymer blends of PSA and PLA exhibit a microphase-separated morphol-ogy of bulgy circular patches in an otherwise flat matrix (Figure 6.A1). The SPR and AFM measurements reveal a three-stage degradation process, consisting of a rapid degradation of the pure PSA domains that coincides with the dissolution of the bulgy patches, an intermediate step where the degradation proceeds at a slower pace compared to the PSA but faster than the rate of PLA, and a final slow step where the domains of pure PLA are dissolved (Figure 6.A2). The dissolution rate of the intermediate step has been found to depend on the presence of PLA, decreasing as the concentration of PLA in the blend increases.

SPM + ELLIPSOMETRY

This combination has been coined *scanning near-field ellipsometry microscopy* (SNEM), and it has received moderate interest in the last decade. The first report on such a combination dates from 2001 (Karageorgiev et al. 2001) and was only fol-lowed by a second study 10 years later (Tranchida et al. 2011).

The reported SNEM setups are very similar and are based on the combination of two commercial instruments (see Figure 6.21). The laser and detector arms of the ellipsometer are mounted on a goniometer and set at equal angles of incidence and reflection, respectively. Both arms are arranged so that the light beam hits the back side of the sample, and the reflected beam is thus collected from there. These angles are fixed throughout the whole experiment. The sample rests on a transpar-ent substrate to allow back-side illumination, and the AFM probe is placed over the sample so that the tip is right on top of the illuminated spot. To avoid possible optical misalignment, the AFM instrument has a tip-moving configuration.

The setup works as follows. Both the polarizer and analyzer of the ellipsometer are previously set for null ellipsometry before the AFM cantilever is positioned on top of the sample. The laser arm produces elliptically polarized light that, upon reflection, is converted into linearly polarized light, which is in turn blocked by the analyzer. The states of the polarizer and analyzer as well as the positions of the laser and detector arms are not changed during the whole experiment. The presence of the

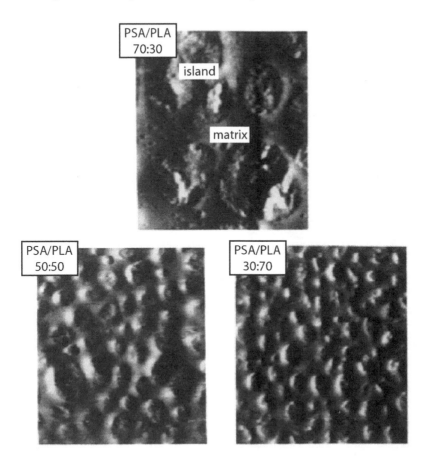

FIGURE 6.A1   AFM images (3×3 mm^2) of the typical morphology of a polymer blend film of PSA and PLA (PSA:PLA, 70:30, 50:50, and 30:70). All of them show circular domains or islands in a flat matrix. (Adapted from Shakesheff et al. 1995. With permission.)

AFM tip, which should be metal-coated, alters the null settings, producing a notice-able increase of the optical signal that is acquired by the detector of the ellipsometer. Topography and SNEM images are thus obtained simultaneously as the tip raster scans the sample.

The modification of the null settings is a consequence of near-field effects whose nature slightly differs from one setup to another. In Karageorgiev's instrument (Karageorgiev et al. 2001), the laser beam is totally internally reflected at the back side of the sample through a prism, in a configuration reminiscent of a surface plasmon resonance setup. The so-created evanescent field is scattered by the metal coating of the AFM probe, thus altering the signal intensity at the optical detector. In Tranchida's instrument (Tranchida et al. 2011), the angle of incidence is smaller than the critical angle and directly illumi-nates the gold-coated AFM tip. This induces field enhancement in a similar way as in the apertureless SNOM setups (see Chapter 4 for details), which in turn alters the reflected beam to produce a change in the signal intensity at the optical detector.

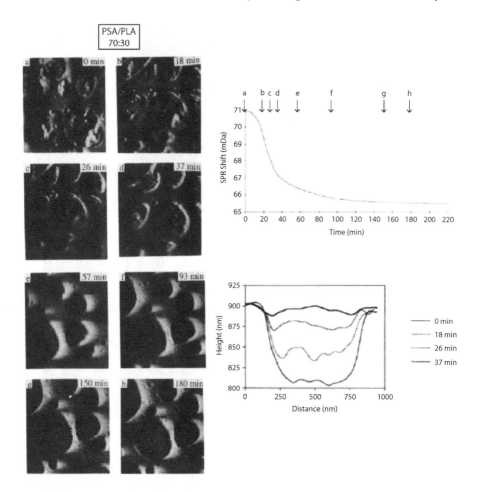

FIGURE 6.A2 AFM and SPR images at different stages of the degradation. *Left*: Images (A–H) are topography images (3 × 3 mm²) as a function of the degradation time. The height profile along an island in the right-hand side (lower graph) shows that material in the island is being depleted as degradation proceeds. The fast dissolution of the islands shows that they are composed of PSA. *Right* (top graph): SPR curve showing the kinetics of film degradation. (Reprinted from Shakesheff et al. 1995. With permission.)

Still at the proof-of-concept stage, SNEM has been mainly applied to test samples of well-defined properties, mainly to quantify the resolution of the technique and the possible mechanisms of optical contrast in the SNEM images (Karageorgiev et al. 2001; Tranchida et al. 2011).

Example B: Testing the Capabilities of a Scanning Near-Field Ellipsometry Microscope (SNEM)

The two available reports on SNEM present the combination as proof-of-concept and discuss the possible sources of optical contrast in the SNEM images on test

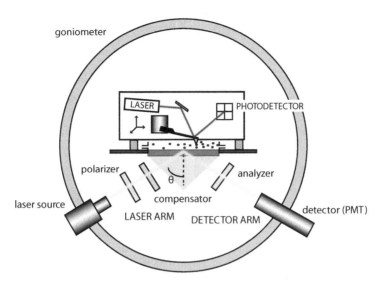

FIGURE 6.21 The combined AFM–ellipsometry technique. The setup highly resembles that depicted in Figure 6.20. A tip-moving AFM is mounted on the same sample platform as the ellipsometer. Null ellipsometry is in this case performed at the underside. Null conditions are initially set and kept unvaried throughout the experiment. The SNEM images are actually light-intensity maps registering deviations relative to the initial conditions of null ellipsometry with the sample position.

samples. Karageorgiev and coworkers made use of a laser-patterned surface grating on an azobenzene-containing film to obtain the surface topography and the corresponding SNEM of the micrometer-sized grating (Karageorgiev et al. 2001). Topography and SNEM images show the grating with reversed contrast; the valleys and ridges in the AFM image correspond to ridges and valleys in the SNEM image at comparable lateral resolution. The nature of the optical contrast in SNEM was further analyzed through another test sample, consisting of a polycrystalline film of a thermotropic liquid crystal. Figure 6.B1 shows the topography and the SNEM of a single crystallite. The SNEM map of the crystallite reveals concentric modulations of the optical signal that cannot be attributed to changes in the refractive index, since the orientation and density of the molecules are constant along the single crystal. The correlation of such modulations with the crystal topology induced the authors to conclude that changes of optical contrast could only be attributed to the crystal shape, namely the thickness variations in the crystallite.

Tranchida and coauthors (2011) go a step further and quantify the different contributions to the overall optical signal in SNEM and extract the near-field intensity out of the parasitic background and interferometry signals, both induced by the vertical movements of the AFM cantilever

$$I(x,y,z) = I_{background} + I_{near\text{-}field} + I_{interferometry} = I_{background} + I_1(x,y,z) + \frac{\partial I_0(x,y,z)}{\partial z} \cdot dz$$

The corrections to the overall intensity were quantified on bare glass (Figures 6.B2a and 6.B2b), which in turn were employed to visualize silver nanoparticles immobilized on a polymer film and a microphase-separated block copolymer film (Figure 6.B2).

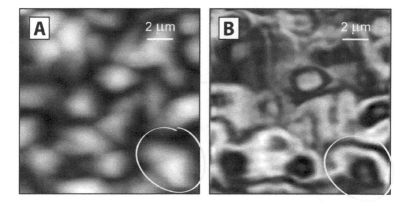

FIGURE 6.B1   Topography (A) and SNEM (B) of a single crystallite, revealing the alterations in optical contrast (a single crystallite is highlighted). Lateral resolution reported =
20 nm. (Adapted from Karageorgiev et al. 2001. With permission.)

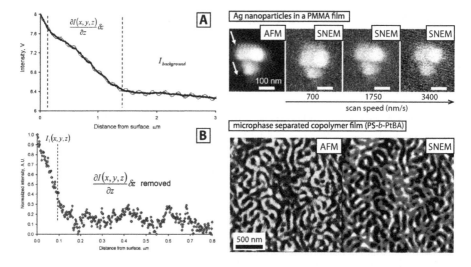

FIGURE 6.B2   Optical curve showing the different contributions to the optical intensity
(a) and the corrected curve (b). The lateral resolution was reported to be better than 80 nm,
irrespective of the scan speed on silver nanoparticles immobilized on a poly(methyl methacrylate) (PMMA) layer. The technique was also tested on a microphase-separated block copolymer film (polystyrene-block-poly(tert-butyl acrylate, PS-PtBA). (All figures reproduced
from Tranchida et al. 2011. With permission of The Royal Society of Chemistry.)

SPM + QCM

Of all the SPM techniques, AFM is the only one that has ever been employed in combination with QCM. Despite its potential utility, only a few attempts have been made
to apply this combination of techniques. The first report on AFM + QCM dates from
1998 (Iwata, Saruta, and Sasaki 1998), when the techniques were simultaneously
used to monitor the electrodeposition of silver on a platinum thin film. In the few

reports that followed, the AFM-QCM technique has been applied as a monitoring tool for the deposition of metals or the adsorption of biomolecules, either onto other metal surfaces or onto organic layers.

Reports describe the QCM-AFM tandem as either a two-in-one or a three-in-one combination of commercial or homemade setups. The *two-in-one* version comprises the QCM and AFM devices, and it has been employed in the adsorption of proteins such as human fibrinogen (Choi et al. 2002) and ferritin (Johannsmann et al. 2008). The *three-in-one* version, which can be envisaged as one step beyond in instrumentation, also allows in situ electrochemistry. The latter has been shown to be especially convenient in the study of electrodeposition of metals such as silver or copper (Iwata, Saruta, and Sasaki 1998; Bund, Schneider, and Dehnke 2002; Friedt et al. 2003).

A schematic of the two-in-one setup is shown in Figure 6.22a. The sensing electrode of the QCM crystal acts as the sample substrate of an AFM. The crystal is mounted either on a tip-moving or on a stage-moving AFM. An alternating current is applied across the electrodes to make the QCM crystal shear-oscillate at frequencies around 5–10 MHz, while the AFM tip scans the surface, usually in intermittent-contact mode, at much lower frequencies (1–10 Hz). The amplitude of the shear-oscillation should be chosen with care, since it has to be small compared to the lateral resolution of the AFM, which implies that it should be just a few nanometers. The three-in-one version includes two to three electrodes to produce electrochemistry and is shown in Figure 6.22b: the working electrode, which is the sensing electrode of the QCM crystal, and the counter electrode and reference electrodes, usually two wires hovering at a close distance over the working electrode. The end of the counter electrode is shaped in a circle around the working electrode, and the reference electrode may or may not be present (Friedt et al. 2003; Bund, Schneider,

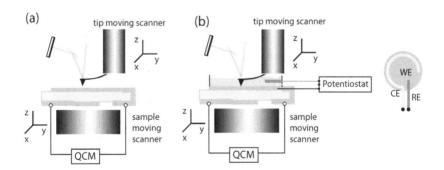

FIGURE 6.22 *(See color insert.)* The combined AFM-QCM technique. (a) The two-in-one version makes use of the exposed side of the QCM sensor as the substrate for AFM studies. The AFM can be either a tip-moving or a sample-moving setup. (b) The three-in-one version includes an electrochemical cell to perform voltammetry. The exposed side of the QCM sensor acts as AFM substrate and additionally as working electrode. The counter electrode (yellow) is a metal wire looped at its end in a three-quarter-of-a-circle shape. Together with the reference electrode (green), they both are immersed in the solution, hovering over the working electrode. The electrodes are connected to a potentiostat that applies the potential in a controlled fashion and measures the current as a function of time or of the applied potential.

and Dehnke 2002). Simultaneous measurements of QCM, AFM, and electrochemistry are possible because the frequency of the electrical signals generated by each technique widely differ: QCM operates at frequencies on the order of megahertz, whereas the AFM operates at scan frequencies of a few hertz and tip oscillations are on the order of kilohertz; electrochemistry sweeps voltages at very low frequencies as well (2 to 50 mV/s; Friedt et al. 2003).

The advantages of such combinations are many: deposited mass can be assumption-free calculated from electrochemical and AFM measurements and hence used to test the adequacy of the model chosen to interpret the QCM data. Indeed, deposited mass can be estimated from cyclic voltammograms by calculating the total charge required to electrodeposit the sensing area, providing that the electrodeposited material is homogeneous. This is obtained by integrating the current administered to the system along half the voltammetric cycle

$$\frac{\Delta m_{\text{EC}}}{A} = \frac{M_{\text{metal}} \int I dt}{Z_{\text{ion}} F} \tag{6.63}$$

where M_{metal} is the atomic mass of the metal, Z_{ion} is the valence of the corresponding metal ion, F is the Faraday constant, A is the sensing area, and I is the electric current (Friedt et al. 2003). Likewise, an estimation of the mass deposited can be extracted from AFM topography images, assuming that the film is homogeneous, and therefore it is possible to extrapolate the deposited mass in a microscale area to the deposited mass in the macroscopic sensing area. In this case,

$$\Delta m_{\text{AFM}} = \rho \int_0^{Ly} \int_0^{Lx} h(x, y) dx dy \tag{6.64}$$

where Lx and Ly are, respectively, the horizontal and the vertical dimensions of the AFM image; ρ is the metal density; and $h(x,y)$ is the distribution of heights in the image, taking the substrate as reference (i.e., $h = 0$). The value of Δm_{AFM} will have dimensions of mass per image area, which corresponds to a surface mass density of $\Delta m_{\text{AFM}}/Lx - Ly$.

Another advantage is that AFM provides a direct means to evaluate the influence of surface roughness or the interaction with the surrounding liquid on the frequency and dissipation shifts of the QCM signals. The model of Urbakh and Daikhin (1994) allows us to calculate the frequency and dissipation shifts brought about by corrugations on the surface, providing a quantitative characterization of such corrugations is available. This characterization can be in the shape of height–height pair-correlation functions that AFM can provide from topography images. On the other hand, surface roughness plays a crucial role in the interaction of the deposited material with the surrounding fluid. Knowing the surface topography eases the way to estimate the amount of trapped liquid in the surface cavities and bumps. Consequently, this might help to evaluate, in turn, the effect of friction between the surface and the bound liquid and its influence on the QCM data (Bund, Schneider, and Dehnke 2002).

Example C: Tracking the Electrodeposition of Metals Using QCM + AFM + EC

Friedt and coworkers applied the three-in-one combination to monitor the electro-deposition of silver and copper (Friedt et al. 2003). They used AFM to visualize the sample topography and QCM to estimate the mass deposition and the extent of shear damping (dissipation) as the metal was being in situ deposited using an integrated electrochemical (EC) setup. Deposition of silver produced smooth films of small granules, which were accompanied by large frequency shifts and small dissipation values. The results were compatible with the Sauerbrey prediction, being the added mass proportional to the normalized frequency, f/M, where M is the overtone number (Figure 6.C1). Conversely, the deposition of copper produced nonuniform granules rather than smooth coatings, which resulted in noticeable dissipation shifts in the QCM signal. The added mass, far from being dependent on the normalized frequency, was found to be proportional to $f/\sqrt{M}$. This dependency is symptomatic of a nonnegligible contribution of viscous drag, most likely caused by the surrounding liquid being trapped in the highly rough metal surface (Figure 6.C2).

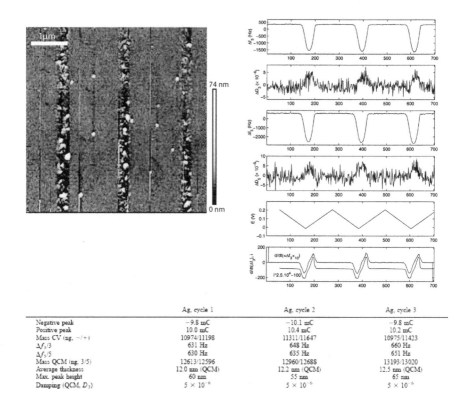

	Ag, cycle 1	Ag, cycle 2	Ag, cycle 3
Negative peak	−9.8 mC	−10.1 mC	−9.8 mC
Positive peak	10.0 mC	10.4 mC	10.2 mC
Mass CV (ng, −/+)	10974/11198	11311/11647	10975/11423
$\Delta f_3/3$	631 Hz	648 Hz	660 Hz
$\Delta f_5/5$	630 Hz	635 Hz	651 Hz
Mass QCM (ng, 3/5)	12613/12596	12960/12688	13193/13020
Average thickness	12.0 nm (QCM)	12.2 nm (QCM)	12.5 nm (QCM)
Max. peak height	60 nm	55 nm	65 nm
Damping (QCM, D_3)	5×10^{-6}	5×10^{-6}	5×10^{-6}

FIGURE 6.C1 *Left*: AFM topography image showing stripes of deposited silver on a QCM sensor. Silver deposition is triggered whenever the saw-tooth-like potential is above the reduction potential of silver. This occurs during scanning and therefore silver is deposited in stripes. Each scan line comprises three voltammetry cycles. *Right* (from top to bottom): Dissipation and frequency shifts of the third and fifth overtones, the electrochemical potential as it is applied, and the resulting current. *Bottom*: QCM results as obtained from the first, second, and third voltammetry cycles. (Table and figure reprinted from Friedt et al. 2003. With permission.)

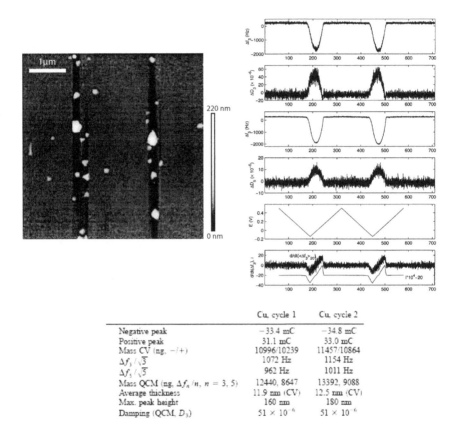

	Cu cycle 1	Cu cycle 2
Negative peak	−33.4 mC	−34.8 mC
Positive peak	31.1 mC	33.0 mC
Mass CV (ng, −/+)	10996/10239	11457/10864
$\Delta f_3 / \sqrt{3}$	1072 Hz	1154 Hz
$\Delta f_5 / \sqrt{5}$	962 Hz	1011 Hz
Mass QCM (ng, $\Delta f_n / n$, $n = 3, 5$)	12440, 8647	13392, 9088
Average thickness	11.9 nm (CV)	12.5 nm (CV)
Max. peak height	160 nm	180 nm
Damping (QCM, D_3)	51×10^{-6}	51×10^{-6}

FIGURE 6.C2 *Left*: AFM topography image showing stripes of deposited copper on a QCM sensor. Copper deposition is triggered whenever the saw-tooth-like potential is above the reduction potential of copper. As for silver, copper is deposited in stripes. Each scan line comprises two voltammetry cycles. *Right* (from top to bottom): Dissipation and frequency shifts of the third and fifth overtones, the electrochemical potential as it is applied, and the resulting current. *Bottom*: QCM results as obtained from the first, second, and third voltammetry cycles. (Table and figure reprinted from Friedt et al. 2003. With permission.)

REFERENCES

Atkins, P., and J. de Paula. 2006. *Physical chemistry*. Oxford, U.K.: Oxford University Press.

Baba, A., W. Knoll, and R. Advincula. 2006. Simultaneous in situ electrochemical, surface plasmon optical, and atomic force microscopy measurements: Investigation of conjugated polymer electropolymerization. *Rev. Sci. Instrum.* 77:064101-1/064101-6.

Barnes, W. L., A. Dereux, and T. W. Ebbesen. 2003. Surface plasmon subwavelength optics. *Nature* 424:824–30.

Bund, A., O. Schneider, and V. Dehnke. 2002. Combining AFM and EQCM for the in situ investigation of surface roughness effects during electrochemical metal depositions. *Phys. Chem. Chem. Phys.* 4:3552–54.

Canit, J. C., and J. Badoz. 1983. New design for a photoelastic modulator. *Appl. Opt.* 22:592–94.

Chen, X., K. M. Shakesheff, M. C. Davies, et al. 1995. Degradation of a thin polymer film studied by simultaneous in situ atomic force microscopy and surface plasmon resonance analysis. *J. Phys. Chem.* 99:11537–42.

Choi, K.-H., J.-M. Friedt, F. Frederix, A. Campitelli, and G. Borghs. 2002. Simultaneous atomic force microscope and quartz crystal microbalance measurements: Investigation of human plasma fibrinogen adsorption. *Appl. Phys. Lett.* 81:1335–37.

Friedt, J. M., K. H. Choi, F. Frederix, and A. Campitelli. 2003. Simultaneous AFM and QCM measurements. *J. Electrochem. Soc.* 150:H229–H234.

Fujiwara, H. 2007. *Spectroscopic ellipsometry: Principles and applications.* Chichester, U.K.: John Wiley & Sons.

Glasmästar, K., C. Larsson, F. Höök, and B. Kasemo. 2002. Protein adsorption on supported phospholipid bilayers. *J. Coll. Int. Sci.* 246:40–47.

Iwata, F., K. Saruta, and A. Sasaki. 1998. In situ atomic force microscopy combined with a quartz-crystal microbalance study of Ag electrodeposition on Pt thin film. *Appl. Phys. A: Mater. Sci. & Proc.* 66:S463–S466.

Jasperson, S. N., and S. E. Schnatterly. 1969. An improved method for high reflectivity ellipsometry based on a new polarization modulation technique. *Rev. Sci. Instr.* 40:761–67.

Johannsmann, D. 2007. Studies of viscoelasticity with the QCM. In *Piezoelectric sensors*, ed. C. Steinem and A. Jahnshoff, 49–109. Heidelberg: Springer Verlag.

———. 2008. Viscoelastic, mechanical, and dielectric measurements on complex samples with the quartz crystal microbalance. *Phys. Chem. Chem. Phys.* 10:4516–34.

Johannsmann, D., I. Reviakine, E. Rojas, and M. Gallego. 2008. Effect of sample heterogeneity on the interpretation of QCM(-D) data: Comparison of combined quartz crystal microbalance/atomic force microscopy measurements with finite element method modelling. *Anal. Chem.* 80:8891–99.

Karageorgiev, P., H. Orendi, B. Stiller, and L. Brehmer. 2001. Scanning near-field ellipsometric microscope: Imaging ellipsometry with a lateral resolution in nanometer range. *Appl. Phys. Lett.* 79:1730–32.

Kretschmann, E., and H. Raether. 1968. Radiative decay of non radiative surface plasmons excited by light. *Z. Naturforsch. A* 23:2135–36.

Lucklum, R., and F. Eichelbaum. 2007. Interface circuits for QCM sensors. In *Piezoelectric sensors*, ed. C. Steinem and A. Jahnshoff, 3–47. Heidelberg: Springer Verlag.

Mason, W. P. 1948. *Piezoelectric crystals and their applications to ultrasonics.* Princeton, NJ: Van Nostrand.

Mecea, V. M. 2005. From quartz crystal microbalance to fundamental principles of mass measurements. *Anal. Lett.* 38:753–67.

Motschmann, H., and R. Teppner. 2001. Ellipsometry in interface science. In *Novel methods to study interfacial layers,* ed. R. Miller and D. Moebius, 1–42. Amsterdam: Elsevier.

Otto, A. 1968. Excitation of nonradiative surface plasma waves in silver by the method of frustrated total reflection. *Z. Phys.* 216:398–410.

Otzen, D. E., M. Oliveberg, and F. Höök. 2003. Adsorption of a small protein to a methyl-terminated hydrophobic surface: Effect of protein-folding thermodynamics and kinetics. *Colloid. Surf. B Biointerfaces* 29:67–73.

Pitarke, J. M., V. M. Silkin, E. V. Chulkov, and P. M. Echenique. 2007. Theory of surface plasmons and surface-plasmon polaritons. *Rep. Prog. Phys.* 70:1–87.

Rodahl, M., F. Höök, C. Fredriksson, et al. 1997. Simultaneous frequency and dissipation factor QCM measurements of biomolecular adsorption and cell adhesion. *Faraday Discuss.* 107:229–46.

Rodahl, M., and B. Kasemo. 1996. A simple setup to simultaneously measure the resonant frequency and the absolute dissipation factor of a quartz crystal microbalance. *Rev. Sci. Instrum.* 67:3238–41.

Röseler, A. 1990. Infrared spectroscopic ellipsometry. Berlin: Akademie-Verlag.

Sauerbrey, G. 1959. The use of quartz oscillators for weighing thin layers and for microweighing. *Z. Physik* 155:206–22.

Schmidt, R. F., J. W. Allen, J. F. Vetelino, J. Parks, and C. Zhang. 2001. Bulk acoustic wave modes in quartz for sensing measured and induced mechanical and electrical property changes. *Sens. Actuators B Chem.* 76:95–102.

Shakesheff, K. M., X. Chen, M. C. Davies, et al. 1995. Relating the phase morphology of a biodegradable polymer blend to erosion kinetics using simultaneous in situ atomic force microscopy and surface plasmon resonance analysis. *Langmuir* 11:3921–27.

Tranchida, D., J. Diaz, P. Schön, H. Schönherr, and G. J. Vancso. 2011. Scanning near-field ellipsometry microscopy: Imaging nanomaterials with resolution below the diffraction limit. *Nanoscale* 3:233–39.

Urbakh, M., and L. I. Daikhin. 1994. Roughness effect on the frequency of a quartz crystal resonator in contact with a liquid. *Phys. Rev. B* 49:4866–70.

Zayats, A. V., I. I. Smolyaninov, and A. A. Maradudin. 2005. Nano-optics of surface plasmon polaritons. *Phys. Rep.* 408:131–314.

7 Scanning Probe Microscopy to Measure Surface Interactions
The Nano Push–Puller

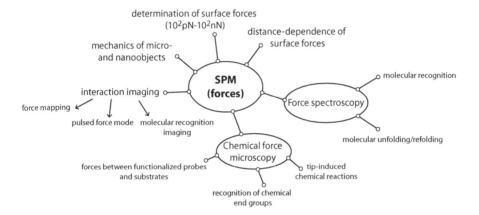

determination of surface forces
(10^2pN-10^2nN)

distance-dependence of surface forces

mechanics of micro- and nanoobjects

molecular recognition

SPM (forces)

interaction imaging

Force spectroscopy

force mapping

pulsed force mode   molecular recognition imaging

molecular unfolding/refolding

Chemical force microscopy

forces between functionalized probes and substrates

tip-induced chemical reactions

recognition of chemical end groups

FORCE CURVES: SURFACE FORCES AND MORE

The principle behind scanning probe microscopies (SPMs) is the measurement of a property that is dependent on the distance between the probe (the meter) and the specimen. This property can either be or embody the interaction between the probe and the sample. In the first case we refer to forces and in the second case to tunneling currents. In this chapter we focus on the measurement of forces. Forces of different nature can be measured as the probe is traveling toward or away from the sample. On the other hand, the mechanics of nano- and microobjects can be locally or globally probed as the probe pushes or pulls them. The latter application has developed into a field of knowledge of its own called *force spectroscopy*, with the main goal of characterizing the interaction landscape between two surfaces.

The discussion in the following background section will help the reader to understand the concepts introduced in this chapter.

Background Information

SCALARS AND VECTORS

Magnitudes that are fully characterized by a number are called *scalars*. Examples of these are dimensional magnitudes like height, width, depth and volume, energy, or temperature. However, a number is not enough for some other types of magnitudes. These are called *vectors*, and they are only fully defined if their direction, together with their extent or magnitude, is given. A typical example is the motion of a body; saying that an object moves at a speed of 20 m/s is not enough to get the complete picture. We need to know something else: Does it move upwards or downwards? To the left or to the right? Toward us or away from us? More generally, the question is: In which direction? The answer is as important as the number. The same applies to other magnitudes such as spatial position, force in any of its forms, or acceleration.

To distinguish vectors from scalars, we represent the former by bold fonts. Forces will be denoted as **F** while temperature is referred to as T. Only in those cases where problems can be reduced to one dimension would the vector magnitude be represented as a scalar.

MECHANICS: NEWTON'S LAWS OF MOTION

Isaac Newton settled the basics of classical mechanics with no more than three laws.

The first law states that an object stays either at rest or moves at constant velocity as long as there is no net force that acts on it. In other words, if the sum of all forces **F** acting on an object is zero, the object is in *mechanical equilibrium* and therefore the velocity of the object **v** is constant. This means that it moves either uniformly and straightforward or not at all.

$$\Sigma\mathbf{F} = 0 \Rightarrow \frac{d\mathbf{v}}{dt} = 0 \tag{7.1}$$

The second law states that the force acting on an object is parallel and directly proportional to the acceleration and the mass of the object. In other words, a force acting on an object produces a parallel acceleration that is proportional to the magnitude of the force.

$$\mathbf{F} = m\cdot\mathbf{a} = m\cdot\frac{d\mathbf{v}}{dt} \tag{7.2}$$

The third law states that for every (force) action there is an equal and opposite (force) reaction. In other words, the forces between two bodies on each other are always equal and directed in opposite directions.

$$\mathbf{F}_{12} = -\mathbf{F}_{21} \tag{7.3}$$

A glass (body 1) resting on a table (body 2) exerts a force on the latter, $\mathbf{F}_{12}$, that is equal and opposite to the force that the table exerts to the glass, $\mathbf{F}_{21}$.

HOOKE'S LAW FOR ELASTIC BODIES

An elastic body can change shape under the action of a force and immediately recover its original shape if the force is no longer applied. If the object is a rod of length L_0 and negligible cross-sectional area and it is pulled by a force applied on its end along the rod axis, it will stretch. In parallel, an *elastic force* evolves that opposes the pulling and tends to recover the object's original shape. This restoring force is proportional to the extension of the rod $L - L_0$

$$F_{\text{elastic}} = k \cdot (L - L_0) = k \cdot \Delta x \tag{7.4}$$

This is the expression of *Hooke's law*, which can be applied in the case of purely elastic bodies that do not undergo permanent or nonlinear deformations. In real life this is not the case, since all materials may have non-Hookean behavior under certain conditions, especially at high forces or when the forces, though small, are applied for a prolonged period of time. Nevertheless Hooke's law is a reasonable model for rigid metals under small forces.

MEASURING FORCES WITH A DYNAMOMETER

A dynamometer consists of either a deflecting or stretching device that can measure forces at mechanical equilibrium. Let us consider a simple case: an object of unknown weight hanging from a spring. The forces acting on the object are the gravitational force, in other words, the object's weight F_w, and the elastic force exerted by the spring F_s. The corresponding force diagram is shown in Figure 7.1. The object's weight can only be measured at equilibrium, in which case both forces should be equal and opposed to one another, $F_w - F_s = 0$. If the spring can only elongate or contract along the vertical dimension and its spring's constant k is known, the weight of the object can be obtained from that relation and Hooke's law

$$F_w = mg = k \cdot \Delta x \tag{7.5}$$

Dynamometers are basically Hookean springs of known spring constant that can measure normal forces from the extent of their elongation or from the extent of their bending or deflection. Springs that can stretch and/or bend are spiral wires like the one shown in Figure 7.1. Trampolines are made of such springs. On the other hand, there are springs that can just deflect, like beams or cantilevers. Examples of commonplace cantilevers are the jump boards at swimming pools or fishing rods.

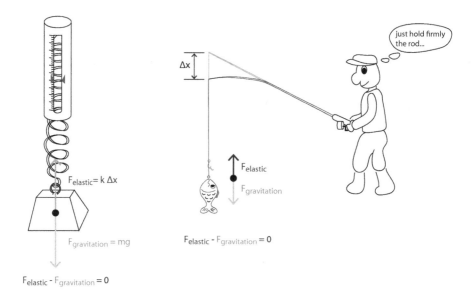

FIGURE 7.1   Measuring forces with a dynamometer. *Left*: The weight of an object is measured from the restoring force of the dynamometer spring. At mechanical equilibrium, both the gravitational and the elastic forces should be equal. *Right*: An everyday example of a bending cantilever. The fishing rod, when held tightly by the fisher bends due to the weight of the fish (the tension along the thread is neglected).

INTERACTION FORCES BETWEEN SURFACES

Surfaces or interfaces are the boundaries of objects through which the latter interact with their surroundings, which can be a fluid or another object. Here we will focus on the interaction between two solid surfaces immersed in a fluid. The forces may tend to keep the objects apart (*repulsive*) or together (*attractive*), and they can evolve at long (*long-range*) or at short distances (*short-range*). According to these characteristics, the forces may be classified in different types.

ELECTROSTATIC "DOUBLE-LAYER" FORCES

When two surfaces are immersed in an aqueous electrolyte solution, they are often charged due to dissociation of their surface chemical groups, ion adsorption, or ion exchange (Israelachvili 2011). In order to balance the surface charge, oppositely charged ions in the solution accumulate in the vicinity alongside those surfaces, defining a layer. This layer is formed by two oppositely and equally charged planes; one is the solid charged surface, and the other is the fluid and rather diffuse layer of *counterions* in the solution. The latter is generally modeled by a sublayer of *bound ions* in the *Stern-Helmholtz* layer at a few angstroms from the solid surface and a diffuse sublayer of *free*-moving ions in the *diffuse electric double layer* farther away. When two similarly charged surfaces come closer, a repulsive force arises in most cases. This so-called *electrostatic double layer* force increases exponentially as the surfaces are brought together, and it is of entropic nature (Israelachvili 2011). The Debye length λ_D is usually identified with the thickness of the diffuse electric double layer and dictates the range of the interaction

$$F_{dl} \propto \exp\left(-\frac{r}{\lambda_D}\right) \quad \text{with} \quad \lambda_D = \sqrt{\frac{\varepsilon\varepsilon_0 k_B T}{\Sigma_i c_i e^2}} \tag{7.6}$$

where k_B is the Boltzmann constant and T the temperature. The variable λ_D depends on the dielectric constant of the fluid, $\varepsilon\varepsilon_0$, and on the concentration of each ionic species i in the solution, c_i. The proportionality parameter in Equation (7.6) depends on the geometry of the interacting surfaces and on their respective surface potentials through the *interaction parameter Z* (Israelachvili 2011). Equation (7.6) is generally valid for separations beyond one Debye length ($r > \lambda_D$). For smaller separations, the exact formulation of the electrostatic force requires the knowledge of the ion distribution in the solution, the surface charge density, and the surface potential at any given distance between the two surfaces. This can only be accomplished by resolving the Poisson-Boltzmann equation, which does not provide an analytical solution.

Usually, two limiting cases with exact numerical solutions are often considered: those of *constant charge* and *constant potential*, and all the real-life cases fall between these two limits. In the case of two planar surfaces with surface potentials Ψ_1 and Ψ_2, the expression for the double-layer force per unit area at the limiting cases is

$$F_{dl} = \frac{2\varepsilon_0\varepsilon\kappa^2}{\left(e^{2\kappa r}-1\right)^2}\cdot\left[\Psi_1\Psi_2 e^{\kappa r}\cdot\left(e^{2\kappa r}+1\right)\pm\left(\Psi_1^2+\Psi_2^2\right)\cdot e^{2\kappa r}\right] \tag{7.7}$$

where the "+" applies to constant charge and the "−" to constant potential. The variable κ is the inverse of the Debye length, $1/\lambda_D$. The electrostatic forces are usually recognized in practice as long-range repulsive interactions, which are highly dependent on electrolyte concentration. Increasing c reduces the interaction range according to Equation (7.6).

Van der Waals Forces

These types of forces are always present and nonnegligible at both large and short separations, which makes them of fundamental importance. They arise from the interaction of dielectric dipoles, either permanent or induced, and they increase at close separations according to a power law of the form

$$F_{vdW} \propto -A\cdot r^{-n} \tag{7.8}$$

where n can be 2, 3, 5/2, according to the geometry of the interacting surfaces. As for the proportionality constant, it also depends on the geometry of the interacting surfaces. There is an additional characteristic parameter, called the *Hamaker constant, A*. The general definition of A is apparently simple

$$A = \pi^2 C \rho_1 \rho_2 \qquad (7.9)$$

where ρ_1 and ρ_2 are the respective atomic densities of the interacting objects, and C is the coefficient in the atom–atom pair potential. However, obtaining a precise value of A for condensed phases is a cumbersome task. To obtain an accurate and ever-valid expression for C it is necessary to consider all the atom–atom pair interactions that simultaneously occur in the material, taking into account that each atom may influence the pair interactions of its neighbors and vice versa. Physics has still not found the analytical solution for the three-body problem, and therefore any attempt to accurately calculate C at either the atomic or molecular level is still out of reach. Instead, attempts to calculate A in the framework of continuum theories do exist that disregard the molecular and hence discrete nature of matter.

Lifshitz's theory, for example, provides an approximate value of A as a function of dielectric permittivity for the interaction of two distinct objects, 1 and 2, across a third medium, 3. In the nonretarded regime, the expression is[*]

$$A \approx \frac{3}{4} k_B T \left(\frac{\varepsilon_1 - \varepsilon_3}{\varepsilon_1 + \varepsilon_3} \right) \cdot \left(\frac{\varepsilon_2 - \varepsilon_3}{\varepsilon_2 + \varepsilon_3} \right)$$

$$+ \frac{3h\upsilon_e}{8\sqrt{2}} \frac{\left(n_1^2 - n_3^2\right) \cdot \left(n_2^2 - n_3^2\right)}{\left(n_1^2 + n_3^2\right)^{1/2} \cdot \left(n_2^2 + n_3^2\right)^{1/2} \cdot \left[\left(n_1^2 + n_3^2\right)^{1/2} + \left(n_2^2 + n_3^2\right)^{1/2}\right]} \qquad 7.10)$$

From the expression of A, it is inferred that van der Waals force between two identical bodies ($n_1 = n_2$) in a medium is always negative, thus attractive. Conversely, van der Waals forces between two distinct bodies in a medium can be either attractive or repulsive.

THE DLVO THEORY: BRINGING DOUBLE LAYER AND VAN DER WAALS FORCES TOGETHER

The DLVO theory is named after its authors (Dejarguin, Landau, Verwey, and Overbeek), who considered both double layer and Van der Waals forces in characterizing the total interaction between two surfaces. They found that the interaction between two planar surfaces is the sum of double-layer and van der Waals forces (Verwey and Overbeek 1948).

---

[*] In the nonretarded regime, the speed of the electromagnetic radiation to reach the second interacting object is large compared with the speed of the fluctuating dipoles within the objects. When this is not the case, especially at large separations or in a fluid where the speed of light is slower, retardation effects should be taken into account (Israelachvili, 2011).

Interactions Arising from Molecular Ordering at Surfaces

Hydrophobic Forces

These are short-range, attractive forces that arise between water-repellent (hence hydrophobic) surfaces immersed in an aqueous medium. Water molecules cannot keep their orientational disorder when they are adjacent to hydrophobic molecules with which they cannot form H-bonds. Consequently, the water molecules tend to orient themselves around the hydrophobic regions so that they can maximize their favorable interactions with other neighboring water molecules. This imposes a new and more ordered structure that is done at the expense of entropy, and it thus costs energy.

The hydrophobic surfaces interact with one another to oppose the entropically unfavorable orientation of water molecules around them. The result is an attractive force that increases with the hydrophobicity of the surfaces and that is even stronger in water than in air (Israelachvili 2011). To date, there is no satisfactory theory on the hydrophobic interaction, which is usually described through empirical expressions. The hydrophobic force has been found to decay exponentially with distance between the surfaces

$$F_h = -2 \cdot \gamma \cdot \exp\left(-\frac{r}{\lambda_h}\right) \tag{7.11}$$

where γ is the *interfacial tension* of the surface with the water, and λ_h is the characteristic length of the interaction that amounts to 1–2 nm (Lekka, Laidler, and Kulik 2007).

Solvation (Hydration) Forces

When liquid is squeezed between two approaching surfaces, solvation forces may develop, which are strongly dependent on the properties of the liquid molecules as well as on the physicochemical properties of the intervening surfaces. They are important at small distances, where continuum theories of van der Waals or electrostatic interactions fail to describe such interactions, and they can either be repulsive, attractive, or even oscillatory with distance. The main type of solvation forces are the oscillatory forces that appear when liquid is confined between two hard, smooth surfaces. The liquid should be composed of molecules that are not too irregular, e.g., rodlike such as hydrocarbons, that exhibit layerlike ordering in regions adjacent to hard, smooth walls and that can easily be exchanged with the reservoir outside the confined volume. Repulsive regimes alternate with attractive regimes as the two surfaces approach, layers of liquid molecules are successively disrupted one by one, and the molecules from the disrupted layer flow away to the reservoir. The oscillation period is equal to the thickness of one layer, and the amplitude often dampens with distance.

The oscillatory profile of solvation forces is highly sensitive to the geometry of the liquid molecules as well as the roughness of the surfaces. Indeed, liquids confined between rough or structured surfaces are forced to arrange differently:

The layerlike order is often lost and, consequently, solvation forces are no longer oscillatory with distance, but they may be instead monotonically attractive or repulsive (Israelachvili 2011).

If the liquid is water or an aqueous solution, solvation forces are regarded as *hydration forces*. They are mainly monotonically repulsive and of very short range—a few nanometers. When two hydrophilic surfaces are immersed in water, their interaction decays with the separation distance between them in an exponential fashion

$$F_{hydr} \propto \exp\left(-\frac{r}{\lambda_{hydr}}\right) \tag{7.12}$$

with $\lambda_{hydr} = 0.6–1.1$ nm for 1:1 electrolytes.

Steric Forces

Polymer coatings define a rather diffuse interface with the surrounding medium, since fragments of the relatively large molecules can dangle into the liquid phase with relative freedom. When two such surfaces approach each other, they experience a repulsive force as the respective loose segments overlap. This occurs when the separation distance is comparable to the molecular dimensions of the polymer molecules, usually a few times the *radius of gyration,*[*] R_g, or the molecule's length, L. Confining polymers into small volumes has thus an entropic cost that leads to the *steric* or *overlap* repulsion. The force is roughly exponential with the separation distance

$$F_{steric} \propto \exp\left(-\frac{r}{\lambda_{steric}}\right) \tag{7.13}$$

The decay parameter λ_{steric} can take different values in two limiting cases: when the density of polymer molecules per unit surface area is small or when it is large. We refer to the first case as the low-coverage regime and the second as the high-coverage regime. The value for λ_{steric} is equal to R_g in the first case and to L in the second case.

SOME CONCEPTS ON THE MECHANICS OF VISCOELASTIC BODIES

The response of objects to an external perturbation that may change their shape determines their mechanical properties. An object is said to be rigid or deformable, elastic or fluid, from such a response. The relation between perturbation and response is a *constitutive equation*. In mechanics, this equation relates a mechanical stress, σ, to a mechanical strain, ε.

---

[*] The radius of gyration is the size parameter of a polymer molecule, when the latter adopts the conformation of a *random coil* in the absence of any perturbation.

In the case of purely elastic bodies, the relation is the following:

$$\sigma = \mathbf{X} \cdot \varepsilon \tag{7.14}$$

The expression in its most generalized form relates the stress tensor, **σ**, and the strain tensor, **ε**, through a mechanical modulus tensor, **X**. The tensor formulation accounts (a) for the distribution of stress and strains along the three dimensions and (b) for the fact that the object may not exhibit the same response along each direction (*anisotropic*). In the case of homogeneous, isotropic objects subjected to a one-directional perturbation, the expression can be simplified into its scalar form

$$\sigma = X \cdot \varepsilon \tag{7.15}$$

This expression is valid for different types of mechanical stress: if the stress is tensile or compressive, the deformation will be also tensile or compressive, and X is referred to as *tensile* or *Young's modulus, E*. If the perturbation is a shear stress, the response will be a shear deformation, and consequently X is referred to as *shear modulus, G*. The tensile and shear moduli are related by the following expression:

$$E = 2 \cdot G \cdot (1 + \nu) \tag{7.16}$$

where ν is the *Poisson's ratio*, which is a measure of the Poisson effect. When an object is compressed or stretched along one direction, it usually thins or expands in the perpendicular directions. The magnitude is a material's property and it can take a value between −1 and 0.5. A perfectly incompressible material that is deformed elastically at small strains would have a Poisson's ratio of 0.5.

Before continuing, let us consider the definition of stress and strain. Stress is the mechanical force per unit contact area. This "contact area" upon which the force is applied arises because the object, to be mechanically perturbed, must be in contact with the force actuator, which is another object that generates the stress. Figure 7.2 shows the area in the case of a rectangular body subjected to a tensile (a) and a shear stress (b), respectively. Strain is the ratio of the object's dimension after the stress is applied to its initial value, before any stress is applied.

Equation (7.15) is not applicable to fluids though. In these cases, the constitutive relation is

$$\sigma = \eta \cdot \frac{d\varepsilon}{dt} \tag{7.17}$$

where η is the fluid's viscosity.

Most of the real-life objects are neither purely elastic nor purely viscous. They are thus said to be *viscoelastic*. The viscoelasticity of materials can thus be characterized by one (or more) elastic moduli and one (or more) viscosities.

We will only consider here the case of *linear viscoelasticity*, i.e., when the characteristic viscoelastic parameters do not depend on the extent of the perturbation.

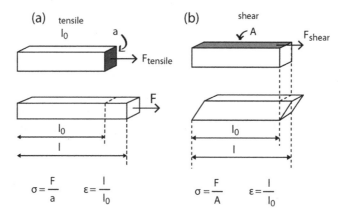

FIGURE 7.2 Mechanical stretch and shear. (a) Object subjected to a tensile force F. The latter is applied to the small area a and produces the object's elongation as well as thinning. The extent of thinning is determined by the Poisson's ratio of the material. (b) Same object subjected to a shear force F, producing a shear deformation. The shear force F is applied to the large area A.

MECHANICAL EQUIVALENTS OF VISCOELASTICITY

To account for the viscoelasticity of materials, mechanical equivalents are often used. These consist of a graphical set of interconnected elements. These elements can be of two types: a spring that represents the purely elastic behavior, and a dashpot that represents the purely viscous behavior. The two simplest models are composed of one spring and one dashpot, either connected in series or in parallel, as Figure 7.3 shows. The first arrangement is referred to as the *Maxwell* model (Figure 7.3a) and the second as the *Kelvin-Voigt* model or simply the *Voigt* model (Figure 7.3b).

EXPERIMENTS TO EVALUATE VISCOELASTICITY

As mentioned previously, the mechanics of materials is tackled by studying their response to a mechanical perturbation. The latter can be applied in mostly two different ways, either suddenly or sinusoidally at a defined frequency. In the first case we have the so-called *transient* experiments; in the second, the *dynamic* experiments.

Transient Experiments: Stress Relaxation and Creep

In these types of experiments, the material is subjected to a sudden strain or a sudden stress, and the corresponding stress or strain is respectively monitored as a function of time. The first case refers to a stress relaxation; the second case refers to creep.

When a sudden strain ε is applied to the Maxwell model, the stress decays with time in an exponential fashion to negligible values at long times (see Figure 7.3a)

$$\sigma(t) = \varepsilon \cdot X \cdot e^{-t/\tau_{\text{relax}}} \tag{7.18}$$

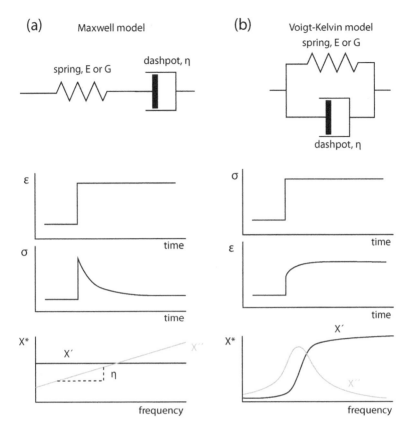

FIGURE 7.3   Two basic viscoelastic models. (a) The Maxwell model is a spring connected in series with a dashpot that predicts stress relaxation but no creep. If a sudden strain is applied, the stress decays exponentially according to Equation (7.18). If the strain is a sinusoidal signal, the real part of the complex elastic modulus is constant and the imaginary part is proportional to the frequency of the sinusoidal signal. (b) The Voigt model is a spring connected in parallel with a dashpot. It predicts material's creep, unlike the Maxwell model. If a sudden stress is applied, the strain increases exponentially to an asymptotic value according to Equation (7.19). If the stress is sinusoidal with time, both real and imaginary parts are functions of the frequency as the plots show.

Conversely, if a sudden stress σ_0 is applied to a Voigt model, the strain increases exponentially as it approaches to an asymptotic value at long times (see Figure 7.3b)

$$\varepsilon(t) = \frac{\sigma_0}{X} \cdot \left(1 - e^{-t/\tau_{creep}} \right) \tag{7.19}$$

In both cases the magnitudes relax with a characteristic time constant, τ, that is the ratio of the viscosity to the elastic modulus X (either E or G), η/X.

Dynamic Experiments: The Complex Modulus

In this case, the perturbation is a sinusoidal signal of frequency $\omega = 2\pi f$. The resulting response is also sinusoidal and varies at the same frequency. However, the amplitude and the phase of the response may not coincide with those of the perturbation. Actually, the response lags behind the perturbation. This is due to the viscoelasticity of the material, which under these conditions is defined by a complex elastic modulus

$$\sigma(\omega) = X^*(\omega) \cdot \varepsilon(\omega) \tag{7.20}$$

where $X^*(\omega) = X'(\omega) + iX''(\omega)$. The real part accounts for the instantaneous response of the material to the perturbation, whereas the imaginary part accounts for its delayed response. Both magnitudes may depend on the frequency.

The complex elastic modulus of a Maxwell model has the following expression:

$$X^*(\omega) = X + i\eta\omega \tag{7.21}$$

The real part is constant with frequency and equal to the real elastic modulus, either E or G. The imaginary part is proportional to the frequency and the slope is equal to the viscosity.

The Voigt model predicts the following expression for the complex elastic modulus:

$$X^*(\omega) = \frac{1}{\dfrac{1}{X} + i(\eta\omega)^{-1}} \tag{7.22}$$

which yields the following expressions for the real and imaginary part of X^*:

$$X' = \frac{X\eta^2\omega^2}{\eta^2\omega^2 + E^2}$$

$$\tag{7.23}$$

$$X'' = \frac{X^2\eta\omega}{\eta^2\omega^2 + E^2}$$

In this case, both real and imaginary parts are functions of the viscosity, the real elastic modulus, and the frequency.

MEASURING THE PROBE–SAMPLE INTERACTION AS A FUNCTION OF THE RELATIVE DISPLACEMENT

In this section we will temporarily set aside the SPM's capability to laterally scan a sample along the *x-y* plane and concentrate on what happens when the scan occurs along the z-axis, at a fixed sample position. By scanning along the z-axis, it is understood that either the probe or the sample moves upwards or downwards, which results in the tip and sample approaching one another or withdrawing from one another. During approaching and withdrawal, the probe may undergo deflections or detect currents due to interactions that arise at particular distances. Though the latter are straightforwardly measured through amperometers, probe deflections and forces are only acquired after proper conversion of the instrumental signal. The conversion of deflection into force requires the knowledge of two parameters: the instrument's sensitivity and the probe's spring constant.

The principle of force measuring is based on the fact that the probe is acting as a force transducer. The probe, a cantilever with a sharp tip or a blunt sphere attached to its free end, may then deflect up or down due to repulsive or attractive forces that arise between the latter and the sample. As mentioned in Chapter 2, if the deflections are small compared to the cantilever's length (typically <0.5%–1%), the *force F is proportional to the cantilever deflection* Δx according to Hooke's law (Equation 7.4). In this case, k is the spring constant of the cantilever, which can be determined by different methods, and it is one of the crucial steps of the instrument's calibration, as will be shown later.

Figure 7.4a shows the typical behavior of a cantilever probe when a sample is displaced toward (1–3) and away from it (3–5). The setup is an atomic force microscope (AFM) with an optical beam detection system (see Chapter 2 for details). As long as the tip and the sample are far away and there is no long-range interaction, the cantilever does not deflect, and consequently the laser spot reflected by the cantilever and registered by the position sensitive detector does not shift its initial position, which is usually set at the center of the position-sensitive detector (PSD, 1). When the tip and sample are close enough for any interaction to occur, the cantilever will deflect. In the case depicted in Figure 7.4a, the cantilever bends downward due to attractive forces (2). As the probe deflects, the laser spot shifts its position toward the lower part of the PSD. If the sample is further moved toward the tip, contact occurs, where the tip touches the sample and starts pressing it. Consequently, the cantilever deflects upwards as repulsive forces develop to resist the penetration of the tip into the substrate (3). The laser spot in turn shifts to the upper half of the PSD (3). If the sample is moved away from the tip at this point, the latter may remain attached to the sample even after a considerable sample displacement (4). Finally, the tip suddenly snaps off when the displacement is large enough and regains its initial condition (5). The laser spot consequently shifts back to the center of the PSD.

Plotting the cantilever deflection as a function of the sample's (or piezo) displacement, one obtains the so-called *approach* and *withdraw* curves or generically *force curves*, as seen in Figure 7.4b. The approach curve is obtained as the tip and sample are brought to contact; it is read from right to left as the approximation occurs. The withdraw or retract curve is registered as the tip and sample are brought apart, and it is read from left to right together with the extent of the separation. In terms of forces and tip–sample distance, which are magnitudes related to those plotted in the figure, the curves can be qualitatively interpreted as follows. When the tip is far apart from the sample, the cantilever does not bend, and a horizontal line at zero deflection (the so-called *zero force line*) is recorded. At shorter

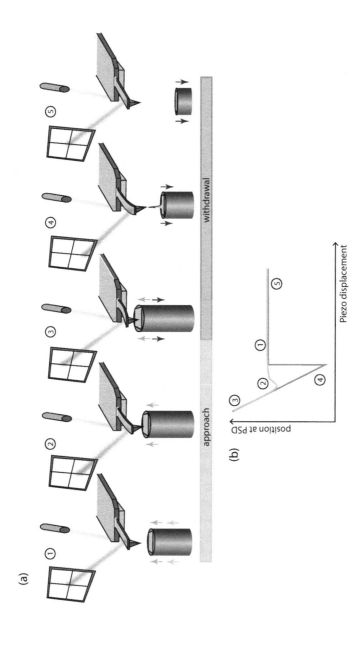

FIGURE 7.4 *(See color insert.)* Force curves measured with the AFM. (a) The position of the laser beam that is reflected at the backside of the cantilever is registered at the PSD at the same time as the sample is moved toward (1–3) and away from (3–5) it. In this case, the sample rests on the top of a tubular piezo actuator, or simply *piezo*, that moves the sample up and down. (b) The position at the PSD is thus plotted as a function of the piezo displacement during approach (light orange curve) and during withdrawal (green curve).

distances, where interactions start to set in, the cantilever may bend upwards (if the interaction is repulsive) or downwards (if the interaction is attractive). Bending occurs in a certain fashion, which is defined by the nature of the forces between the tip and the sample. In this region, the restoring force of the cantilever is balanced by the interaction, providing a means to measure it. On some occasions the balance may be disrupted when the distance between the tip and the sample is very small (i.e., <5 nm). In these cases, surface interactions change with distance too rapidly to be counterbalanced by the restoring force of the cantilever. Consequently, the cantilever is not in mechanical equilibrium and thus accelerates toward the sample in a sort of "snap-in" until it regains equilibrium when it contacts the sample. It is then said that the AFM tip "jumps" into contact. At this point the contact region starts, where the tip is pressed against the sample. If sample deformation does not occur, the cantilever deflection linearly increases with the piezo displacement until a certain value of deflection is attained.

Upon separation from the value of maximum deflection, the cantilever may or may not follow the same trend as during the approach. When it does not, the approach and separation curves do not overlap; the difference between the curves is called *hysteresis*. The major reasons for this behavior are, on the one hand, the adhesion between the tip and the substrate that extends beyond the contact region and, on the other hand, plastic deformations in the sample. The plot shows hysteresis due to adhesion forces that extend the linear behavior of the cantilever deflection beyond the contact region. When the restoring force of the cantilever overcomes the adhesion force, the cantilever snaps off the sample in a sudden jump back to the zero-force line. The vertical jump allows one to calculate the force required to separate tip and sample, which is called the *pull-off or adhesion force*.

For any interaction to be sensed by the cantilever and accurately registered by the instrument, two requirements must be fulfilled. On the one hand, the cantilever should be in equilibrium; on the other hand, the deflection should be detectable. Both requirements should simultaneously occur at each value of the tip-to-sample relative displacement, Z.

For the cantilever to be in mechanical equilibrium, the restoring force exerted by the latter should be compensated by the tip–substrate interaction. This means that the elastic force should always oppose and be equal to the tip–substrate interaction. The first condition is characteristic of restoring forces, and it is always fulfilled; the second condition, however, can only be met if the force gradient—the variation of the interaction F with Z—is smaller than the cantilever's spring constant

$$\frac{\partial F}{\partial Z} < k \tag{7.24}$$

If the condition described in Equation (7.24) does not apply, the cantilever is mechanically unstable and swiftly moves until a new equilibrium position is attained. The transition from one equilibrium position to the next one is shown in the curves as a sudden step. Within the step, force is not properly measured, and therefore the points of the curve in this region are meaningless. Examples of steps that usually appear in the force curves are the *snap-in* and *snap-off* jumps.

When probing viscous samples or dynamics of processes such as molecule unfolding, the tip–sample interaction is time dependent, and proper mechanical equilibrium is never attained at the available approach and retract rates. In these cases, interaction parameters can only have full meaning at zero rate, where true equilibrium holds. These conditions are unattainable in practical terms, but they can be extrapolated if approach and retract curves are done at different speeds. Obtaining force curves at different rates is thus key in addressing dynamic interaction parameters, and they define the experimental ground of *force spectroscopy*.

But according to Equation (7.4), there is no measurable force if there is no measurable deflection. By measurable it is understood that the bending of the cantilever, and hence the displacement of its free end, should be large enough and hence clearly discernible out of the background noise, optical interference, or instrumental drift. The latter effects impose a lower limit of detection below which forces cannot be measured. On the other hand, the deflection cannot be as large as one may desire; indeed, cantilever deflections that are too large produce huge deviations of the reflected laser spot that may fall outside the range of the PSD. Besides, the cantilever may not behave linearly under those conditions, invalidating Hooke's law and impairing the calculation of forces. This makes the overall force range inaccessible for a single cantilever of given k. Choosing the right cantilever, and hence the right k, is key to obtaining reliable force measurements.

Since larger deflections are always easier to detect than small deflections at a particular value of the force, soft cantilevers of low spring constants are more appropriate to measure small forces. Conversely, hard cantilevers are more suitable to measure large forces to keep the signal within the measurable range of the detector.

NOISE AND ARTIFACTS

The first effect is mainly thermal noise that is manifested as ever-present, nondirectional deflections of small amplitude. The change in position of the laser spot at the PSD, which is an optical magnification of the cantilever deflection, is then affected by erratic fluctuations. The second effect is brought about by interference between light reflected by the back side of the cantilever and light reflected by the sample. The laser spot, though focused on the free end of the cantilever, is usually larger than the cantilever dimensions, and part of the light unavoidably reaches the sample. If the latter is reflective, such as a gold surface, light is thus reflected and can in turn interfere with the light reflected by the cantilever. Interference is manifested in the shape of wavy zero-force lines of considerable amplitude and frequency that depend on the relative position of the cantilever and the sample. The third effect can be caused by environmental variables such as temperature gradients or fluid viscosity. When the tip and sample are motionless and do not interact with each other, drift is clearly recognizable as a very slow, large-amplitude oscillation in the position of the laser spot. Drift usually fades away with time.

QUANTITATIVE DETERMINATION OF FORCES: INSTRUMENT AND CANTILEVER CALIBRATION

Raw data of approach and retract curves consist of two voltage signals. One of those signals is applied to the piezoelectric actuator that displaces vertically either the tip or the sample. The other is the signal derived from the position of the reflected laser spot at the PSD, which is proportional to the cantilever deflection and hence to the force. The first of the signals is converted into displacement by the calibration parameters of the scanner, which are obtained from the manufacturer and updated regularly through calibration protocols with reference samples; these protocols are instrument dependent, and they will not be discussed here. The second signal, however, can only be converted into force through a *force calibration procedure* that must be done for each cantilever. The procedure requires a hard, nondeformable substrate as reference sample and consists of two steps. The first one involves the conversion of the voltage signal from the PSD into cantilever bending in length units, and the second one involves the conversion of the cantilever's bending into force.

$$\text{Force (N)} = \text{raw signal (V)} \times \text{sensitivity (m/V)} \times k \text{ (N/m)} \qquad (7.25)$$

VOLTAGE TO DEFLECTION: DETERMINATION OF THE SENSITIVITY

Voltage signals coming from the PSD must be converted into displacements of the free end of the cantilever. The amplitude of the voltage signal is proportional to the displacement of the laser spot in the PSD from its initial position, which is usually at the center of the PSD (zero volt), as seen in Figure 7.5a. This magnitude, Δ_{PSD}, is in turn proportional to the displacement of the free end of the cantilever due to its bending Δx. In fact,

$$\Delta x = \frac{\Delta_{\text{PSD}} l}{3L} \qquad (7.26)$$

where l is the length of the cantilever and L the distance between the position of the laser spot at the back side of the cantilever and the PSD. The proportionality factor is a parameter that depends on the instrument configuration, the type of cantilever, and laser alignment, and it is called *sensitivity*. In practical terms, the sensitivity can be graphically obtained from force curves between the tip and a hard substrate. In particular, the sensitivity is the inverse of the slope of the contact region, a straight line. Under these conditions, the displacement of the free end of the cantilever is equal to the extent of the piezo displacement, Z (see Figure 7.5a). The sensitivity thus has units of meters per volt (m/V).

DEFLECTION TO FORCE: DETERMINATION OF SPRING CONSTANT

The accurate determination of cantilever spring constants is still challenging yet necessary. The manufacturing processes are capable of fabricating wafer-scale

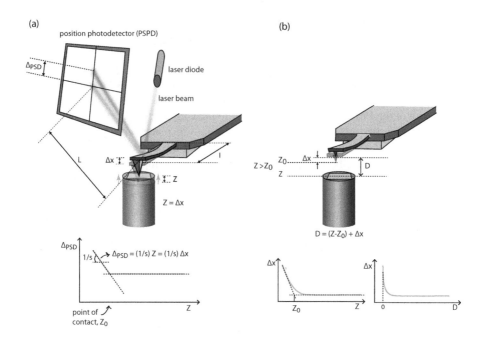

FIGURE 7.5 Conversion of piezo displacements (Z) to tip–sample distance (D). Calibration. (a) Calculation of the sensitivity. That is a required parameter that converts the voltage signal at the PSD into a measurable distance. The obtainment of this magnitude requires a hard substrate; the tip of the cantilever is pressed against this surface and its deflection measured at the PSD. Under these conditions the piezo displacement Z is equal to the cantilever deflection Δx, $Z = \Delta x$. The sensitivity is thus calculated from the slope of the contact region of the resulting force curve as shown below. The sensitivity is usually expressed as $1/s$, which has units of distance/volt. (b) Relating Z and D. Tip–sample distances and relative displacements of the piezo actuator ($Z - Z_0$) coincide as long as the cantilever does not bend due to tip–sample interactions. In the latter case, the piezo displacement must be corrected by the cantilever deflection Δx. This magnitude should be added when the cantilever is bent upwards and subtracted when the cantilever is bent downwards. Z_0 is the point of contact. Some simple criteria to define Z_0 include the minimum Z at which the deflection is greater than 3 times the standard deviation of the fluctuations at the baseline or the point at which the deflection is 1% or some other fraction of the maximum value (Melzak et al. 2010). These definitions or a simple visual inspection of the data give reasonable values of Z_0 just in few cases, especially when long-range interactions do not occur. The variable Z_0 is depicted in the approach curve in the case of a hard substrate exhibiting long-range interactions. For $Z < Z_0$ the relative displacement and the cantilever's deflection are equal in magnitude and have opposite sign, giving $D = 0$.

cantilevers with excellent lateral resolution, but they still lack the ability to produce cantilevers of equal thickness. Moreover, the cantilever materials often present variations in their properties that contribute to altering the value of the spring constant in cantilevers coming from different production batches. Thickness and material properties are two of the parameters that largely affect the cantilever's stiffness and hence its spring constant, and therefore calibration at a single-cantilever basis is required. On the other hand, the nonsimple geometry and composite nature of

TABLE 7.1

Summary of the Theoretical Methods to Calculate the Cantilever's Spring Constant

	Theoretical		
	Bernoulli's Model	**Sader's Method**	**FEA**
Uncertainty	E and t, 80% (V-shaped)	E and t; 16%, 2%, or smaller (errors from FEA results)	10%
Maths	$k = \dfrac{Ewt^3}{2L^3}$ w = width t = thickness L = length	$k = \dfrac{Ewt^3}{2L^3}\cos\alpha\left[1 + \dfrac{4w^3}{t^3}(3\cos\alpha - 2)\right]^{-1}$ α = half-opening angle w = (arm) width; t = thickness	Finite element Numerical treatments
Advantages	Parallel beams or V-shaped beams (parallel beam approx.)	V-shaped cantilevers	Accurate
Disadvantages	Requires t and E Geometry dependent Coating not considered	Requires t and E Geometry dependent Coating not considered	Requires t and E Time consuming Complex and expensive programs

cantilevers—often back-side coated with a thin layer of metal to enhance their reflectivity—impose difficulties in the theoretical calculation of their stiffness. Many methods have been reported in the literature of differing complexity and accuracy. A good account of these methods is presented by Cumpson et al. (2008) and Butt, Cappella, and Kappl (2005), from which Tables 7.1 and 7.2 show a brief summary.

The methods can be categorized as *theoretical* and *pseudo-theoretical*, among which are included both *dynamic* and *static* experimental as well as thermal methods. The theoretical methods calculate spring constants from the cantilever geometry and material properties such as the Young's modulus (E) or Poisson's ratio (ν). These methods include mathematical models of varying level of approximation and refinement and numerical methods such as finite element analysis (FEA). Their uncertainty is often large, since both cantilever thickness and stiffness are not accurately known. Yet, FEA provides the most accurate result. The pseudo-theoretical methods combine a theoretical method with an experimental measurement of a cantilever's property, such as the resonance frequency in the dynamic experimental methods or the thermal mechanical vibrations in the thermal methods. Static experimental methods use a reference cantilever or spring of known spring constant to calibrate other cantilevers by pressing them against the reference. Alternatively, hanging masses on the cantilevers produces deflections or shifts in the resonance frequency that can be used to calculate spring constants.

TABLE 7.2
Summary of the Experimental and Pseudo-Experimental Methods to Calculate the Cantilever's Spring Constant

	Static Experimental		Dynamic Experimental		
	Mass Hanging	**Cantilever On a Ref. Cantilever or Spring**	**Resonance Frequency**	**Thermal Noise**	**Mass Attachment**
Uncertainty	15%–25%	10%	15%–20%	15%–20%	15%–25%
Maths	$k = \dfrac{F}{\Delta x}$ Δx = cantilever deflection	$k = k_{ref}\dfrac{Z_p - \Delta x}{\Delta x}$ k_{ref} = stiffness of the ref. cantilever or spring Z_p = piezo position	$k = 0.1906\rho_f w^2 L Q \Gamma_i(\text{Re})\omega_0^2$ ρ_f = density of the fluid w = cantilever's width L = cantilever's length Γ_i = imaginary part of the hydrodynamic function, dependent on Reynold's number $\text{Re} = \dfrac{\rho_f \omega_0 w^2}{4\eta}$ η = fluid viscosity Q = quality factor of the resonance ω_0 = (angular) resonance frequency	$k = \beta\dfrac{k_B T}{\langle\Delta x^2\rangle_{OBL}}$ $\langle\Delta x^2\rangle_{OBL}$ = apparent cantilever deflection (detected through OBL) β = correction factor $0.817 \rightarrow$ rectangular cantilever $0.764 \rightarrow$ V-shaped cantilever	$k = \dfrac{4\pi^2 M}{\left(\dfrac{1}{v_0'}\right)^2 - \left(\dfrac{1}{v_0}\right)^2}$ M = added mass v_0 = resonance frequency without M v_0' = resonance frequency with M
Advantages	E or geometry not needed Nondependent on geometry	E or geometry not needed Nondependent on geometry	E or geometry not needed Nondependent on geometry Surrounded medium = viscous fluid	Easily implemented in an AFM Geometry or E is not needed	Geometry or E not required
Disadvantages	Requires sphere attachment Low reproducibility Precalibration of spheres and deflection	Requires an accurate value for k_{ref}	Expression valid for rectangular cantilevers Requires Γ_i, which depends on cantilever's cross section	Equipartition theorem, cantilever as harmonic oscillator Correction factors for different geometries and surrounding fluids	Requires sphere attachment Not reproducible Requires determination of spheres' mass

THE ISSUE OF GETTING ABSOLUTE DISTANCES

To fully characterize tip–sample interactions it is necessary to know the dependence of the force on the distance between both surfaces. SPM can provide the piezo position, Z, of either the tip or the sample. However, it cannot give a direct means of obtaining the separation distance between these surfaces. Instead, this magnitude must be derived from Z through a conversion relation that is not free of assumptions. If Z increases in the retract direction, this relation is the following:

$$D = (Z - Z_0) + \Delta x + \delta_{sample} \qquad (7.27)$$

The separation distance can thus be obtained by adding the cantilever deflection Δx and the sample deformation δ_{sample} to the piezo displacement $(Z - Z_0)$, as depicted in Figure 7.5b. The latter is referred to the position where the tip and sample are brought into contact, the so-called contact point, Z_0, where the distance is zero. The value of Δx is, in this case, positive when bending is upwards and negative when bending is downwards, whereas δ_{sample} is positive if the sample is indented and negative if the sample is stretched. In order to apply Equation (7.27), one parameter has to be known: the zero distance.

The main difficulty is to ascribe the zero distance to a particular point, Z_0 (the contact point), or to a portion of the assumed contact region in the force curves. The task is usually done by visual inspection of the data. Though not strictly rigorous, the method gives reasonable results, but only in few cases. Indeed, the assessment of the contact point may be straightforward in the case of hard, nondeformable materials where no long- or short-range interactions occur. However, the uncertainty in the determination of Z_0 is high in the case of deformable materials and even higher in the presence of long-range interactions (Butt, Cappella, and Kappl 2005).

The simplest case is depicted in Figure 7.6a. Here the contact region is a straight line with slope equal to the sensitivity or, in calibrated devices, with slope equal to 1, since in this region $Z - Z_0 = -\Delta x$ and hence $D = 0$. Out of contact, the force is zero, $\Delta x = 0$, and hence $D = Z - Z_0$. The contact point is then the Z value of the intersection between two straight lines: the zero force line and the contact region line. In practical terms, a fairly correct value for Z_0 can be thus found by just visual assessment of the curves. The same approach for the contact point can be used in the presence of surface interactions, as in Figure 7.6b. In this case, the contact point is obtained from the intersection between the *extrapolated* zero force and contact-region lines.

Deformable materials depict a more complicated scenario, since in this case $\delta_{sample} \neq 0$, which makes the assignment of the contact point especially troublesome (see Figure 7.6c). The contact region may not be linear with Z, and it can be wrongly assigned to a noncontact interaction, misinterpreting the curve and hence the value of Z_0. If the piezo further moves beyond the point of contact, it will bring the tip deeper into the sample, causing sample deformation, even though $D = 0$. Under these conditions, the extent of deformation δ_{sample} (also called *indentation*), rather than distance D, is the characteristic parameter of the force curve for the contact region. Therefore, in the noncontact region, the force depends on *distance* D, defined as

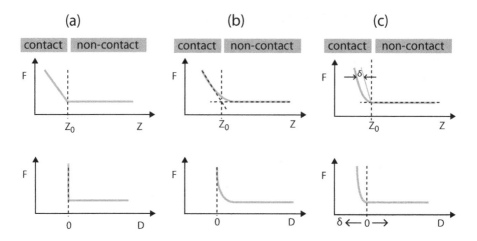

FIGURE 7.6 Conversion of piezo displacement (Z) to tip–sample distance (D). Same case examples. (a) Nondeformable substrates in the absence of long-range interactions. (b) Nondeformable substrates in the presence of long-range interactions. (c) Deformable substrates in the absence of long-range interactions.

$(Z - Z_0) + F_{interaction}/k$, whereas in the contact region, $D = 0$, and the force is sensitive to the indentation, $\delta_{sample} = -(Z - Z_0) - \Delta x_{contact} = -(Z - Z_0) - F_{contact}/k$.

Correct assignment of Z_0 in force curves exhibiting long-range repulsions remains a challenging task, although there have been attempts to develop a systematic, assumption-free method to detect it. The authors have suggested a method based in computing a sample's relative deformations as a function of the applied load (Melzak et al. 2010). The value of Z_0 is obtained by extrapolating such data to zero load. The method is illustrated in Figure 7.7 and applied to films of glycopolymer brushes in water. The force curves between these films and a silica bead exhibit long-range ionic repulsions, which makes it a suitable system to test the method. Force curves are performed within a range of loads, L, from a minimum L_{min} to a maximum value L_{max}. However, instead of inspecting the force curves, we rather focus on the plots of the displacement Z as a function of time. As the approach and retract rates are constant, the curves have a typical V-shape (Figure 7.7a,b). The minimum is the value of Z at which the value L is attained. The relative deformation is thus obtained by subtracting the minima of each curve from that of the curve at the lowest load, L_{min}, $Z(L) - Z(L_{min})$. When this procedure is done on a bare, hard substrate (Figure 7.7a), the results are actually relative cantilever deflections. If these results are plotted as a function of the relative load, $L - L_{min}$, we obtain a straight line whose slope is the inverse of the cantilever's spring constant, $1/k$ (Figure 7.7c). When applied to a soft sample such as to the glycopolymer brush film (Figure 7.7b), the data points $Z(L) - Z(L_{min})$ (Figure 7.7d, black dots) vary linearly as a function of L with a slope higher than $1/k$. After subtracting the cantilever deflections to these data points (Figure 7.7d, grey dots) and extrapolating to zero load, we obtain the sample's contact point Z_{cp}^{sample}.

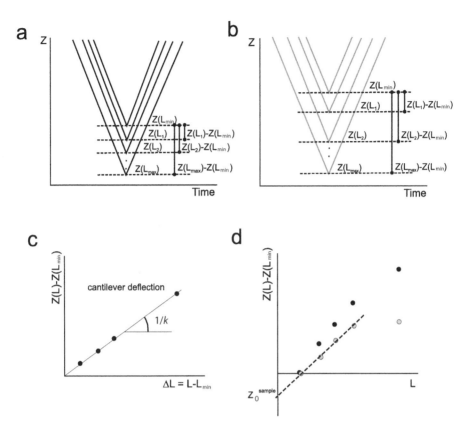

FIGURE 7.7 Method to obtain the point of contact Z_0 from relative deformations. Raw data are Z versus time from force curves at different loads, from the smallest (L_{min}) to the highest (L_{max}) for a hard substrate *a*) and the sample. (b) The relative displacements $Z(L_i) - Z(L_{min})$ are calculated as the plots show. (c) On the hard substrate, $Z(L_i) - Z(L_{min})$ varies linearly with the added load, $L - L_{min}$, the slope being the inverse of the spring constant, $1/k$. (d) On a soft sample, the slope should be higher and eventually equal to $1/k$ at sufficiently high loads, when the sample cannot be further deformed. It thus acts as a nondeformable material (black dots). By subtracting the cantilever deflection and linear extrapolating to zero load (grey dots), the contact point is obtained as the *y*-intercept. (All figures adapted from Melzak et al. 2010. With permission.)

This value was used to calculate sample indentations as a function of load, as shown in Figure 7.8. The values (filled symbols) were compared to those obtained from force–distance curves, where the contact point has been detected by visual inspection of the data (open symbols). Only the data obtained using the method described here display a reasonable dependence with load, showing a linear region at small loads ($L < 1$ nN), where the brush film is compressible, and a plateau region at higher load ($L > 1$ nN), where the film cannot be further compressed. The value of the sample indentation at this plateau region is comparable to the film thickness.

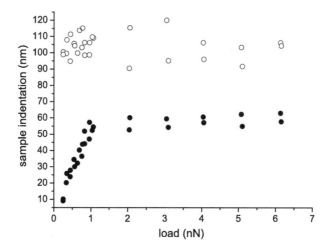

FIGURE 7.8 Method to obtain the point of contact Z_0 from relative deformations. Sample indentation. The sample indentation of a film of glycopolymer brush with a colloidal probe is calculated according to the expression $d_{sample} = -(Z - Z_0) - F_{contact}/k$ and plotted as a function of the applied load. Filled symbols result from using in the expression the value of Z_0 obtained from the method described in the text and in Figure 7.7. Open symbols are the results from a value of Z_0 extracted from visual inspection of the force curves. Only the black dots show a reasonable dependence of the sample indentation with load; indentation increases until it is no longer possible, where it reaches a plateau. The presence of the plateau may be either due to entropic repulsion that makes the film fairly incompressible, or due to the probe pressing the substrate, since the film has a finite thickness. The value of the sample indentation at the plateau can thus be approximated to the film thickness. (Reprinted from Melzak et al. 2008. With permission.)

Another approach is to use an independent instrumental source that can assess absolute distances in parallel to the performance of force curves. A promising way is to combine microinterferometry, a technique that can accurately provide the distance between two reflecting surfaces, and AFM as a tool to calculate forces. Such a fruitful combination is described in detail in Chapter 8.

QUALITATIVE INTERPRETATION OF FORCE CURVES

A force curve is a rich source of information about all kinds of surface interactions. Figure 7.9 shows some typical force curves. By just visually inspecting these curves, it is already possible to make some valuable remarks on the matter. For example, the shape of the out-of-contact region already hints at the nature of noncontact forces (electrostatic, dispersive, steric, solvating) between two interacting surfaces. The contact region of the same curve is indicative of the sample's stiffness or compliance as a function of the applied load. The irreversibility of the sample's deformation— the sample's plasticity—can be estimated from the hysteresis of the approach and the retract curves. Adhesion forces are extracted from the retract curves, and these are indicative of the surface affinity, which may be attributed to a specific or nonspecific

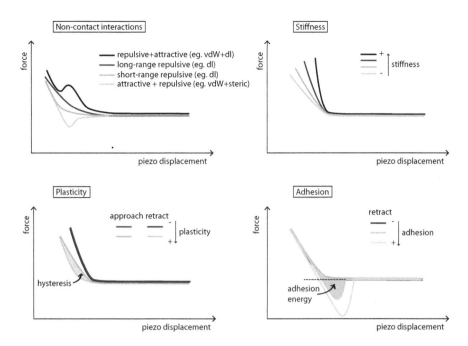

FIGURE 7.9   *(See color insert.)* The information in a force curve. From the approach curves, it is possible to infer the nature of non-contact interactions (top-left graph) or qualitatively address the sample stiffness (top-right graph). The hysteresis between the approach and retract curves in the contact regime is indicative of the degree of plasticity in the sample (bottom-left graph), whereas the adhesion is a measure of the affinity between the sample and the probe (bottom-right graph).

molecular interaction or may have implications in the tribological behavior of two surfaces in contact.

However, the potential of the force curves is not limited to this. The results have also inspired the development of SPM-based techniques and new fields of knowledge, as we will see now.

CHEMICAL FORCE MICROSCOPY

The chemical force microscopy (CFM) technique allows identifying sample constituents from the type and the magnitude of their interactions with the probe. The latter is *chemically modified* with molecules of such type that can distinctively interact with various other molecules present in the sample. Adhesion and friction forces are most typically addressed to quantify the extent of the probe–sample affinity. The more favorable the interaction, the stronger is the adhesion or the stronger the friction. Although unambiguous chemical identification cannot be assigned from adhesive forces alone, the technique can certainly throw light on the spatial distribution of constituents in samples of known chemical composition, thus providing information about the extent of the component miscibility or separation. The proof-of-concept of

CFM appeared in 1994, when Frisbie and coworkers showed that it was possible to use a tip functionalized with carboxylic COOH groups or with methyl CH_3 groups to probe the presence of CH_3 and COOH groups on the surface (Frisbie et al. 1994). This opened the way for chemical imaging as well as for accurate studies of surface interactions with the scanning force microscope (SFM).

A typical example to illustrate the capabilities of CFM is the probing of hydrophilic (water loving) and hydrophobic (water hating) chemical groups in mixed self-assembled monolayers (SAMs). SAMs are monomolecular thick films of densely packed molecules that are able to organize in an ordered manner on top of a hard substrate, usually a metal. They have a well-defined surface chemistry, determined by the outer functional groups of the self-assembling molecules. An archetype of a SAM is that formed by alkanethiols on gold, which is also most commonly used as sample and functionalizing coating for tips in CFM. Mixed SAMs of hydrophobic and hydrophilic groups are readily formed by mixtures of alkanethiols with methyl and hydroxyl groups, respectively. It has been found that adhesion forces between such layers and hydrophobic tips—coated with methyl-terminated alkanethiols—in aqueous solutions are particularly sensitive to the presence of hydrophobic end groups (Sinniah et al. 1996; Alsteens et al. 2009) or to their relative accessibility (Moreno-Flores et al. 2006). Indeed, adhesion increases with the concentration of hydrophobic end groups on the surface or when the hydrophobic end groups protrude from equally numbered hydrophilic end groups. In both case studies it could be proved that adhesion was connected to the macroscopic wettability of the surfaces by water.

Likewise, CFM has proved to be useful in probing the ionization state of acidic or basic groups in the sample when the tip is functionalized with either an acidic or a basic SAM in water. That is the case of the interactions between acid-terminated SAMs. Performing force curves throughout a wide range of pH values, it could be observed how the interactions changed from electrostatic repulsive (pH 10), as inferred between deprotonated, negatively charged COO^- groups in both tip and sample, to van der Waals attractive (pH 3) between uncharged COOH groups (Ahimou et al. 2002).

THE SCIENCE OF PULLING SINGLE MOLECULES OR LIGAND–RECEPTOR PAIRS

Ligand–receptor interactions and the mechanical stability of entangling biomolecules play key roles in all physiological processes that sustain life. Ever since the first successful attempt to characterize such forces, those systems have been the focus of interest and inspired a new SPM-based discipline, so-called *force spectroscopy*.

SPM together with other physical techniques such as optical tweezers or the bio-membrane force probe can quantitatively address a wide range of interactions, ranging from the weakest entropic forces (10^{-15} N or femtonewtons) to covalent bonds (10^{-9} N or nanonewtons). The experimental approach is common to all these techniques. They apply a pulling force to either the ligand–receptor complex or to the coiled protein until the bond or bonds break. The disruption of such bonds appears as sudden discontinuities or snap-offs in the force-displacement curves in the shape of one or more consecutive adhesion *events* (see Figure 7.10). These events take place at a particular value of the piezo displacement or pulling *extension* in this case.

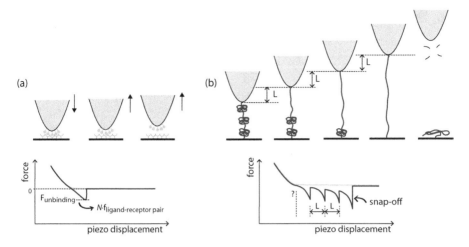

FIGURE 7.10 Pulling at a molecular scale. (a) Ligand-receptor pairs. The observed unbinding force is usually observed as a spike in the retract curve and it is proportional to the number of ligand–receptor interactions that are disrupted. (b) Single molecules. The scheme depicts a molecule composed of repetitive domains (i.e., a polyprotein) being pulled by an AFM tip. The disentanglement of each individual domain is seen as a spike in the retract curve.

The nature of unbinding or unfolding forces is stochastic, and it is mainly governed by thermal fluctuations that may alter the strength of the interactions. It is thus necessary to acquire a large number of force curves *under the same experimental conditions* and perform a proper statistical analysis to accurately obtain the magnitude of such interactions. The results are often presented as force histograms that represent the empirical probability function of a particular interaction or interactions. The maximum of such distributions is the most probable force value that characterizes the interaction, and it is usually referred to as the *interaction strength* at that particular experimental condition.

But the whole description does not end here. To evaluate the mechanisms of unbinding or unfolding, the changes of molecular energy states and conformations, and to finally obtain energy landscapes of the interaction, *the time scale of the pulling experiments must be systematically varied*. This is done by performing force curves at different pulling rates. Thus combining statistical analysis at each pulling rate results in a set of force distributions. The maxima of such force distributions when plotted as a function of the pulling rate constitute an invaluable tool to infer the energy landscape and the possible pathways involved in the dynamics of the processes involved.

STATISTICAL DESCRIPTION OF BOND RUPTURE: TWO-STATE MODEL

In our description, we will consider that the interactions between ligands and receptors and those that keep the three-dimensional (3-D) scaffold in proteins can be described by a common model. The model evolves about the transition between two states, the *bound* and the *free* states, separated by an energy barrier and characterized

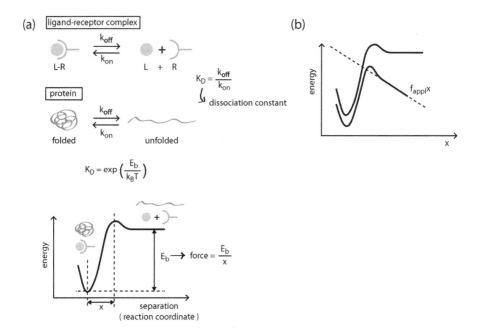

FIGURE 7.11 Two-state model. (a) A ligand–receptor bond or a single molecule entanglement can be treated with the same formalism. Both can be considered as transitions characterized by a dissociation constant. Assuming that the Boltzmann Ansatz applies, K_D can be related to the binding energy E_b, the depth of the energy minimum. The binding (or unbinding) force is the ratio of E_b to the reaction coordinate, x, the half width of the well. (b) In the presence of an applied force, the energy landscape is tilted. To obtain the parameters in the absence of forces it is required to perform force pulling experiments at different speeds.

with a value of a characteristic interaction parameter, in this case the separation or extension (see Figure 7.11a). The former corresponds to the ligand–receptor complex or the entangled (folded) protein and the latter to the detached components or the disentangled (unfolded) protein (Kienberger, Gruber, and Hinterdorfer 2007)

$$S_{bound} \leftrightarrow S_{free}$$

for ligand-receptor interactions

$$L - R \leftrightarrow L + R$$

for protein unfolding

$$P_{folded} \leftrightarrow P_{unfolded}$$

the equilibrium constant K_D can be expressed as a function of the constants for the forward and backward transitions, namely k_{on} and k_{off}, respectively

$$K_D = \frac{k_{off}}{k_{on}} = \exp\left(-\frac{E_b}{k_B T}\right) \tag{7.28}$$

where E_b is the energy difference between the bound and the free states, k_B is the Boltzmann's constant, and T is the temperature. If one assumes a one-step kinetic mechanism, k_{on} and k_{off} are also the kinetic constants of association/entangling and dissociation/disentangling, respectively. Classical mechanics describes detaching forces f as the derivative of the energy with respect to the interaction coordinate. In this case, a fairly good estimation would be $f = E_B/x$, where x is the distance of the energy barrier from the energy minimum according to Figure 7.11a. The scenario described here occurs under thermodynamic equilibrium and thus, to experimentally obtain the energy profile depicted in Figure 7.11a, the pulling techniques should in theory "pull" with infinitely low forces and at infinitely low speeds, which is practically unrealizable. Under these conditions, proteins and ligands will disentangle and entangle, associate and dissociate without the application of any force. The lifetime of the *bond*, $\tau(0) = 1/k_{off}$, will be short if compared with the experimental time frame. In practice, pulling experiments are done at finite pulling rates, where the experimental time frames may be small compared with $\tau(0)$, and in this case the bond requires a force to break.

If thermal fluctuations govern the associating and dissociating processes, the lifetime of those bonds can be expressed according to the Boltzmann's Ansatz

$$\tau(0) = \tau_{osc} \cdot \exp\left(\frac{E_b}{k_B T} \right) \tag{7.29}$$

where τ_{osc} is the inverse of the natural oscillation frequency. Applying a force to break the bond, f_{appl} reduces the energy barrier by an amount proportional to the magnitude of the force (Evans and Ritchie 1997)

$$\tau(f) = \tau_{osc} \cdot \exp\left(\frac{E_b - f_{appl} \cdot x}{k_B T} \right) = \tau(0) \cdot \exp\left(\frac{-f_{appl} x}{k_B T} \right) \tag{7.30}$$

and tilts the energy landscape (see Figure 7.11b). As mentioned previously, if one applies the Boltzmann's Ansatz in combination with the stochastic description of the process leading to bond disruption, one experimentally obtains a series of force distributions at different pulling rates. Authors refer rather to loading rate, r, defined as the time gradient of the applied force, which is obtained by multiplying the pulling velocity by the spring constant of the SPM probe (Evans and Ritchie 1999). Each distribution is characterized by a maximum, $f^*(r)$, that is related to the dissociation constant k_{off} by the following expression:

$$f^*(r) = \frac{k_B T}{x} \cdot \ln\left(\frac{r \cdot x}{k_B T k_{off}} \right) \tag{7.31}$$

In a logarithmic representation of the force versus the loading rate, Equation (7.31) will predict a linear dependence for a single barrier. If there were more barriers of different magnitudes, the plot would show several linear regimes, each

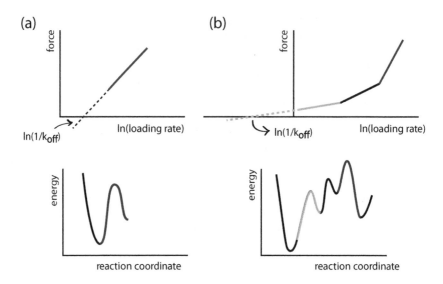

FIGURE 7.12 Force versus the natural logarithm of the loading rate in the case of a single barrier (a) and of three barriers (b). The value of k_{off} is obtained from linear extrapolation at zero force. In the case of a single barrier, there is a single linear regime, whereas the presence of more than a barrier results in different linear regimes according to the loading rate. It is thus important to produce experimental data in a wide range of loading rates to obtain a fairly correct value of k_{off}. AFM provides a fairly narrow range of loading rates (10^4–10^5 pN/s) and therefore may be insufficient. The experimental range can be considerably expanded if other complementary techniques are employed, such as the bead molecular probe (BMP) or optical tweezers (OT). The first technique provides intermediate loading rates (10–10^3 pN/s) whereas the second accounts for the smallest loading rates (1–10 pN/s; Evans and Ritchie 1997).

one characterized by a distinct slope. The value for k_{off} can thus be calculated by extrapolation at $f*(r) = 0$. Figure 7.12 illustrates the shape of the f–r plots and the corresponding energy profile along the interaction coordinate in the case of a single barrier (Figure 7.12a) and of a multibarrier (Figure 7.12b). *Force spectroscopy* can thus unveil the energy landscape of interactions at the single-molecule level.

PARTICULARITIES OF MOLECULAR RECOGNITION SPECTROSCOPY

Negative Controls

The assignment of adhesion events to unbinding between a ligand and a receptor is tricky. There are many likely phenomena behind a sudden jump in a force curve that may lead to data misinterpretation. A usual and yet necessary procedure to avoid this is performing negative control experiments. They consist of performing pulling experiments with the same receptor-functionalized tip and a ligand-coated surface that has been previously incubated in a receptor-containing solution. Under these circumstances, the receptors bind the ligands on the surface until saturation, where there are no more available binding sites. The tip will thus not specifically interact with the saturated surface, and hence the force curves will not exhibit ligand–receptor unbinding events. Still, adhesive events due to unspecific interactions may occur, although to a much reduced extent. As a result, adhesion events that appear in force curves from noncontrol experiments and do not appear in force curves of the negative controls can be reasonably assigned to ligand–receptor interactions.

Inferring Single Ligand–Receptor Interactions from Adhesion Events

The number of ligand–receptor unbinding events that can occur between the tip and sample is defined by the number of binding molecules per unit area in both surfaces, i.e., the surface density of binding groups. But this is not the only parameter to take into account. Before pulling, the tip has to be brought close to the surface to induce any specific interactions. The area of the sample where the tip surface has been as close as or even closer than x, the ligand–receptor interaction parameter (see Figure 7.10a), is the other important parameter, and it is called the *contact area, a_c*. Knowing the contact area, the surface density σ, and the magnitude of the adhesion force of a specific event, it is possible to infer the interaction of a ligand–receptor single pair if one assumes that the observed unbinding force is directly proportional to the number of ligand–receptor pairs

$$F_{unbinding} = N \cdot f_{l-r} = \sigma \cdot a_c \cdot f_{l-r} \tag{7.32}$$

in the absence of nonspecific forces. However unbinding events are stochastic processes, and the number of ligand–receptor pairs may still vary from one pulling experiment to another; moreover, both the contact area and the surface density of ligands are most frequently unknowns. The performance of many such experiments has produced force histograms, as mentioned previously, whose shape depends on the case studied and the number of molecules present in the contact area. One peak-force histogram usually evidences a single bond rupture or one most frequent number of involved ligand–receptor pairs, whereas a multiple peak force histogram is symptomatic either of simultaneous rupture of two or more bonds or of two or more most frequent numbers of ligand–receptor pairs. A usual approach is to assume that the histogram peak (or peaks), represented by $F_{unbinding}$, follows a Poisson's distribution (Lekka, Laidler, and Kulik 2007) whose variance, $\sigma^2_{F_{unbinding}}$, has the following expression:

$$\sigma^2_{F_{unbinding}} = f_{l-r} \cdot F_{unbinding} - F_{unbinding} \cdot F_{nonspec} + \sigma^2_{F_{nonspec}} \tag{7.33}$$

as long as the $F_{unbinding}$ can be described as follows:

$$F_{unbinding} = N \cdot f_{l-r} + F_{nonspec} \tag{7.34}$$

where $F_{nonspec}$ is the mean value of nonspecific forces that are, in turn, assumed to be stochastic.

Plotting $\sigma^2_{F_{unbinding}}$ versus $F_{unbinding}$ gives a straight line for a particular type of ligand–receptor pair. According to Equation (7.33), the slope is the force of a single

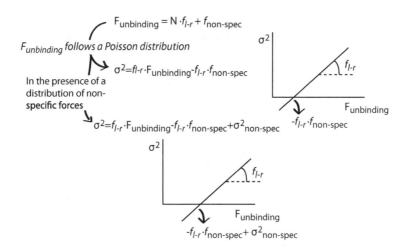

FIGURE 7.13 Obtaining the force of a single ligand–receptor pair from the statistical distribution of unbinding forces. If the unbinding force follows a Poisson distribution, the plot of the variance versus the unbinding force is a straight line. The slope of the line is directly the force of a single ligand–receptor pair.

ligand–receptor pair, whereas the *x*-intercept is the sum of two terms containing the mean and the variance of the nonspecific forces (see Figure 7.13).

PARTICULARITIES OF MOLECULAR UNFOLDING

Pulling experiments unambiguously unveil the mechanical resistance of molecular scaffolds as long as they can be disrupted in a repetitive manner. This means that the target molecules should contain two or more fragments or *domains* of a very similar or identical nature that can unfold one after another upon being pulled. The resulting adhesion events appear as successive peaks in a kind of sawtooth pattern, which can be viewed as characteristic of the single molecule disentanglement (see Figure 7.10b). The number of peaks coincides with the number of repeated domains, a structural parameter that is usually known or characteristic of a specific intermolecular arrangement. An example of the former is the synthesis and use of polyproteins as target molecules in unfolding studies (Rief et al. 1997). These molecules are usually generated from genetically engineered DNA fragments or constructs that contain two or more encoding sequences of a whole protein or a protein fragment (i.e., *domain*). The construct is inoculated into a living microorganism (e.g., a bacterium) that in turn fabricates the encoded molecule as a builder would mount a piece of furniture from a set of instructions. The process is called *expression*, which results in the generation of many of such molecules, each one consisting of a linear array of whole proteins or protein domains covalently linked to one another.

An example of an intermolecular arrangement may be the conformation of a large membrane protein in a lipid bilayer, a ubiquitous system in cells (see Figure 7.14). One of the possibilities for a membrane protein to be inserted in a lipid membrane is in a series of consecutive loops like thread stitches in a fabric. Dismantling the

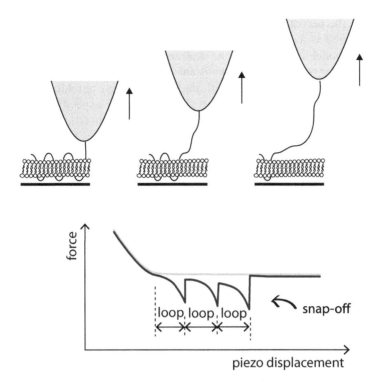

FIGURE 7.14 Disentangling repetitive structures in nature. Membrane proteins are inserted in membranes in a series of consecutive loops that can be in turn unraveled one by one by pulling.

membrane protein from its membrane would thus be equivalent to disentangling or "unstitching" the (almost) identical loops, one after the other. In practice, the poly-proteins or fragments of native membranes are immobilized on rigid substrates at very low surface densities and probed with the AFM tip. The latter will repetitively approach and withdraw the surface in search of molecules. A pulling experiment is said to be successful when a molecule is "trapped" between the incoming tip and the substrate and subsequently stretched as the tip retracts from the surface. The force curve will then show the sawtooth pattern as proof of successively occurring unfolding events. In these experiments, not only the force, but the extension at which unfolding occurs is important. The extension, or the piezo displacement, provides valuable information about the molecular dimensions. To obtain this and other struc-tural and mechanical parameters, force-extension curves are usually analyzed in terms of models borrowed from theories on polymers. An archetypal example is the wormlike chain model (WLC), that is able to predict the elastic behavior of single-polymer strands up to forces of several hundred piconewtons (Bustamante et al. 1994). Briefly, the stretching force F of a WLC polymer can be described as follows:

$$F = \frac{k_B T}{P} \cdot \left[\frac{1}{4}\left(1 - \frac{x}{L}\right)^{-2} - \frac{1}{4} + \frac{x}{L} \right] \tag{7.35}$$

where P is the persistence length, L is the contour length, k_B is the Boltzmann's constant, and T is the temperature. The variable P is the only adjustable parameter of the model and describes the molecule's stiffness. The model works reasonably well for proteins whose persistence length does not change much within the range of extension forces or with the extent of their unfolding. Alternatives consist of defining a persistence length of the protein for a specific range of forces or of extending the model by adding more WLC elements (Rief, Fernandez, and Gaub 1998).

Another important aspect of the phenomenology associated with protein unfolding has to do with the reversibility of the process. The question to address here is whether the unfolded molecules are capable of *refolding* and thus regaining their native conformation. The answer can be provided by a SPM-based procedure called *force clamp*.

In a force-clamp experiment, the tip pulls a molecule and unfolds it either partially or totally. The tip then stops for a certain time at a given extension value or at a given stretching force. When this period of time is over, the tip resumes pulling the molecule. The curves before and after the so-called equilibration time are then compared. Any unfolding event appearing in both curves is associated with a reversible unfolding–refolding process. Force-clamp experiments at different extents of unfolding or equilibration times can shed light on the dynamics of the refolding (Crampton and Brockwell 2010).

THE SCIENCE OF PUSHING: CONTACT NANOMECHANICS

There are several SPM-based methods to measure local viscoelastic properties, and they can be divided into static and dynamic (Szoszkiewicz and Riedo 2007). The *static* methods rely on force curves, where tip and sample are brought into contact and further pressed due to the application of a normal force or *load*, which is released (or unloaded) immediately afterwards or after a certain period of time. The static methods are usually done at constant loading-unloading rates. In the *dynamic* methods, the sample is either laterally or vertically oscillated as the AFM tip is pressed against it. The oscillation amplitudes are usually very small, just a few angstroms, and the oscillation frequencies are on the order of a few kilohertz (in the case of modulated nanoindentation or SPM-driven microrheology) or on the order of megahertz (in the case of acoustic force atomic microscopy). In this section we will mainly focus on the static methods.

A Short Note on Probes for Nanomechanics

In nanomechanics, the probe geometry especially matters. To derive an accurate mathematical treatment and thus reliably obtain mechanical parameters, the geometry of the probe or indenter must be known with the greatest accuracy. Though widely used in these types of experiments, the microfabricated AFM-cantilevered tips have conical or pyramidal shapes that in some cases may have ill-defined heights, sharpness, or opening angles. The exact geometry can thus only be accurately determined by independent imaging methods such as scanning electron microscopy (SEM). A widespread method to obtain probes of better-defined geometry consists of attaching a spherical microbead of known diameter at the free end of the cantilever. The force curves are then performed with the attached bead and not with the tip. However, a drawback of the so-called *colloidal probe* method is the lack of lateral resolution due to the relatively large size of the bead, usually between 1 and 50 µm in diameter. In practice, both sharp tips and colloidal beads are used to mechanically probe samples more or less locally.

CONTACT REGION OF THE FORCE CURVE: BEYOND THE POINT OF CONTACT

As mentioned previously, the force steadily increases when tip and sample are further pressed to one another beyond the initial contact. The corresponding part of the force curve, the so-called *contact region*, can adopt various shapes. These shapes are intimately related to the mechanical properties of both probe and sample. In most cases the probe is much stiffer than the sample, and the contact region is thus determined by the sample's mechanical properties. Under these conditions, a linear contact region represents the behavior of a rigid, nondeformable surface, whereas nonlinear curves are typical of compliant, deformable surfaces. In the latter case, the tip either penetrates or deforms the sample's surface as the tip and sample are further pressed toward one another. The extent of the deformation is called *indentation*, and it is calculated from the piezo displacement $(Z - Z_0)$ after subtracting the cantilever deflection

$$\delta = (Z - Z_0) - \frac{F}{k} \qquad (7.36)$$

Another valuable piece of information is the degree of overlapping between the approach and the retract curves in the contact region. Figure 7.15 illustrates this point. If the sample deforms reversibly, the retract curve will follow exactly the same path of the approach curve, and thus both curves coincide; if, however, the sample deforms irreversibly or *plastically*, the retract curve will "lag" behind the approach curve. The difference between the approach and the retract curves is called *loading-unloading hysteresis*, and it constitutes a measure of the sample's plasticity.

Just a quick inspection of the force curves is enough to qualitatively estimate the sample's stiffness or compliance, especially when these curves are compared with those obtained from a rigid sample or from a sample's component, i.e., the reference sample used to calibrate the probe. For a quantitative calculation, however, theoretical modeling is required. The widespread approaches to describe a sample's elasticity are the continuum contact mechanics theories that work reasonably well at the mesoscale, since they do not consider the finite nature—either atomic or molecular—of matter. Instead, the theories treat the objects in contact as homogeneous, continuum media with flat, flawless surfaces. A brief account of the models is shown in Table 7.3.

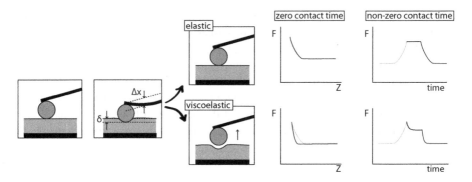

FIGURE 7.15 Testing the mechanics of materials with the AFM. Pressing the material beyond the contact point with a colloidal probe attached to a microcantilever produces a deformation d in the material. This deformation or indentation is brought about by the application of a compressive (in opposite to a tensile) force, $F = k\Delta x$. The application and release of the force can be tracked in the force curves. If the material is elastic, the deformation is reversible, and both approach and retract curves coincide. If the probe remains in contact for a certain time, the force does not vary, the material "stores" the elastic energy. If the material is viscoelastic, the deformation is partly irreversible, and the approach and retract curves exhibit hysteresis. If the probe remains in contact with the sample, the force decays with time, which means some of the energy is dissipated and lost in permanently deforming the sample.

From the models, one can obtain the Young's modulus, E, a characteristic parameter of the sample elasticity and the contact area, which depends on the tip or indenter's geometry. Indeed, the expression that relates the sample stiffness k_s and the E is the following:

$$k_s = 2aE' \tag{7.37}$$

where a is the tip–sample contact radius, and E' is the reduced Young's modulus (Butt, Cappella, and Kappl 2005).

The former quantity is not experimentally accessible through SPM techniques alone, and it is normally assumed to follow the theoretical predictions. E' is related to the Young's moduli of the tip and sample, E_t and E_s, respectively, through

$$\frac{1}{E'} = \frac{1 - v_s^2}{E_s} + \frac{1 - v_t^2}{E_t} \tag{7.38}$$

where v_s and v_t are the Poisson's ratio of sample and tip, respectively. If the tip is much stiffer than the sample ($E_t \gg E_s$), E' can be approximated by the first term of Equation (7.38), and the expression for the sample's stiffness reduces to

$$k_s = 2a\left(\frac{E_s}{1 - v_s^2}\right) \tag{7.39}$$

TABLE 7.3

Summary of the Available Theories That Describe the Contact Mechanics of Elastic Bodies

MODELS	GEOMETRY	FORCE-INDENTATION ($F=f(\delta)$, $\delta=f(F)$)	APPLICATION
Hertz		$F = \dfrac{8GR^{1/2}}{3(1-\nu)}\delta^{3/2}$	- continuum theories - mesoscopic scale: van der Waals (vdW), adhesion and capillary forces are negligible
Sneddon		$F = \dfrac{4Ga\delta}{1-\nu}$ $F = \dfrac{4G\cot\alpha}{\pi(1-\nu)}\delta^2$ $F = \dfrac{8G}{3(1-\nu)}a\delta$ $a=$ contact radius	- soft-elastic surfaces - the flat surface is a half-infinite space - tip-sample forces outside the contact area are **not** considered
Bilodeau		$F = \dfrac{G\alpha}{(1-\nu)\beta}\left[(a^2+\beta^2)\log\dfrac{\beta+a}{\beta-a} - 2a\beta\right]$ $a=$ contact radius $F = \dfrac{1.4906\,G\cot\alpha}{(1-\nu)}\delta^2$	
DMT		$F - F_{adh} = \dfrac{8GR^{1/2}}{3(1-\nu)}\delta^{3/2}$	vdW forces in the contact area; large probes; high adhesion; stiff samples tip-sample forces outside the contact area are **not** considered
JKR		$a = \sqrt[3]{\dfrac{3R(1-\nu)}{8G}\left[F+\dfrac{3}{2}F_{adh}+(3F_{adh}F+(\dfrac{3}{2}F_{adh})^2)^{1/2}\right]}$ $\delta = \dfrac{2}{3}\sqrt{\dfrac{(1-\nu)F_{adh}a}{8GR}}$	short-range forces in the contact area; large, low-stiffness probes; high adhesion; soft samples

Maugis and Pollock		$\bar{F} = \bar{A}^3 - \lambda\bar{A}^2(\sqrt{m^2-1}+\arctan\sqrt{m^2-1})$ $\bar{\delta} = \bar{A}^2 - \dfrac{4}{3}\lambda\bar{A}\sqrt{m^2-1}$
		
		$\bar{F} = \dfrac{2F}{F_{adh}}$ $\lambda = 4-\dfrac{2.06}{D_0}\sqrt[3]{\dfrac{9F^2_{adh}(1-\nu)^2}{4\pi^2RG^2}}$
		
		$\bar{\delta} = \dfrac{4\delta}{\sqrt[3]{\dfrac{9F^2_{adh}(1-\nu)^2}{4RG^2}}}$ $\bar{A} = \dfrac{2a}{\sqrt[3]{\dfrac{3RF_{adh}(1-\nu)}{2G}}}$
			- generalized, parametric expressions
			- large and small probes, large and small adhesion forces
			- tip-sample forces outside the contact area are considered
			- $m=$ ratio of contact radius a to an annular region around that contributes to the adhesion parameter λ
			  $\lambda \to$ infinite => JKR
			  $\lambda \to 0$ => DMT

As mentioned previously, the theories have limited applicability when surface roughness is considered. Certainly, most real surfaces are not infinitely smooth; rather, they have asperities of various sizes that are randomly distributed. The existence of these asperities complicates extremely the calculation of contact areas, since contact between two rough bodies occurs wherever their asperities meet. Real contact areas can thus be either larger or smaller than those calculated from purely elastic and geometric criteria. However, it is possible to get a better approximation of the real contact area by replacing a single asperity by a simple geometrical body, calculate the contact area for that single asperity, and average over all asperities as long as the distribution of asperity heights is known (Buzio and Valbusa 2006).

Another important aspect about the theories listed in Table 7.3 is that they all assume that there is no plastic deformation and thus that the sample behaves as a purely half-infinite elastic body. However, this is not the case for compliant samples, or even for elastic samples, when the approach speed, the indentation, or the load is too high (Szoszkiewicz and Riedo 2007). In this case, permanent deformations occur, which result in loading–unloading hysteresis. In these cases, viscoelastic models should be applied, which take into account the elastic and the viscous nature of the sample as well as finite depth effects (Dimitriadis et al. 2002; Mahaffy et al. 2004).

Several approaches have been tried out in both the SPM-based static and dynamic methods. In the static methods, a widespread procedure consists of modeling the sample's behavior by a mechanical equivalent. The mechanical representation is usually a series of springs and dashpots connected to one another in a specific way. The springs account for the sample's elasticity, whereas the dashpots represent the sample's viscous behavior. A widespread configuration for highly compliant samples such as living cells is the so-called three element model, consisting of a spring connected in series with a single Voigt element (Darling et al. 2007; Radmacher et al. 1996). Applying high loads on cells may affect cell constituents with differing viscoelastic properties, and the model may need to be replaced by a multi-element version (Moreno-Flores et al. 2010). In the dynamic or *rheological* methods, where the indentation is sinusoidally oscillated, the usual approach is to employ a continuum theory of contact mechanics (usually the Hertz theory or similar) that has been extended to account for viscoelastic effects (Mahaffy et al. 2004; Alcaraz et al. 2003).

Alternatively, it is always possible to estimate the extent of plastic deformation by calculating the area between the approach and retract curves: This area is related to the energy that is lost or *dissipated* in the process (see Figure 7.9).

MAPPING INTERACTIONS

FORCE CURVE MAPS

See Figure 7.16. If one combines the capability of the AFM to laterally displace the probe with force curves, it is possible to obtain the spatial distribution of forces together with a sample's topography. In other words, it is possible to obtain *force* images as well as height images.

In force mapping, force displacement curves are performed in predefined positions in the sample, distributed in an orderly fashion within a regular area. This

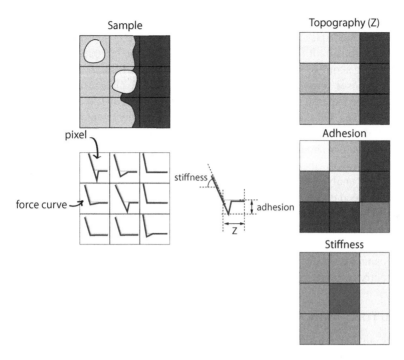

FIGURE 7.16 Force maps. If force curves are systematically performed in a square or rectangular area within the sample, it is possible to produce three-dimensional maps of different magnitudes. The area is in turn divided in small squares or pixels at which a force curve is performed. The slope of the contact region, which is related to the sample's stiffness, or the minimum of the retract curve, i.e., the adhesion force, can be extracted and mapped parallel to the sample's topography. The resulting maps are usually pixilated, the lateral resolution of the maps depends on the size of the pixel but it never outperforms the lateral resolution of conventional AFM imaging.

regular area is usually a square or a rectangle, and the positions are the centers of small squares or *pixels* that result from dividing the area into *a* rows and *b* columns. The number of these pixels determines the lateral resolution of the force image: the higher the number, the better the resolution. Force curves will thus be done at the center of each little square, from the first to the last position, resulting in $a \times b$ pixilated height and force images. The height image reveals the spatial distribution of heights at constant force; this means that the intercepts of the force curve (either the approach or the retract curve) correspond with a horizontal line positioned at a particular force value. The force image shows the distribution of surface interactions at constant height, and it is thus obtained from the intercepts of the force curve with a vertical line positioned at a particular height. The most commonly displayed parameters are the minimum of the retract curves giving out the so-called *adhesion* maps, and the slope of the force region at that position is characteristic of—though not identical to—the sample *stiffness*.

Other types of maps that can be extracted after proper data treatment are elasticity and relaxational multiparametric maps. The former show the spatial distribution of Young's moduli (Radmacher et al. 1996), while the latter show the spatial distribution of load relaxation times and amplitudes (Moreno-Flores et al. 2010). The images are thus built for visualization purposes, but the complete force curves and positions are acquired in a single data set. This is quite advantageous, but at the same time, it imposes serious limitations to the technique. A high-resolution force image is done at the expense of increasing the number of pixels and, therefore, the number of force curves. On the other hand, each force curve must contain a relatively high number of sample points to accurately characterize short-range interactions. This inevitably leads to longer acquisition times, and the images may be distorted due to instrumental drift, thermal drift, or sample transformations. At the same time, the storage and analysis of such high-resolution maps requires high computer memory and high-speed computer processing, which may impair data handling.

In practice, a compromise is reached between lateral resolution and data volume. Pixels cannot be as small as one may want, and in no case should they be smaller than the tip–sample contact area in order to avoid oversampling. On the other hand, pixels cannot be as large as one may want if the sample characteristic features cannot be properly resolved (undersampling).

PULSED FORCE MODE

See Figure 7.17. This technique goes a step further in acquiring adhesion and stiffness maps at lateral resolutions comparable to those typically encountered in AFM imaging. In pulsed-force mode, the tip–sample relative position is sinusoidally oscillated along the normal direction. The frequencies of oscillation are usually low, off-resonance frequencies, usually in the range 100 Hz–2 kHz, and the amplitudes are in the range of 20–500 nm. The extent of the amplitude should be large enough to acquire the complete shape of the force curve (i.e., contact and noncontact regions). The cantilever tip simultaneously scans the area along the X-Y plane in a raster manner. The scanning speed should be low enough to let the tip oscillate a certain number of times at each sample position before moving to the next one. The scanning speed can thus be regulated so that the following relation always holds:

$$\frac{1}{2 \cdot f_{\text{fast axis}}} \geq N_{\text{pixel}} \cdot \frac{1}{f_{\text{osc}}} \tag{7.40}$$

where $f_{\text{fast axis}}$ is the frequency at which two image lines (the forward and backward lines) are scanned, N_{pixel} is the number of pixels per image line, and f_{osc} is the frequency of oscillation of the tip–sample position.

Unlike in force maps, in the so-called analog pulse force mode, only certain values of the force curves are stored instead of the whole curves. Some of these values are indicated in Figure 7.17. This means that part of the information about surface interactions and surface deformations is lost. However, this limitation is overcome by the recent development of the digital pulse force mode, with the capability of storing the complete force curves without loss of resolution.

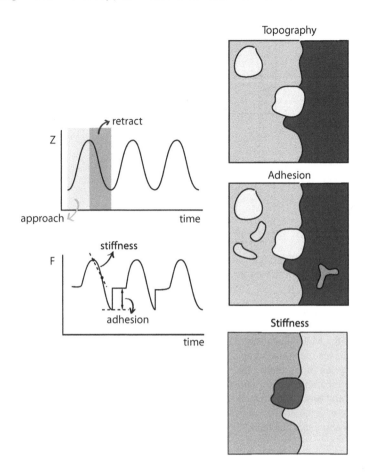

FIGURE 7.17   Pulsed force mode. The piezo position Z is varied sinusoidally as the cantilever scans the sample's surface. The amplitude of the oscillation is such that a complete force curve can be obtained for every cycle as the figure shows. Images of topography, adhesion and stiffness (the latter may be approximate values) can be simultaneously obtained as in force maps. The lateral resolution of such images is however much better and comparable to that of conventional AFM imaging.

MOLECULAR RECOGNITION IMAGING

See Figure 7.18. Although molecular recognition events may be recognizable through conventional force mapping, they can be easily misinterpreted or disguised by non-specific interactions, mainly derived from contact between tip and sample. A way to circumvent the interference consists of applying an off-resonance oscillation at the probe along the normal direction. The frequency is usually lower than the resonance frequency and the amplitude such that the lower half of the cycle should touch the surface and the upper half of the cycle should keep the tip and sample at a distance comparable to x, the characteristic parameter for specific interactions (see

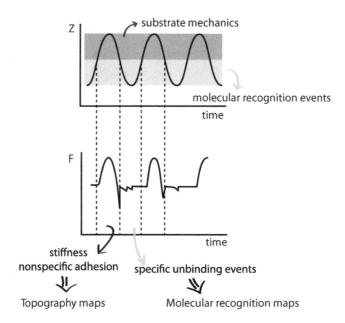

FIGURE 7.18   Molecular recognition imaging. The piezo position Z is varied sinusoidally as the cantilever scans the sample's surface. The amplitude of the oscillation is such that the lower cycles probe molecular recognition events, where the distance between tip and sample is comparable to the characteristic interaction parameter (x). The upper cycle, where the sample and probe are pressed together, is sensitive to substrate mechanics. In this way it is possible to separate the contributions of specific interactions from those of nonspecific interactions or contact mechanics.

Figure 7.18). Decoupling the cantilever deflections that occur on the lower half cycle from those occurring at the upper half cycle allows one to simultaneously obtain topography and molecular recognition maps (Stroh et al. 2004a, 2004b).

SUMMARY

We have seen that SPM is a most convenient and versatile tool to study the interactions between two surfaces. Table 7.4 shows a brief account of the features of the technique. Forces of a most diverse nature—as low as a few tens of piconewtons—can be detected and mapped. Nevertheless, the technique is not exempted of artifacts and limitations. Instrumental drift and noise derived from external perturbations, such as vibrations or acoustic noise, can have a significant effect on the force sensitivity and induce misinterpretation of the results. One of the main limitations of the technique is the determination of the absolute distance between tip and sample. Indeed, the magnitude depends in turn on the contact point. The correct assignment of contact points from force curves is still a most challenging, even controversial task. On the other hand, contact areas are key magnitudes in the characterization of the mechanics of materials that are not accessible from SPM measurements. Instead, theoretical values are often employed, which derive from diverse theories on contact mechanics

TABLE 7.4

Summary of the Features and Limitations of SPM as a Measuring Tool of Surface Interactions

CHARACTERISTICS	SPM (forces)		Force imaging		
	Force curves and force spectroscopy	Chemical Force Microscopy	force mapping	pulsed force mode	molecular recognition imaging
Lateral resolution	tip contact area (10 nm² - 10² μm²)		pixel size (≥ tip contact area)	tip contact area (10 nm² - 10² μm²)	
Vertical resolution	< 1 nm (f(piezoactuator along Z))				
Sample requirements	samples supported on solid substrates				
	entangled (macro)molecules with repetitive patterns (unfolding) samples softer than the cantilever and the solid support (mechanics)	recognizable/reactive chemical groups functionalized tips with equal or different chemical groups			specifically recognizable molecules (i.e. receptors) functionalized tips with specifically recognizable molecules (i.e. ligands)
Force sensitivity	>10 - 10² pN (f(cantilever spring constant))				
Local/global	local				
Contrast	chemical (specific and non-specific interactions) mechanical (mechanics)	chemical and mechanical	chemical (non-specific)		chemical (specific & non-specific)
Other features	quantitative determination of non-contact interactions distance-dependence determines the nature of surface forces mechanical properties of materials: elastic moduli, viscosities unbinding/unfolding constants	- local recognition of chemical end groups - determination of interfacial tensions - local triggering chemical reactions		simultaneous obtention of topography, adhesion and stiffness maps	decoupled specific and non-specific interactions simultaneous obtention of topography and molecular recognition events
Limitations	the absolute tip-sample distance cannot be unambiguously determined contact areas are not accessible				
	reduced dynamic range (force spectroscopy)	applicable to few types of chemical groups		rough estimation of stiffness	multiadhesion events are not considered

of dubious applicability, especially in nonelastic materials. Still, the technique is powerful, and some of its limitations can be readily overcome if combined with the proper techniques, as we will see in Chapter 8.

REFERENCES

Ahimou, F., F. A. Denis, A. Touhami, and Y. F. Dufrene. 2002. Probing microbial cell surface charges by atomic force microscopy. *Langmuir* 18:9937–41.

Alcaraz, J., L. Buscemi, M. Grabulosa, et al. 2003. Microrheology of human lung epithelial cells measured by atomic force microscopy. *Biophys. J.* 84:2071–79.

Alsteens, D., V. Dupres, E. Dague, et al. 2009. Imaging chemical groups and molecular recognition sites on live cells using AFM. In *Applied scanning probe methods XII—Characterization,* ed. B. Bhushan and H. Fuchs, 32–48. New York: Springer-Verlag.

Bustamante, C., J. F. Marko, E. D. Siggia, and S. Smith. 1994. Entropic elasticity of lambda-phage DNA. *Science* 265:1599–1600.

Butt, H.-J., B. Cappella, and M. Kappl. 2005. Force measurements with the atomic force microscope: Technique, interpretation and applications. *Surf. Sci. Rep.* 59:1–152.

Buzio, R., and U. Valbusa. 2006. Morphological and tribological characterization of rough surfaces by atomic force microscopy. In *Applied scanning probe methods III—Characterization,* ed. B. Bhushan and H. Fuchs, 261–98. New York: Springer-Verlag.

Crampton, N., and D. J. Brockwell. 2010. Unravelling the design principles for single protein mechanical strength. *Curr. Op. Struct. Biol.* 20:508–17.

Cumpson, P. J., C. A. Clifford, J. F. Portoles, J. E. Johnstone, and M. Munz. 2008. Cantilever spring-constant calibration in atomic force microscopy. In *Applied scanning probe methods VIII—Scanning probe microscopy techniques*, ed. B. Bhushan, H. Fuchs, and M. Tomitori, 289–314. New York: Springer-Verlag.

Darling, E. M., S. Zauscher, J. A. Block, and F. Guilak. 2007. A thin-layer model for viscoelastic, stress-relaxation testing of cells using atomic force microscopy: Do cell properties reflect metastatic potential? *Biophys. J.* 92:1784–91.

Dimitriadis, E. K., F. Horkay, J. Maresca, B. Kachar, and R. S. Chadwick. 2002. Determination of elastic moduli of thin layers of soft materials using the atomic force microscope. *Biophys. J.* 82:2798–810.

Evans, E., and K. Ritchie. 1997. Dynamic strength of molecular adhesion bonds. *Biophys. J.* 72:1541–55.

———. 1999. Strength of a weak bond connecting flexible polymer chains. *Biophys. J.* 76:2439–47.

Frisbie, C. D., L. F. Rozshayai, A. Noy, M. S. Wrighton, and C. M. Lieber. 1994. Functional group imaging by chemical force microscopy. *Science* 265:2071–74.

Israelachvili, J. 2011. Intermolecular and surface forces, 3rd ed. San Diego, CA: Academic Press.

Kienberger, F., H. Gruber, and P. Hinterdorfer. 2006. Dynamic force microscopy and spectroscopy. In *Applied scanning probe methods II—Scanning probe microscopy techniques*, ed. B. Bhushan and H. Fuchs, 143–64. New York: Springer-Verlag.

Lekka, M., P. Laidler, and A. J. Kulik. 2007. Direct detection of ligand-protein interaction using AFM. In *Applied scanning probe methods VI—Characterization,* ed. B. Bhushan and S. Kawata, 165–203. New York: Springer-Verlag.

Mahaffy, R. E., S. Park, E. Gerde, et al. 2004. Quantitative analysis of the viscoelastic properties of thin regions of fibroblasts using atomic force microscopy. *Biophys. J.* 86:1777–93.

Melzak, K., S. Moreno-Flores, K. Yu, J. Kizhakkedathu, and J. L. Toca-Herrera. 2010. Rationalized approach to the determination of contact point in force-distance curves: Application to polymer brushes in salt solutions and in water. *Microsc. Res. Techn.* 73:959–64.

Moreno-Flores, S., R. Benitez, M. dM. Vivanco, and J. L. Toca-Herrera. 2010. Stress relaxation and creep on living cells with the atomic force microscope: A means to calculate elastic moduli and viscosities of cell components. *Nanotechnology* 21:445101-1/445101-8.

Moreno-Flores, S., A. Shaporenko, C. Vavilala, et al. 2006. Control of surface properties of self-assembled monolayers by tuning the degree of molecular asymmetry. *Surf. Sci.* 600:2847–56.

Radmacher, M., M. Fritz, C. M. Kacher, J. P. Cleveland, and P. K. Hansma. 1996. Measuring the viscoelastic properties of human platelets with the atomic force microscope. *Biophys. J.* 70:556–67.

Rief, M., J. M. Fernandez, and H. E. Gaub. 1998. Elastically coupled two-level systems as a model for biopolymer extensibility. *Phys. Rev. Lett.* 81:4764–67.

Rief, M., M. Gautel, F. Oesterhelt, J. M. Fernandez, and H. E. Gaub. 1997. Reversible unfolding of individual titin immunoglobulin domains by AFM. *Science* 276: 1109–12.

Sinniah, S. K., A. B. Steel, C. J. Miller, J. E. Reutt-Robey. 1996. Solvent exclusion and chemical contrast in scanning force microscopy. *J. Am. Chem. Soc.* 118:8925–31.

Stroh, C. M., A. Ebner, M. Geretschläger, et al. 2004a. Simultaneous topography and recognition imaging using force microscopy. *Biophys. J.* 87:1981–90.

Stroh, C., H. Wang, R. Bash, et al. 2004b. Single-molecule recognition imaging microscopy. *Proc. Natl. Acad. Sci. USA* 101:12503–7.

Szoszkiewicz, R., and E. Riedo. 2007. New AFM developments to study elasticity and adhesion at the nanoscale. In *Applied scanning probe methods V—Scanning probe microscopy techniques*, ed. B. Bhushan, H. Fuchs, and S. Kawata, 269–86. New York: Springer-Verlag.

Verwey, E. J. W., and J. Th. Overbeek. 1948. *Theory of the stability of lyophobic colloids.* Amsterdam: Elsevier.

8 Tidying Loose Ends for the Nano Push–Puller

Microinterferometry and the Film Balance

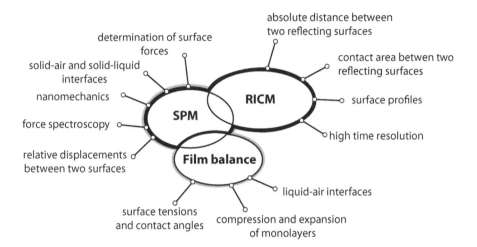

absolute distance between two reflecting surfaces

determination of surface forces

contact area betwen two reflecting surfaces

solid-air and solid-liquid interfaces

nanomechanics

RICM

surface profiles

SPM

force spectroscopy

high time resolution

relative displacements between two surfaces

Film balance

liquid-air interfaces

surface tensions and contact angles

compression and expansion of monolayers

MICROINTERFEROMETRY

We have seen in the previous chapter that surface interactions cannot be accurately characterized without knowing the separation distance between the interacting surfaces. Likewise, the mechanics of contact between two surfaces can only be properly described if the geometry as well as the extension of the contact area is unambiguously known. However, absolute distances and contact areas are not directly measurable by any of the available scanning probe microscopy (SPM) techniques. Instead, they have been the result of assumptions that may lack applicability in some cases. Help is needed, and it readily comes in the shape of an additional optical technique: reflection interference contrast microscopy (RICM). This technique can infer the absolute distance between two nearby surfaces or the geometrical profile of two surfaces in contact from interference patterns.

As Figure 8.1 shows, the interference pattern caused by the reflection of light between two surfaces is a well-known phenomenon, which was studied independently by Robert Hooke and Isaac Newton (Hecht 2002). Coined *Newton rings* after the latter scientist, they appear when monochromatic light impinges onto two

RICM

1664 -- Robert Hooke describes the interference pattern of two reflected surfaces.

1717 -- Isaac Newton studies the interference pattern produced when light is reflected by two surfaces in contact at one point.

1958-1960 -- Van den Tempel and Vasicek respectively present inspiring works on thin-layer optics.

 Van den Tempel, M. 1958. Distance between emulsified oil globules upon coalescence. *J. Colloid Sci.* 13: 125-133

 Vasicek, A. 1960. *Optics of thin films*. Amsterdam: North Holland Publishing Company

1964 -- Curtis applies interference reflection microscopy (IRM) to investigate cell-substrate interactions. The techniques was later coined reflection interference contrast microscopy (RICM).

 Curtis, A.S.G. 1964. The mechanism of adhesion of cells to glass. A study by interference reflection microscopy. *J. Cell Biology* 20:199-215

1975 -- Ploem invents the reflection contrast, also called *antiflex* technique.

 Ploem, J.S. 1975. Reflection contrast microscopy as a tool for investigation of the attachment of living cells to a glass surface. In *Monomolecular phagocytes in immunity, infection and pathology*. Ed. R.V. Furth. Oxford: Blackwell Scientific Publications

1979 -- Gingell and Todd develop the finite aperture interferometry theory: a mathematical framework for IRM data applicable to finite illumination apertures.

 Gingell, D., Todd, I. 1979. Interference reflection microscopy. A quantitative theory for image interpretation and its application to cell-substratum separation measurements. *Biophys. J.* 26:507-526

1993 -- Rädler and Sackmann apply RICM to determine optical thickness, contact profiles and separation distances of supported lipid monolayers and giant lipid vesicles.

 Rädler, J., Sackmann, E. 1993. Imaging Optical thicknesses and separation distances of phospholipid vesicles at solid surfaces. *J. Physique II* 3:727-748

FIGURE 8.1 History of reflection interference contrast microscopy. Referred to as interference contrast microscopy (IRM) in the first years, the technique was developed in the early 1960s and readily applied to one of the most complex systems, such as living cells.

transparent surfaces that are brought together at a point. However, the application of interferometry in microscopy came much later, partly inspired by the works of Van den Tempel and Vasicek on the optics of thin layers at the edge of the 1960s. Curiously, RICM was born and readily applied on one of the most complex systems to tackle: eukaryotic cells. Indeed, the first application of RICM appeared in 1964 as a tool to study cell–substrate interactions by measuring the distance between cells and the supporting substrate (Curtis 1964). Since then, the applicability of RICM has considerably broadened hand in hand with the development of ever better methods for data analysis and interpretation.

The discussion in the following background section will help the reader to understand the basics of microinterferometry.

Background Information

THE SUPERPOSITION OF REFLECTED LIGHT BEAMS

The role of light interference in optics has been explained in Chapter 3 together with some basic relations, such as that between the wavelength of light and the optical path to produce constructive or destructive interference. There we mentioned that two light beams of equal wavelength constructively interfere when the optical path difference is an integer of the light wavelength and destructively interfere when the optical paths differ in an integer of half the light wavelength (see Equations [3.4 and 3.5]). Here we will focus on the interference between two *reflected* light beams of equal wavelength.

Reflection alters the relation between the wavelength of light and the optical path in order to produce either constructive or destructive interference. In particular, this occurs between an *internally* reflected light beam and an *externally* reflected light beam.

We have also seen in Chapter 3 that external reflection occurs when the refractive index of the incident medium is *smaller* than that of the transmitting medium, such as a light beam impinging on a glass slide from the air. Conversely, internal reflection takes place when the refractive index of the incident medium is higher than that of the transmitting medium, i.e., when the light beam impinges the air–glass interface from the glass side. Both internal and external reflection can thus be interpreted as two reverse processes.

The phase difference between an internally reflected wave and an externally reflected wave is 180° (or π) (Hecht 2002). This has an effect on the conditions of constructive and destructive interference. Both beams can be thought as though there were propagating waves with an *initial* phase shift of λ/2. In this case, constructive interference occurs when the optical path difference Δ is an odd integer of half the wavelength

$$\Delta_{\text{consructive}} = (2m+1) \cdot \frac{\lambda}{2}, \quad m = 0, 1, 2, \ldots \tag{8.1}$$

and destructive interference when Δ is an integer of the wavelength

$$\Delta_{\text{destructive}} = m \cdot \lambda, \quad m = 0, 1, 2, \ldots \tag{8.2}$$

These conditions determine the interference pattern in reflection interference contrast microscopy, since interference occurs between externally and internally reflected beams, as we will later see.

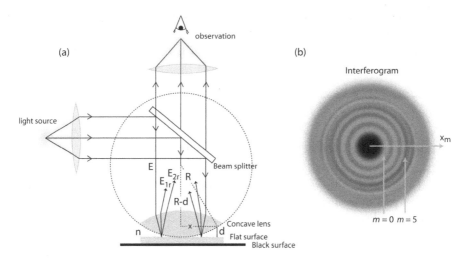

FIGURE 8.2 A setup to observe Newton rings. (a) A quasi-monochromatic source illuminates the optical arrangement consisting of a concave lens in contact with a flat surface at a point. The radius of curvature of the lens is R. The point is focused by the observation lens. Light travels different paths when the distance between the substrate and the lens, d, is nonzero, thus producing interference between the reflected beams. In the scheme only the first two, E_{1r} and E_{2r} are depicted. (b) A typical set of Newton rings seen on a reflection microscope. The contact area appears dark surrounded by various concentric annuli of high (bright) and low (dark) intensity. The order of interference increases and the fringes narrow with the distance to the central dark spot, x. The shape and arrangement of these fringes determine the quality of optical lenses.

NEWTON RINGS

Newton rings are a series of concentric, nearly circular fringes that are formed around the point of contact between two transparent surfaces (Hecht 2002). These can be spheres, objects of concave shape such as a lens, or even flat glasses. The fringe arrangement, or *interferogram*, is the result of the interference between the rays reflected by the surface of the concave lens (internal reflection) and those reflected at the surface of the flat substrate (external reflection). An arrangement to visualize the Newton rings is shown in Figure 8.2a in the case of a lens placed on a flat glass, while Figure 8.2b shows a typical interferogram where the Newton rings are visible. The lens is illuminated at normal incidence with quasi-monochromatic light of wavelength λ_0. Geometrical considerations relate the radius of curvature of the lens, R, with the distance, x, and the separation, d, as follows:

$$x^2 = R^2 - (R-d)^2 \approx 2Rd \quad \text{if } R \gg d \tag{8.3}$$

On the other hand, the mth-order interference maximum occurs when the separation d obeys the expression

$$2nd_m = \left(m + \frac{1}{2} \right) \cdot \lambda_0 \qquad (8.4)$$

Combining the last two expressions, it is possible to obtain the lateral position of the bright and the dark fringe of the mth-order interference

$$x_m^{\text{bright}} = \left[\left(m + \frac{1}{2} \right) \cdot \lambda_0 \cdot R \right]^{\frac{1}{2}} \qquad (8.5)$$

$$x_m^{\text{dark}} = \left(m\lambda_0 R \right)^{\frac{1}{2}}$$

and hence to obtain d by applying Equation (8.3).

If the two surfaces have a good, clean contact, the central fringe should be dark, if observed in reflection.

FUNDAMENTALS OF REFLECTION INTERFERENCE CONTRAST MICROSCOPY

Let us consider an object hovering in a fluid at a certain distance from a glass substrate, as depicted in Figure 8.3a. When illuminated with monochromatic light, part of the beam is reflected at the glass–fluid interface, and part of the transmitted light is reflected at the fluid–object interface. At a certain plane below the glass substrate, both reflected beams will have traveled along paths of different length. Consequently, an interference pattern is created at this plane, which consists of a series of concentric dark and bright fringes whose intensity and sharpness decrease outwards in the radial direction, as shown in Figure 8.3a. If the refractive index of the medium between the substrate and the object is smaller than those of the two, then light at the glass substrate is internally reflected, while surface light at the object is externally reflected. Dark fringes arise due to destructive interference of the reflected beams, and the length difference between their respective paths is an integer of the wavelength of light ($p \cdot \lambda$). On the other hand, bright fringes are the result of constructive interference of the reflected beams, and the path-length difference is thus an odd integer of half the wavelength ($[2p + 1] \cdot \lambda/2$) (Hillner, Radmacher, and Hansma 1995). The integer p determines the order of interference.

As Figure 8.3b intuitively shows, the shape and relative intensity of the interference fringes are strongly dependent on the distance between the object and the glass substrate as well as on the profile of the object's surface that is closest to the substrate. However, extracting such information from the interference patterns is not an easy task. Three main approaches have been developed throughout the years as attempts to mathematically describe the intensity profile along the interference pattern. The first to appear, the *simple or normal-incidence theory* (Curtis 1964), was extended to the *finite-aperture theory*, gaining in applicability and rigor (Gingell and Todd 1979). The more recent *nonlocal theory* (Kühner and Sackmann 1996) appears to be the most comprehensive

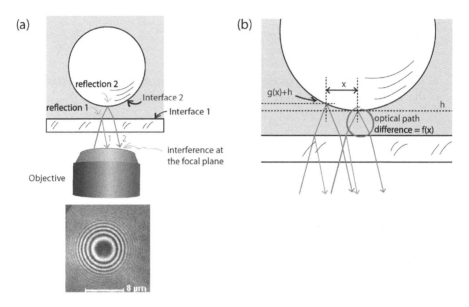

FIGURE 8.3   Basics of RICM. (a) A sphere hovering on a flat substrate. Light is reflected at the interfaces 1 and 2 and collected by the objective. The reflected beams have traveled a different path length and interfere either constructively or destructively. A RICM image is shown below, consisting of a concentric arrangement of bright and dark fringes. The intensity at the centre of the image varies sinusoidally with the distance to the substrate: bright at $\lambda/4$, dark at $\lambda/2$. (RICM photograph reprinted from Dubreuil, Elsner, and Fery 2003. With permission.) (b) Details of the optical path difference and its dependence with the lateral distance, x. The shape of the object $g(x)$ determines the shape of the fringes and thus the radial dependence of the intensity.

of the three. Far from being mutually exclusive, each method has its niche of applicability as long as certain experimental conditions and optical settings are met.

The object of study is common to all three: it has refractive index n_2 and hovers above a glass slide of refractive index n_0 at a height h in a fluid medium of refractive index n_1. The incident light, of intensity I_0, is partly reflected at the glass–medium interface and partly reflected at the glass–object interface. The intensities of the reflected light beams are I_1 and I_2, respectively.

Briefly, the *simple* theory (Kühner and Sackmann 1996) assumes that the illumination has a zero aperture angle. This basically means that the light source produces a beam whose propagation direction is normal to the glass substrate (see Figure 8.4a). In this case, no illumination "cone" is created but, rather, an illumination "cylinder." Additionally, the object's surface is assumed to be parallel to the glass substrate, which means that normal incidence is preserved in all cases. Under these conditions, the intensity profile across the interference pattern can be described as

$$I(g(x,y),\lambda) = I_1 + I_2 + 2\sqrt{I_1 I_2}\,\cos\left[\frac{4\pi n_1(g(x,y)+h)}{\lambda}+\phi\right] \qquad (8.6)$$

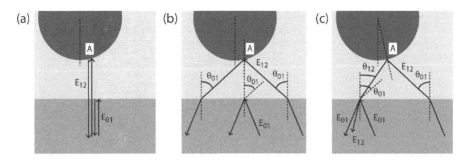

FIGURE 8.4 Models in RICM: 2-D representation. (a) Normal incidence, the simplest and first to appear, considers that parallel rays of light approach the interfaces at the normal direction to the substrate. (b) The finite aperture theory takes into account the angle of incidence θ due to a finite illumination aperture. These models assume that all the interfaces are parallel to the glass substrate. (c) The nonlocal theory is a generalization of the previous two, where the interfaces may have different orientations. (Redrawn from Kühner and Sackmann 1996. With permission.)

In this expression, $g(x,y)$ denotes the geometrical profile of the bottom surface of the object at each coordinate pair (x,y), and h is the distance of the nearest point of the object's surface to the glass slide. The variable φ is the phase shift of the light reflected from the object; this magnitude is equal to π only if the refractive index of the object is higher than that of the medium ($n_2 > n_1$) and the object is thick enough to consider only the contribution of the bottom surface to the interference (Rädler and Sackmann 1992). In the case of stratified or very thin objects, I_2 is the contribution of multiple reflections at the different interfaces within the object, which may lead to an effective phase shift that is unequal to zero or π. This introduces an uncertainty in the determination of h and $g(x,y)$ that can be overcome experimentally, as we will show later.

In the particular case where the object is axisymmetric, such as a sphere of radius R, only one coordinate is necessary to characterize the intensity profile, i.e., x. The sphere profile can be described as $g(x) = R - (R^2 - x^2)^{1/2}$, and Equation (8.6) results in

$$I(x) = I_1 + I_2 + 2\sqrt{I_1 I_2}\, \cos\left[\frac{4\pi n_1}{\lambda}\left(R - \left(R^2 - x^2\right)^{\frac{1}{2}}\right) + \phi\right] \qquad (8.7)$$

However, microscopes work at finite and even high illumination numerical apertures, which means that they produce illumination cones where the light beams hit the glass and the object at different angles (see Figure 8.4b). The cone is sustained by a solid angle Ω or by α in a two-dimensional representation, which in this case is nonzero. Another consequence of the illumination with high apertures is the decaying contrast of the higher-order fringes and the resulting loss of visibility (Rädler and Sackmann 1993). The finite-aperture theory takes all of this into account to produce the following expression for the intensity:

$$I(g(x,y),h) = 4\pi\sin^2\frac{\alpha}{2}\left[I_1 + I_2 + 2\sqrt{I_1 I_2} \cdot \frac{\sin\left(2k(g+h)\sin^2\frac{\alpha}{2}\right)}{2k(g+h)\sin^2\frac{\alpha}{2}}\right.$$

$$\left.\cos\left[2k(g+h)\cdot\left(1-\sin^2\frac{\alpha}{2}\right)+\phi\right]\right] \tag{8.8}$$

Alternatively, the second term of the sum, I_2, and the multiplying term preceding the cosine function can be approximated to Gaussians. In particular, the Gaussian expression for I_2 describes the spatial bending of the background intensity due to diffuse reflections coming from regions farther away from the glass slide, i.e., the upper interface of the object (Rädler and Sackmann 1992).

Equation (8.8) is obtained under the following assumptions. The intensity is constant for all incidence angles within the illumination cone; the reflectivity of the interfaces does not depend on the incidence angle; and the incident light is circularly polarized (Rädler and Sackmann 1993).

Both normal-incidence and finite-aperture theories rely on light reflection at interfaces parallel to the glass slides. The real picture may be different, though, and reflection usually occurs at curved interfaces. In this case, nonlocal theory is to be applied (Kühner and Sackmann 1996) (see Figure 8.4c), and we have two light cones: that of the illumination (sustained by an angle α_{IA}) and that of the detection (sustained by an angle α_{DA}). The expression for the intensity of the interference pattern along the coordinate x has the form

$$I(x) \propto 4\pi\sin^2\left(\frac{\alpha_{IA}}{2}\right)\cdot I_1 +$$

$$\int_{0,0}^{2\pi,\alpha_{DA}} \Theta\left(\alpha_{IA} - f_\vartheta(\Omega_2)\right)\cdot\left[I_2 + 2\sqrt{I_1 I_2}\cos\left(k\delta(\Omega_2,x)\right)\right]\cdot d\Omega_2 \tag{8.9}$$

where Θ and δ are the Heaviside and delta functions,[*] respectively. The Heaviside function ensures that each reflected light wave is caused by an illuminating light wave. The Ω_2 denotes the pair of polar angles of reflection at the medium–object interface, $\Omega_2 = (\varphi_2, \vartheta_2)$, and the function $f_\vartheta(\Omega_2) = \vartheta_1$ correlates the angles of reflection at the glass and at the object surfaces. The integral of Equation (8.9) can only

---

[*] The Heaviside function centered at zero ($Q(0)$ or $H(0)$) is a unit step function whose value is zero for negative values and 1 elsewhere. Within the theory of distributions or generalized functions (Edwards 1995), the derivative of the Heaviside function is the Dirac delta function. The latter, $d(0)$, is the unit impulse function, which is formally defined as a singular distribution that is also a measure. However, such distribution is usually considered to be zero for all values except for zero, where it takes the value of infinity; it can be regarded as the weak limit of a sequence of functions that have a peak in the origin (e.g., Gaussian functions).

be evaluated analytically (Kühner and Sackmann 1996), though it is not exempted from assumptions. Indeed, Equation (8.9) is a result of a series of simplifications. The reflectivity coefficients do not depend on the angle of incidence; the amplitude of the incident beam depends neither on the angle of incidence nor on position; and both incident and reflected light is assumed to be polarized in the same direction (Kühner and Sackmann 1996).

The normal-incidence theory works reasonably well when small illumination numerical apertures (i.e., illumination numerical aperture, INA = 0.42), are used and the objects are thick ($\approx$1 µm or higher; Rädler 1992). In this case, the influence of upper interfaces is minimal, and the rays may be considered to impinge normally to the only contributing interface. For thinner films or smaller objects, the contribution to the interference image of upper interfaces is negligible as long as large illuminating apertures are employed (i.e., INA = 1.2); in this case, the finite-aperture theory proves to be appropriate (Gingell and Todd 1979). This seems to be the case of nonplanar interfaces such as wedge-shaped liquids even for small illumination numerical apertures (INA = 0.42, 0.75) (Wiegand, Neumaier, and Sackmann 1998). The nonlocal theory, also called the *theory of nonplanar interfaces*, seems to respond to a more realistic picture of optical reflections, but it is complex and yields nonanalytical solutions. However, the theory was further applied to wedge-shaped and spherical interfaces, which yielded the best agreement between theory and experiment (Wiegand, Neumaier, and Sackmann 1998). Indeed, the surface microtopography of liquid droplets was reconstructed from the fringe patterns, giving reasonable values of contact angles.

LATERAL AND VERTICAL RESOLUTION IN RICM

As suggested previously, most objects of interest exhibit curved surfaces, which means that the fringes of the interference pattern narrow when the gradient of the object's profile is long, that is to say, $\partial g(x,y)/\partial x$ increases along the x-axis. That is easy to understand if the object is a sphere, whose contour is such that the gradient is zero at the most proximal point to the glass slide and gradually increases as the position is moved toward the sphere side. For small objects and steep gradients, the fringe pattern will consequently reach the limit of the spatial resolution of the microscope, where the fringes will no longer be distinguishable. The limiting gradient or slope is given by (Rädler and Sackmann 1993)

$$\left.\frac{\partial h}{\partial x}\right|_{\text{limiting}} = \frac{n_1}{n_0}\cdot\left(\text{NA} - \text{INA}\right) \tag{8.10}$$

where NA is the numerical aperture of the objective and INA is the illumination numerical aperture.

Some numbers for lateral and vertical resolutions of RICM have been reported in the literature. Lateral resolutions are on the order of 200–300 nm, whereas vertical resolutions are found to vary within 1–4 nm (Stuart and Hlady 1999; Rädler and Sackmann 1993; Kühner and Sackmann 1996).

RICM Setup

An experimental RICM setup is shown in Figure 8.5a. A powerful source of monochromatic light is required, which in most cases consists of a mercury lamp (100–200 W) and a narrow band filter, usually a band-pass or an interference filter that selects the green or the blue emission line of mercury, at 546 nm or 436 nm, respectively. Alternatively, these filters may be located before the detection system. An additional ultraviolet (UV) and infrared (IR) filter can be placed between the mercury lamp and the filter to block the UV and IR radiation that may damage the sample. The light beam reaches a beam splitter and is directed toward the sample (the glass slide and the proximal object) through a special type of objective, the so-called *antiflex* objective. The intensity of the reflected monochromatic light is usually very low, on the order of 0.1%–1% of the intensity of the incident light and is very easily obscured by stray light coming from unwanted reflections at surfaces other than that of the object. The reflection at the glass–air interface between the glass slide and the objective can be easily removed by oil immersion; however, other weak reflections such as those arising within the objective produce enough stray light to greatly alter the RICM image.

One of the ways to minimize stray light consists of reducing the field-stop diaphragm to a small size so that most of the stray light is cut off. However, this does not reduce the reflections inside the objective, and there is always some remnant stray

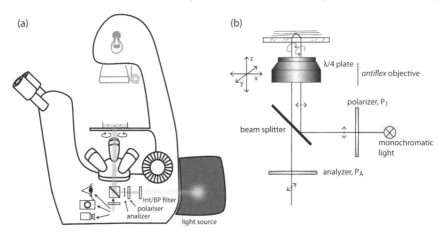

FIGURE 8.5   RICM setup. (a) An inverted microscope is equipped with a strong light source that is rendered monochromatic by an interference (Int) or a band-pass (BP) filter. Light traverses a polarizer and an antiflex objective to produce circularly polarized light. Only the reflected light produced by the object and the substrate is collected by the same objective and traverses the analyzer, which in turn delivers the optical signal to the oculars, a photo camera, or a video camera. (b) The antiflex principle to remove stray light: A polarizer P_1 converts the beam from the monochromatic source into linearly polarized light, which in turn is converted into a circularly polarized light by a quarter-wave (λ/4) plate situated at the objective before reaching the sample. The light is reflected and remains circularly polarized but in the reverse direction. After traversing the beam splitter, only the reflected light coming from the interfaces of interest traverse the crossed analyzer, P_A, to the display device.

light that cannot be removed unless the field-stop diaphragm is narrowed to imprac-ticably low sizes. Alternatively, one can use optical objectives with ring-shaped aper-tures, since most of the stray light comes from the central parts of the lens, but the use of a central beam stop is not recommended in RICM (Verschueren 1985).

The solution came in the shape of the *antiflex* illumination technique, also called *reflection contrast*, developed by Ploem (1975). The principle of the antiflex tech-nique is shown in Figure 8.5b: A polarizer, P_1, is placed in the illumination path, and it produces linearly polarized light out of the incident monochromatic beam. The light beam, in turn, becomes circularly polarized after passing the microscope objective, which is covered with a quarter-wave ($\lambda/4$) plate at the top. Reflection at both interfaces of the sample reverses the direction of polarization of the circularly polarized light, and when it passes the $\lambda/4$ plate a second time, the reflected light becomes linearly polarized again but rotated 90° with respect to the polarization of the incident beam. A second crossed polarizer, called analyzer P_A, is placed in the path of the reflected light. P_A lets pass only the reflected light coming from the object while it blocks the stray light reflected from interfaces below the $\lambda/4$ plate. Therefore, the objectives used in RICM setups are all antiflex type, oil-immersion objectives with typical magnifications of 63× (though 100× and 40× have also been employed) and relatively high numerical apertures (usually 1.25–1.3). The $\lambda/4$ plate is oriented at 45° with respect to the polarizer's direction to get maximum signal intensity (Bereiter-Hahn, Fox, and Thorell 1979).

To further increase the sensitivity and image contrast of objects in contact with the glass substrate, a layer of dielectric material may be deposited on top of the glass slide (Rädler and Sackmann 1993). Indeed, a thin membrane or a sphere in con-tact with a glass slide appear indistinguishable from the background intensity in an RICM image. A dielectric layer, usually of magnesium fluoride (MgF_2), introduces a defined path difference and thus shifts the interference pattern so that the objects become clearly distinguishable from the background. The change in intensity ΔI depends on the thickness of the layer L and can be expressed as

$$\left.\frac{\Delta I}{I}\right|_{d=0} = \frac{2r_{12}r_{23}\sin 2kL}{r_{01}^2 + r_{13}^2 + 2r_{01}r_{13}\cos 2kL} \cdot 2k\Delta d \tag{8.11}$$

where the subscripts 0, 1, 2, and 3 refer to the glass slide, the dielectric layer, the object of study, and the buffer, respectively; k is the wave vector; and Δd is the thickness of the object. According to Equation (8.11), the maximum intensity gain is attained when the thickness of the dielectric coating is equal to $\lambda/8$.

The RICM image is thus formed at the image plane of the objective and acquired by a charge-coupled device (CCD) camera. Especially convenient is the ability to capture the optical images at high time resolutions if one wants to track in real time the varia-tions in position or in shape of the objects under study, e.g., due to Brownian motion or thermal fluctuations. For that purpose, video-rate CCD cameras equipped with image-processing software can acquire and process one RICM image every 10–70 ms (Rädler and Sackmann 1992; Hlady, Pierce, and Pungor 1996; Schilling et al. 2004).

Dual-Wavelength RICM (DW-RICM)

Using two wavelengths instead of one makes it possible to unambiguously determine absolute heights h or the direction of local deflections in undulating membranes, i.e., inward versus outward deformations in vesicles (Schilling et al. 2004). We have already mentioned that the mathematical treatment of the intensity of an interferogram depends on the phase shift, φ, an unknown parameter in the case of stratified media or very thin films (see Equations [8.6–8.8]). If RICM images are simultaneously acquired at two different wavelengths, namely λ_α and λ_β, the phase shift can be eliminated from Equations (8.6 and 8.7). In the case of Newton rings, a unique value of h can be assigned for every pair of values I_α and I_β and the respective intensities of bright and dark fringes, I_{max} and I_{min}

$$I_\beta = \left(I_\alpha^{max} - I_\alpha^{min}\right) - \left(I_\beta^{max} + I_\beta^{min}\right)\cdot\cos\left(\frac{\lambda_\alpha}{\lambda_\beta}\right)\cdot\cos^{-1}\left(\frac{I_\beta^{max} - I_\beta^{min} - I_\alpha}{I_\alpha^{max} + I_\alpha^{min}}\right) \quad (8.12)$$

Equation (8.12) is a parametric equation in h, which yields characteristic plots as that shown in the left-hand side of Figure 8.6. Each point designs a defined height. In the case of giant vesicles where membrane undulations can occur, DW-RICM

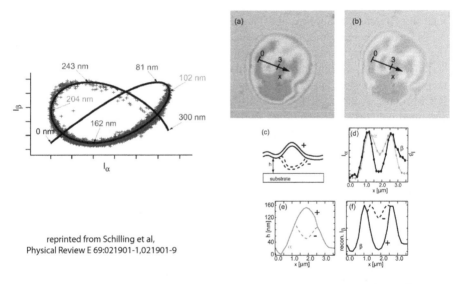

reprinted from Schilling et al,
Physical Review E 69:021901-1,021901-9

FIGURE 8.6 DW-RICM. (*Left*): Parametric plot of the intensity at a wavelength λ_β versus the intensity at the wavelength λ_α. Each point of the graph unambiguously determines a value of the height. (*Right*): (a) and (b) RICM images of buckling vesicles in contact with a glass substrate at $\lambda_\alpha = 546$ nm and $\lambda_\beta = 436$ nm, respectively. (c) The possibilities of an inward (+) or an outward (−) fold. (d) Lateral dependence as obtained from the experiments. (e) Calculated height profile for a (+) and a (−) fold at the wavelength λ_α calculated from Equation (8.12). (f) Reconstructed intensity profiles for both folding scenarios at the wavelength λ_β. Only the profile predicted by the (+) fold agrees with the measured profile at (d). (All figures reprinted from Schilling et al. 2004. With permission.)

can help in distinguishing whether a cavity or a pouch is formed as a result. The right-hand side of Figure 8.6 shows RICM images of a giant vesicle in contact with a glass substrate taken at two wavelengths (546 and 446 nm in Figures 8.6a and 8.6b, respectively). The images show a bright loop in the contact area that is attributed to a likely membrane fold. The image intensity as a function of the coordinate x depicted in the interferogram images at both wavelengths is shown in Figure 8.6d, where both I_α and I_β show a two-peak profile. To find out whether the fold occurs inwards (for a cavity, +) or outwards (for a pouch, −, see Figure 8.6c), the experimental profile is compared to the calculated ones for the respective cases, shown in Figures 8.6e and 8.6f at both wavelengths. Match between the experiment and the calculation only occurs for an outward (+) fold.

APPLICATIONS OF RICM

As we mentioned at the beginning of the chapter, the first applications of RICM focused on the study of the adhesion of living cells on substrates. Cells are particularly challenging, not only due to their structural and optical complexity, but also to the fact that they show very poor contrast when viewed under the microscope. Quantitative RICM was first used to estimate the gap distance between the cell basal membrane and the underlying substrate by applying the normal-incidence theory (Curtis 1964). The study assumed that the contributions of other interfaces, such as the apical or cytoplasmic membrane, were negligible regardless of the cell position, which may not hold true when the cell thickness is below 100 nm (as in pseudopodia) or the illumination aperture is close to zero, in which case the normal-incidence theory is no longer valid. Further attempts to study cell–substrate contact abound in the literature, where the limitations of the technique and thus the validity of the results are also discussed (Verschueren 1985 and references therein). Yet semiquantitative and even qualitative interpretation of the RICM results is possible when the cytoplasmic layer is larger than 1 µm.

Another important application of RICM deals with the determination of optical constants and contact profiles of molecularly thin films or giant lipid vesicles in contact with a dielectric coated-glass substrate (Rädler and Sackmann 1993). To measure both layer thickness and refractive index of an adsorbed layer, a so-called contrast variation experiment is carried out, where the refractive index of the medium is gradually varied by adding another component, i.e., glycerol. The thickness is obtained from the intensity profile in the case where the sample is composed of two reflecting interfaces, the bottom and the upper one. The intensity I_2 that appears in Equations (8.6 and 8.7) depends now on the complex reflection amplitude R, as $I_2 = |R|^2 I_0$, and the phase shift φ, as

$$\phi = \arctan\left(\frac{\mathrm{Im}(R)}{\mathrm{Re}(R)}\right) \approx \frac{\pi}{2} + 2kd,$$

where d is the thickness of the layer. The variable R can be expressed in terms of the reflectivity constant of the layer–fluid interface, r_2, as follows:

$$R = r_2 \cdot \frac{1 + e^{2kd}}{1 + r_2^2 \cdot e^{2kd}} \qquad (8.13)$$

For thin membranes, it is possible to calculate the thickness from the maximum and minimum intensities of the bright and dark fringes, as well as from the background intensity of the membrane interference pattern

$$\frac{I_{max} - I_{min}}{I_1} = \frac{8r_2}{r_1} \sin(kd) \qquad (8.14)$$

This approach has been proved to be more sensitive when the membrane thickness is calculated from giant vesicles, as from supported membranes (Rädler and Sackmann 1993).

RICM can also be used to calculate interaction potentials between surfaces from thermal motions of spheres in close proximity to transparent substrates. The technique can obtain in real time the out-of-plane fluctuations of the particles' position as a function of time, $h(t)$. With this information, it is possible to obtain the height correlation function $<h(\tau) \cdot h(0)>$ and the relative probability density of finding the particle at a certain position, $p(h)$. The function $<h(\tau) \cdot h(0)>$ in turn allows calculation of the effective viscosity of the medium into which the particle is immersed, whereas the probability density function leads to evaluation of the height dependence of the interaction potential around the minimum $U(h)$.

This approach has been validated with latex microbeads in close proximity with a glass substrate (Rädler and Sackmann 1992; Schilling et al. 2004). In these studies, repulsive interaction potentials have been obtained, which can be subsequently analyzed according to the available theories on interaction forces (see Chapter 7 for details). In this way, it is possible to probe repulsive interaction forces between different surfaces and various media, provided that the surfaces are transparent and the coatings used, if any, do not have a significantly different refractive index from the glass underneath (Kühner and Sackmann 1996). The sensitivity in the determination of forces through this method has been determined as long as the mass of the investigated objects was known (Rädler and Sackmann 1992), since the restoring force that opposes the repulsive interaction potential is the gravitational force. For latex spheres, the balancing forces are on the order of 10^{-11}–10^{-12} N, whereas for the lighter giant vesicles the forces are as small as 10^{-15} N.

RICM: Summary

Table 8.1 compiles the characteristics of RICM in its two versions: the conventional or single wavelength-RICM and the dual wavelength-RICM. As an optical technique, RICM shares most of the features and limitations of other optical microscopies (e.g., bright field). The interference-based contrast makes the technique unique in the determination of very small vertical distances, attaining a resolution much beyond the diffraction limit (i.e., $\lambda/200$) without need of sample labeling, as in the case of high-performance fluorescence microscopies. However, the technique is

TABLE 8.1

Summary of Reflection Interference Contrast Microscopy

CHARACTERISTICS	REFLECTION INTERFERENCE CONTRAST MICROSCOPY (RICM)	
	Single-wavelength (SW)	Dual-wavelength (DW)
Lateral resolution	$\lambda/2$ (200 - 300 nm)	
Vertical resolution	$\lambda/200$ (4 - 1 nm)	
Sample requirements	transparent and reflective (coated or uncoated glass, thin mica sheets)	
Illumination volume	10^{18} nm^3	
Local/global technique	laterally: non local; vertically: local	
Contrast	interference - extent of constructive and destructive interference optimal for pure reflecting, non-absorbing materials antiflex technique improves the contrast–reduction of stray light	
Other features	quantitative analysis of RIC micrographs yields: separation distance between an object and a sphere object profile close to the glass substrate contact areas high time resolution (video rates)	unambiguous determination of absolute distance between object and substrate when objects are stratified media or thin films
Limitations	bound to transparent, microsized objects model-based technique quantitative RICM impracticable for objects of non-simple geometry or when more than two interfaces are involved uncertainty in the determination of the absolute distance between an object and a substrate when the former is a stratified system or a thin film	

restricted to transparent substrates, which means either glass or coated glass. Mica can also be used, as long as it is a thin layer and it is deposited on a glass coverslip through an index-matching glue (Rädler and Sackmann 1993). RICM can provide quantitative results in a straightforward manner as long as it is applied to objects of simple geometry (i.e., spheres) and no more than two interfaces are involved.

THE COMBINED SPM-RICM TECHNIQUE

The great potential of RICM in combination with SPM is the complementarity of both techniques to attain quantitative information, especially in the determination of surface forces and their distance dependence. On the one hand, SPM- and AFM (atomic force microscopy)-based techniques, in particular, can measure forces very precisely; on the other hand, RICM can measure separation distances and contact profiles very accurately as long as the geometry of the surfaces involved is simple and the system of interest comprises no more than two interfaces (see Figure 8.7b).

In this case, AFM is once more the technique of choice (see Figure 8.7a). The AFM probe is generally a tipless cantilever with a microsphere, either of latex or of glass, attached to its free end. The so-called *colloidal probe* acts as the "object" in RICM. The microsphere serves two purposes: the geometry of such beads is well defined, which simplifies considerably the interpretation of AFM data; on the other hand, the geometry sphere-plane produces the well-known Newton rings, which can be as well easily interpreted and characterized through the RICM technique and the

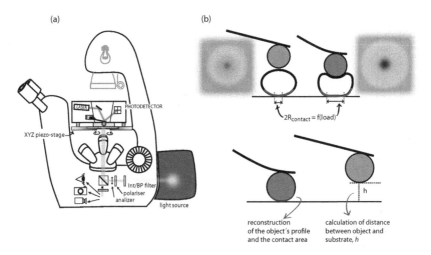

FIGURE 8.7   The combined AFM/RICM. (a) The configuration on an inverted microscope suitable for RICM is compliant to the installation of a sample-moving AFM setup. (b) RICM images can be analyzed to reconstruct contact profiles of simple objects when these are pressed against a hard substrate by a colloidal probe AFM or the contact profile of the colloidal probe itself. The area of contact is one of the parameters that can be obtained. On the other hand, RICM can be used to calculate absolute distance between the colloidal probe and the substrate, which is of particular importance in unveiling surface interactions.

application of the simplest theories. In this case, the object or colloidal probe can be moved upwards and downwards at controlled speed and extent, maintained at a certain height above the substrate, or pressed against the substrate with a certain load. To be visible under the microscope, the bead should be considerably larger than standard AFM tips, which are normally 10–50 μm in diameter (Hillner, Radmacher, and Hansma 1995; Stuart and Hlady 1999; Dubreuil, Elsner, and Fery 2003; Ferri et al. 2008; Erath, Schmidt, and Fery 2010). To avoid losing the optical focus of the object, the AFM is usually of a sample-moving type. A bare, coated-glass slide or any transparent flat support acts as substrate for both techniques. The support rests on the sample stage that is able to move precisely along the three dimensions.

A critical step in integrating both techniques is the synchronization of the vertical movement of the AFM cantilever with the acquisition of RICM images. The speed at which force curves are performed should not be too fast to avoid producing blurred interferograms or compromising the time resolution of the CCD camera and the image processor. Even under controllable speeds, jumps in or out of contact produce sudden changes in the shape of interferograms that cannot be properly tracked, causing discontinuities in the intensity profiles (Hlady, Pierce, and Pungor 1996). At the same time, the speed should not be too slow compared to the acquisition speed of the RICM images in order to avoid the accumulation of redundant images where the vertical position does not change from one image to another (*oversampling*). To determine contact areas and their time variation, the probe should be kept sufficiently long in contact with the substrate in order to acquire the required RICM images. CCD cameras can usually be protected with IR filters that block the laser employed in AFM cantilever alignment (Ferri et al. 2008).

The SPM/RICM combination has been mainly used to test the deformability of microcapsules and microdroplets under the application of forces (Dubreuil, Elsner, and Fery 2003; Ferri et al. 2008) as well as the adhesion forces between beads and substrates of differing surface chemistry and their dependence on the applied load (Erath, Schmidt, and Fery 2010). In the field of force spectroscopy, RICM has been used as an external means to verify the meaning of distance in unbinding events between proteins and ligands (Stuart and Hlady 1999).

Example A: Deformability of Polyelectrolyte Microcapsules Using a Combined AFM/RICM

Dubreuil and coworkers employed AFM and RICM to study the forces required to deform individual capsules made of multilayered polyelectrolytes and to reconstruct the capsules' shape (Dubreuil, Elsner, and Fery 2003). In parallel, they monitored the changes in the shape of the capsule bottom and the contact area with a glass substrate as the capsule was pressed from above with a colloidal AFM probe. Figure 8.A1 shows a typical force-deformation curve as well as RICM images at different stages of the deformation when a single capsule is subjected to an axisymmetric deformation by a colloidal AFM probe. The capsule is immobilized on a PEI-coated glass slide. Three regions are to be observed in the curve. At small deformations, the force is proportional to the deformation, and hence the capsule behaves elastically; at intermediate deformations, the contact area increases which

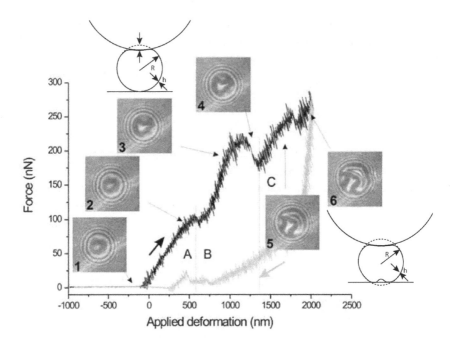

FIGURE 8.A1 Force-versus-deformation curve on a single polyelectrolyte capsule and the accompanying changes in the capsule shape as shown by RICM. The three deformation regions, A, B, and C, are depicted in the figure. Region A comprises the small deformation region where the capsule behaves elastically and Young's moduli can be determined. (Reprinted from Dubreuil, Elsner, and Fery 2003. With permission.)

means that the capsule deforms on top and at the bottom; at even higher deformations, the capsule finally buckles in the contact area. The Young's modulus of the single capsules was determined from the slope (S) of the linear part of the force-deformation curve, in the small deformation regime. The capsules were assumed to be isotropic objects of radius R, thickness h and composed of a homogeneous, continuous material characterized by a Young's modulus (E) and a Poisson ratio (v). Under these conditions, E can be calculated from the following expression:

$$S = \frac{AEh^2}{R} \tag{8.15}$$

where A is the contact area that depends on Poisson's ratio and has the form

$$A = \frac{1}{1+v}\sqrt{6(1-v)}$$

or

$$A = \frac{1}{1-v}\sqrt{6(1-v)}$$

if the capsule is subjected to shear or stretch deformation, respectively. The so-calculated Young's moduli range between 1.3 and 1.9 GPa for 10-μm-diameter capsules.

THE FILM BALANCE AND AIR–FLUID INTERFACES

We hope to have sufficiently probed the high versatility of the SPM techniques as measuring tools of various surface properties under different environmental conditions. However, the techniques are constrained in the type of interface they are able to study, for only solid surfaces or samples firmly attached to solid supports can be properly measured. Air–fluid or fluid–fluid interfaces are particularly challenging to be investigated with cantilever-type probes because they are highly dynamic and fragile in most cases. Fluid surfaces in air can be easily disrupted by capillary forces, which makes their study by conventional SPM techniques fairly impracticable.

The interest in fluid interfaces increased in the last quarter of the eighteenth century (see Figure 8.8). Benjamin Franklin, followed by Lord Rayleigh a century later, concentrated on the spreading of oil on water and suggested a likely structure at the molecular scale. The main experimental work was done by an amateur scientist, a housewife called Agnes Pockels, who developed the predecessor of the film balance and intensively tried to measure surface tensions and create monolayers of diverse substances at the air–water interface. Following Pockels's work, Irving Langmuir improved the apparatus, which carries his name since then. Langmuir, together with his student Katherine Blodgett, developed the technique to transfer monolayers onto solid supports. Today, the film balance and the so-called Langmuir-Blodgett technique of monolayer transfer are of widespread use in surface science.

The discussion in the following background section will help the reader to understand the film balance.

Background Information

INTERFACIAL TENSION

Interfacial tension, also called surface tension, γ, is a magnitude characteristic of the boundary between two homogeneous media and $_{ni}$ is the number of mole species. Thermodynamically, it is defined as follows:

$$\gamma = \left(\frac{\partial G}{\partial A} \right)_{T,p,ni} \tag{8.16}$$

where G is the free Gibbs' energy and A is the area of the interface. In other words, γ is the energy required to maintain the boundary between two phases, and it is defined as the change of free energy when the area of the interface is increased at constant temperature, pressure, and composition. It has units of

Langmuir/film balance

1774 -- Benjamin Franklin describes the spreading of oil films on the water surface

Franklin, b., Brownrigg, W., Farish, M. 1774. Of the stilling of waves by means of oil. Philosophical Transactions 64:445-460

1890 -- Lord Rayleigh suggests that the spreading of oil results in a monolayer of oil molecules.

Rayleigh, F.R.S. 1890. Measurements of the amount of oil necessary in order to check the motions of camphor upon water. *Proc. R. Soc.* 47:364-367

1891-1894 -- Agnes Pockels develops an apparatus to measure the surface tension of substances and applies it to measure the amount of various materials to form a monolayer. She comments on purity and clenliness required to perform accurate measurements of surface tension.

Pockels, A. 1891. Surface Tension. *Nature* 43:437-439

Pockels, A. 1892. On the relative contamination of the water surface by equal quantities of different substances. *Nature* 47:418-419

Pockels, A. 1893. Relations betwwen the surface tension and relative contamination of water surfaces. *Nature* 48:152-154

Pockels, A. 1894. On the spreading of oil upon water. *Nature* 50:223-224

1917 -- Irving Langmuir improves Pockels´ apparatus and describes at the same time as Harkins the orientation of amphiphilic molecules in monolayers

Langmuir, I. 1917. The constitution and fundamental properties of solids and liquids. II.Liquids. *J. Am. Chem. Soc.* 39:856-879

Harkins, W.D. 1917. The evolution of the elements and the stability of complex atoms. *J. Am. Chem. Soc.* 39:856-879

1920 -- Langmuir develops the transfer of amphiphilic films at the water-air interface to solid substrates.   Nobel Prize (1932)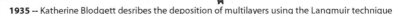

1935 -- Katherine Blodgett desribes the deposition of multilayers using the Langmuir technique

Blodgett, K.B. 1935. Films built by despositing successive monomolecular layers on a solid surface. *J. Am. Chem. Soc.* 57:1007-1022

FIGURE 8.8   History of the film balance in a timeline. Interfacial science starts with observations of everyday-life events like oil films on water, followed by the dedicated work of a housewife and the scientific improvements of Langmuir.

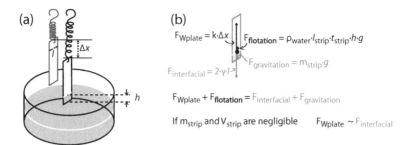

FIGURE 8.9 Measuring surface tension of liquids with a Wilhelmy plate. (a) A strip of light material attached to a spring is immersed in the liquid. The spring's constant is k and the strip has a thickness t_{strip} and a width l_{strip}. The strip has a mass m_{strip} and the length of the strip immersed in the liquid is referred to as h. Due to the surface tension and the gravitational force, the spring elongates a distance Δx. (b) Balance of forces. At mechanical equilibrium, the restoring force, F_{Wplate}, and the flotation force, $F_{flotation}$, balance the interfacial force $F_{interfacial}$, and the gravitational force, $F_{gravitational}$. If the strip is thin and light, flotation and gravitational forces are negligible, and the interfacial tension can be calculated from F_{Wplate} and l_{strip}.

energy per unit area or force per unit length. An alternative interpretation of the surface tension is therefore a force that opposes any attempt to alter the interface. Its magnitude is γ and its direction is normal to the surface pointing downwards to the bulk of the liquid.

The surface tension of liquids is thus a property defined at the air–liquid interface, which may depend on pressure and temperature. A common and simple method to measure surface tension is the so-called *Wilhelmy plate*.

A thin strip of a material that is wetted by the fluid is placed across the interface, as seen in Figure 8.9a. The strip of width l is connected by a spring that measures the normal force in a similar way as a dynamometer does. The strip disturbs the interface, creating a three-phase line of length $2l$ and, due to the surface tension, it will be drawn downwards, opposing the restoring force of the spring. At mechanical equilibrium we have

$$F_{Wplate} = 2\gamma l \tag{8.17}$$

where F_{Wplate} is the force measured by the Wilhelmy plate. This magnitude may be corrected by the presence of gravitational and flotation forces (Figure 8.9b), which may be negligible if the strip is small and made of a light material, such as filter paper.

AMPHIPHILIC MOLECULES: SURFACTANTS AND LIPIDS

Etymologically the term *amphiphilic* comes from the Greek *amphi* meaning "both" and *philia* meaning "love." Which "both" does the word refer to? *Amphi* may implicitly refer to two *opposing* concepts: in this case, two chemically opposing environments, such as water and oil. These media are immiscible, and most

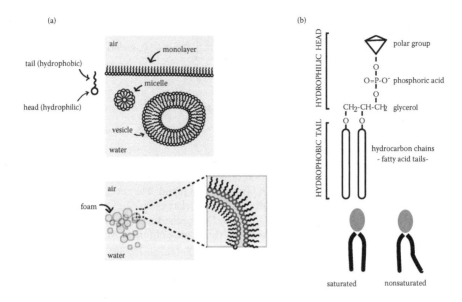

FIGURE 8.10 *(See color insert.)* Amphiphilic molecules. (a) A surfactant is an amphiphilic molecule composed of a hydrophilic head and a hydrophobic tail. Surfactants can form stable structures in both hydrophobic and hydrophilic media, such as micelles, vesicles or foams. At the interface between a hydrophobic and a hydrophilic fluid, such as air and water, they readily form monolayers. (b) Phospholipids are amphiphilic molecules of particular biological value. The chemical structure shows that their head is composed of glycerol, a phosphoric group and a polar (or ionic) group. The hydrocarbon chains are the tails of fatty acids linked to the glycerol liked through ester bonds. If the hydrocarbon chains do not contain double bonds, the lipids are said to be *saturated*. Otherwise, they are nonsaturated. The double bond impairs the free rotation of the carbon–carbon bond, and hence reduces the flexibility of the hydrophobic tail. The presence of the double bond is depicted as a kink in the tail.

of the substances that are "water-loving" (or *hydrophilic*) are not "oil-loving" (*hydrophobic*) and vice versa—except for the amphiphilic molecules.

Amphiphilic molecules are thermodynamically stable in both hydrophobic and hydrophilic media. An archetype of an amphiphilic molecule is a surfactant, which is the main component of soaps and detergents. Amphiphilics are usually organic compounds whose chemical structure is characterized by one or two lengthy hydrocarbon chains: the *hydrophobic tail* and a rather compact group—either polar, ionic, or zwitterionic*—that is the *hydrophilic head*. A surfactant molecule is depicted in Figure 8.10a.

When dissolved in polar or nonpolar liquids at a concentration above the so-called critical micellar concentration (CMC), they may form micelles, vesicles, or foams (Figure 8.10a). In each of these aggregates, tails associate to tails and heads associate to heads in order to maximize the attractive interactions and hence

---

* Zwitterionic = a compound that has one positive and one negative charge.

their thermodynamic stability. Depending on the nature of the liquid, either heads or tails may be exposed (Berg 2010).

These molecules are also surface-active, which means that they readily rest on interfaces between two immiscible fluids. The most typical of all is the water–air interface; on such an interface, these molecules can organize in a way that their tails are oriented toward the air, whereas the heads are exposed to the water. Such a molecular assembly can be huge along the plane of the surface but no more than a molecule thick. Therefore it is called *monolayer.*

A macroscopic consequence of such behavior is a strong decrease of the surface tension of water. Pure water has a large surface tension when exposed to a hydrophobic fluid such as air (71.99 mN/m at 25°; Butt, Graf, and Kappl 2003). This magnitude drops significantly even if only a few amphiphilic molecules are present at the interface. The difference between the new surface tension γ and the surface tension of the pure liquid γ_0 is called surface pressure (Berg 2010)

$$\pi = \gamma_0 - \gamma \qquad (8.18)$$

From all the known surfactants, *phospholipids* are the best-characterized surfactants. Their importance is not small, for they are the main constituents of biological membranes. The tails are hydrocarbon chains, where the carbon atoms are joined by single bonds (*saturated lipids*), although they may contain some double bonds (*nonsaturated lipids*), as seen in Figure 8.10b. The presence of double bonds has dramatic consequences in the flexibility of the molecule as well as in the stiffness of the corresponding aggregates. The chains are joined to a glycerol molecule through one ester bond each. The head is composed of a polar or ionic group joined to the tail through a phosphodiester group, which is deprotonated and hence negatively charged at physiological pH (7.0).

FUNDAMENTALS OF THE FILM BALANCE

The *film balance* is a simple and yet brilliant tool to study monolayers at the air–fluid interface (Butt, Graf, and Kappl 2003). The balance operates under a very intuitive principle, which is shown in Figure 8.11. A trough is filled with a polar liquid onto which a defined number of molecules is deposited. The liquid is referred to as the *subphase*, and it is mostly water or an aqueous solution. The molecules should be surface-active so that they remain at the air–fluid interface instead of diffusing into the liquid. The molecules spread over the available surface of the liquid, which in turn can be expanded or contracted by means of a movable barrier. As the latter moves toward the further end of the trough, the liquid surface gets smaller and the molecules are forced near to each other. The barrier is then said to compress the monolayer. Conversely, if the barrier moves in the opposite direction, the liquid surface becomes larger, and the molecules in the monolayer can loosely move around. The barrier is thus said to expand the monolayer.

The setup is completed with a Wilhelmy plate that measures the surface tension, γ, of the air–fluid interface as the barrier compresses or expands the monolayer.

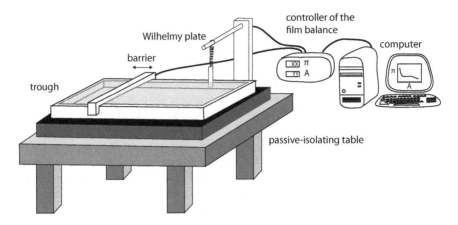

FIGURE 8.11 *(See color insert.)* Setup of a film balance. A trough usually made of polytetrafluoroethylene (PTFE) is filled with a liquid (subphase). The area of the liquid surface is varied by a barrier (or two) that can be laterally moved at a controlled speed. The surface tension is measured at all times by a Wilhelmy plate connected in turn to the controller. The computer registers the values of the surface pressure and the trough area to produce the Langmuir isotherms. To avoid mechanical vibrations the trough is placed on a passive vibration-isolating table and usually enclosed in a protective hood to avoid dirt being deposited on the surface.

The surface pressure, π, is defined as the difference between the surface tension of the bare subphase, γ_0, and the surface tension of the subphase covered with molecules, γ (see Equation [8.18]). If the number of molecules at the air–fluid interface is kept constant during compression and expansion, it is possible to determine the area per molecule a by simply dividing the available area A by the number of molecules at the interface, N^*. The variations of the surface pressure of the air–fluid interface as the area per molecule is increased or decreased characterize the phase behavior of the monolayer. They constitute the *Langmuir* or *pressure isotherm*. Langmuir isotherms can thus be acquired during compression and/or during expansion.

The operation of the film balance is simple yet demanding in the preparations. For an optimal performance of the trough, it must be placed in a quiet, dust-free environment. Mechanical vibrations produce waves at the fluid surface, which can ruin the experiment. Therefore, the film balance is usually placed in a massive, vibration-isolating table. Moreover, any dust particle or impurity in the air or in the balance can sit on the interface and alter the readings, thus compromising the reproducibility of the results. To minimize the influence of dirt or impurities, the film balance is usually enclosed in a transparent hood that is thoroughly cleaned before the experiment. This also applies to the subphase surface, which should be as clean as possible. We have already seen that the surface tension of a liquid is most sensitive to the presence of a few molecules of a foreign substance. The usual way to remove these impurities is to

---

* N is known, since the molecules are dosed by depositing a defined volume of molecule-containing solution of known concentration, c. The solvent of such a solution is volatile and immiscible with the subphase. This ensures that only solute molecules are present at the interface.

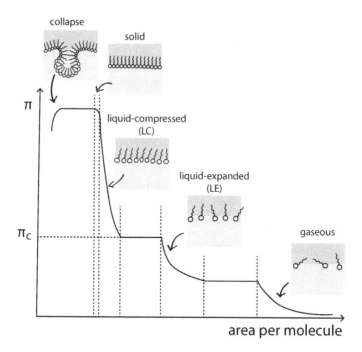

FIGURE 8.12 The Langmuir isotherm. The typical result of a film balance experiment, where the surface pressure is plotted against the area per molecule, either when the film is compressed or when it is expanded. All possible phase transitions that may occur are depicted in the graph, though amphiphiles may not show all. Phases are referred to as gaseous, liquid or solid, separated by plateau regions where phase transitions occur.

suck off the outermost layer of liquid with a small aspirator until the measured value of the surface tension is constant with time and equal to that of the liquid (Berg 2010).

A typical plot of pressure versus surface area per molecule is shown in Figure 8.12. Each discontinuity of the curve represents a possible phase change in the monolayer. In practice, monolayers may not show all phases. The phases are referred to as gaseous, liquid, or solid, in analogy to the three-dimensional case. Gaseous monolayers exist at large molecular areas; the molecules in this state can freely move just as molecules in a gas. Further compression leads to a liquid-like phase, where the surface pressure depends more strongly on the molecular area. Under these conditions, the molecules are forced to be closer to one another, and thus the lateral intermolecular interactions increase. Two types of liquid phase exist: a liquid-expanded phase (LE or L_1) and a liquid-compressed phase (LC or L_2), usually separated by a plateau at a critical surface pressure π_c.[*] In the liquid-expanded phase, there is no molecular order, and the molecules are highly solvated; whereas in the liquid-compressed phase, the molecules, if linear, may start organizing in a tilted manner to the subphase. The chains may subsequently adopt a normal orientation to the subphase as the film is further compressed, and the heads may get less solvated. In

---

[*] More precisely, there are numerous liquid condensed phases that differ in lattice type, chain tilt orientation, and positional correlation length. There are as well many types of solid phases.

the solid phase, the pressure-area isotherm is rather linear, with areas per molecule at zero pressure quite close to the molecular cross sections. In this state, the molecules are tightly packed. If the film is further compressed, the surface pressure may either not further change or suddenly drop. This indicates that the film is collapsing, with the monolayer buckling inwards and with a likely loss of material that may flow down into the subphase (Berg 2010).

THE TRANSFER OF MONOLAYERS ONTO SOLID SUPPORTS

The Langmuir-Blodgett Technique

Monolayers can be transferred to solid supports by the Langmuir-Blodgett technique, shown in Figure 8.13a. The support, a hydrophilic substrate, is initially immersed in the subphase and oriented normal to the air–fluid interface. If the support is vertically moved out of the subphase at a controlled speed, the molecules of the monolayer may adsorb on the support. This produces a drop in the surface tension due to the reducing number of molecules that remain in the monolayer. The decrease in π is readily compensated by compressing the monolayer so that even more molecules can be transferred to the substrate. This coating procedure is thus done at constant surface pressure and results in a layer of molecules adsorbed on the hydrophilic substrate by their hydrophilic parts, whereas the hydrophobic parts are exposed to the air. If the procedure is repeated, multilayers are produced where the molecules of consecutive layers are either head-to-head or tail-to-tail oriented.

The Langmuir-Schaeffer Technique

An alternative way to deposit monolayers on solid supports consists of placing a hydrophobic solid support on top of the monolayer and lifting it afterwards, as depicted in Figure 8.13b. In this case, the monolayer may be horizontally transferred. The result is a monolayer of molecules whose hydrophobic parts are oriented toward

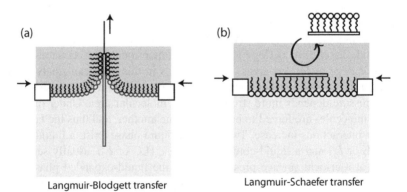

FIGURE 8.13   Transfer of monolayers on solid supports. (a) The Langmuir-Blodgett technique uses a vertically oriented substrate, initially immersed in the subphase. Molecules are transferred to the substrate heads-on as the picture shows. (b) In the Langmuir-Schaeffer technique, the substrate is placed over the monolayer and flipped over to produce a horizontal, tails-on transfer of the molecules.

TABLE 8.2
Summary of the Film Balance

CHARACTERISTICS	FILM BALANCE
Lateral sensitivity	molecular cross sections ($Å^2$)
Vertical sensitivity	molecular lengths (nm)
Sample requirements	surface-active molecules(e.g., surfactants), two immiscible liquids (e.g., oil and water)
Local/global technique	laterally: non local vertically: local (air-fluid interface)
Contrast	surface pressure, π
Other features	- simple, non-destructive - determination of surface tensions of liquids - langmuir isotherms: phase behaviour of monolayers as a function of temperature and area per molecule - measuring and preparative technique: can be combined with monolayer transfer (e.g., Langmuir-Blodgett, Langmuir-Schaeffer)
Limitations	- extremely sensitive to mechanical vibrations and dirt - restricted to monolayers and air-fluid interfaces

the substrate, whereas the hydrophilic parts are exposed to air. Again, if this procedure is repeated at constant surface pressure, multilayers can be deposited one on top of the other in a similar manner as described previously.

THE FILM BALANCE: SUMMARY

Table 8.2 summarizes the characteristics of the film balance, a molecularly sensitive technique that measures macroscopic magnitudes such as surface pressure and surface area. Though restricted to air–fluid interfaces and to the characterization of monolayers, the film balance can be used either as a measuring or as a preparative technique. The latter application refers to the different transfer methods that make use of the film balance to produce mono- or multilayers on solid supports.

THE COMBINED AFM + FILM BALANCE: THE MONOLAYER PARTICLE INTERACTION APPARATUS (MPIA)

Far from what may be expected, AFM has been extensively used to characterize monolayers produced in the Langmuir film balance. However, the monolayers previously had to be transferred to a solid support. Film transfer was usually done by either the Langmuir-Blodgett or the Langmuir-Schaeffer techniques, which may alter the nature of the monolayer. Reorganization, dehydration, or condensation may occur in the film, and in any case, the so-transferred films can usually only be investigated from the air-exposed side, as shown in Figure 8.13. Transfer does not allow study of the deformability or stiffness of the monolayer as it is at the air–liquid interface, since the presence of a hard support largely influences these properties.

This last issue can be tackled with a combined AFM-based technique and a Langmuir film balance. The setup has been reported recently (Graf et al. 2010) and was readily applied to study the interactions of lipids and proteins as well as the deformability of lipid and polymer monolayers (McNamee et al. 2010a, 2010b).

The setup, shown in Figure 8.14, comprises an AFM-type force measurement device and a Langmuir film balance. A polytetrafluoroethylene (PTFE)-coated Langmuir trough equipped with a Wilhelmy system and two movable barriers was mounted above a closed-loop piezo actuator. The latter had an operating range of 12 μm and drove the AFM probe up or down, either toward or away from the liquid–air interface at the subphase side. This was done through a connecting rod made of a chemically inert material (i.e., polyether ether ketone, PEEK). The rod holds, in turn, a small platform made of the same material onto which the AFM probe is fixed in such a way that the cantilever is tilted with respect to the air–liquid interface. The rod traverses the bottom of the trough through an optically transparent window, sealed with an O-ring to avoid leakage of liquid. The purpose of the window is to facilitate the passage of the laser beam in and out of the trough, which is required for the detection of the AFM cantilever deflection. This means that the laser beam enters the trough, reflects upon the back side of the cantilever, and gets back out of the trough to the position-sensitive detector (PSD). The AFM probe is a colloidal particle, usually a microsized sphere made of silica or borosilicate glass. The setup can be complemented with a temperature control unit that allows to keep the subphase at a constant temperature and an upright microscope for optical inspection of the monolayers (Graf et al. 2010).

The combined instrumentation can measure forces between a colloidal particle and a monolayer at the air–liquid interface as a function of temperature, surface pressure, or area per molecule. The MPIA outperforms the standard AFM-based techniques on transferred monolayers in the sense that it allows studying the dynamics of the monolayers, an intrinsic property that is ruined when transferred to solid supports. Indeed, the kinetics of molecular rearrangements within the monolayer can be correlated to the time dependence of the surface interactions as the monolayer is expanded or compressed. Although the range of operative speeds in the recently developed technique is relatively small due to its high sensitivity to interface oscillations and the effect of hydrodynamic forces (McNamee et al. 2010b), it is, to date, the only available technique that can properly address such issues.

Example B: Measuring Interactions of Phospholipid Monolayers at the Air–Water Interface

McNamee and coauthors (2010b) studied the interaction forces between a silica or borosilicate particle and a phospholipid monolayer in water with the MPIA. The monolayers of interest were composed of phospholipids with the same hydrophilic head and hydrophobic tails of differing length. The authors thus focused on the effect of the tail length and the surface pressure on the adhesive forces to a hydrophilic microsphere, which helped in gaining insight into the relation of molecular structure and monolayer stiffness. The stiffness was estimated by comparing the contact regions of force curves obtained when the particle is pressed against the monolayer and when the same particle is pressed against a hard substrate. Jumpins observed in the force curves were attributed to binding of lipids to the bead, which in turn covers the latter with hydrophobic tails. This induces an increase in the hydrophobicity of the sphere and thus the formation of a three-phase contact line (see inset of Figure 8.14 and Figure 8.B1) with a defined contact angle, θ

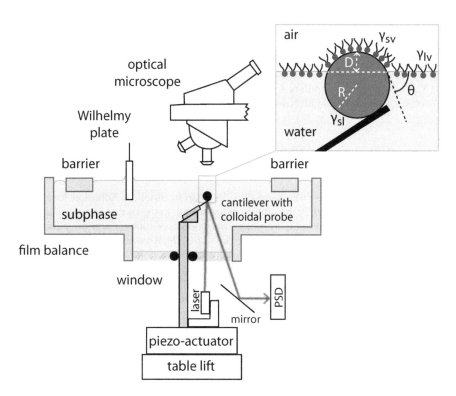

FIGURE 8.14 The particle monolayer interaction apparatus (PMIA). The device allows us to study interaction forces between a colloidal probe and a monolayer at the air–liquid interface. The PMIA combines a film balance (a PTFE trough equipped with two movable barriers and a Wilhelmy plate) and an AFM-like device (a piezo actuator and an optical position sensitive detector [PSD]). Forces are measured from the cantilever deflection as the colloidal probe is approached to the interface. The inset shows a detail of how the sphere may deform the monolayer and consequently produce a three-phase line between the liquid subphase, the air, and the solid sphere. A contact angle, θ, characterizes the wetting of the sphere by the monolayer and it can be obtained from the jump-in distance in the force curves, D, and the radius of the sphere, R. (Redrawn from Graf et al. 2010. With permission.)

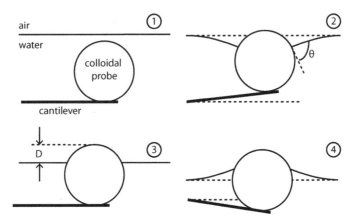

FIGURE 8.B1 The scenarios for a particle before (1) and in contact (2–4) with an air–water interface. A snap-in in the force curve (2) occurs if the monolayer readily wets the particle. The adhesion is the result of attractive interactions between the monolayer and the sphere. (3) The case where no capillary forces occur and (4) when the sphere further presses the monolayer. (Redrawn from McNamee et al. 2010b. With permission.)

$$\cos\theta = \frac{R-D}{R} \tag{8.19}$$

where R is the sphere radius and D is the snap-in distance. Adhesion forces of the retract curves quantify the affinity of the sphere to the monolayer. A summary of the results is shown in Figure 8.B2. The monolayer stiffness and the adhesion forces decrease with increasing surface pressure, which means that the ordering of the molecules leads to a higher deformability of the monolayer and that the hydrophilic heads determine the interaction with the sphere in water. As the hydrophobic tail increases, the adhesion decreases while the stiffness increases. However, the authors claim not to have a satisfactory explanation for such an observation.

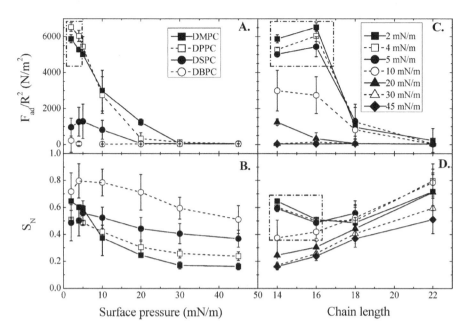

FIGURE 8.B2 Adhesion forces (normalized to R^2) and monolayer stiffness as a function of surface pressure and length of the hydrophobic tails. (Reprinted from McNamee et al. 2010b. With permission.)

REFERENCES

Bereiter-Hahn, J., C. H. Fox, and B. Thorell. 1979. Quantitative reflection contrast microscopy of living cells. *J. Cell Biol.* 82:767–79.

Berg, J. C. 2010. *An introduction to interfaces and colloids: The bridge to nanoscience.* Singapore: Word Scientific.

Butt, H.-J., K. Graf, and M. Kappl. 2003. *Physics and chemistry of interfaces.* Weinheim, Germany: Wiley-VCH.

Curtis, A. S. G. 1964. The mechanism of adhesion of cells to glass: A study by interference reflection microscopy. *J. Cell Biol.* 20:199–215.

Dubreuil, F., N. Elsner, and A. Fery. 2003. Elastic properties of polyelectrolyte capsules studied by atomic-force microscopy and RICM. *Eur. Phys. J. E* 12:215–21.

Edwards, R. E. 1995. *Functional analysis theory and applications.* New York: Dover.

Erath, J., S. Schmidt, and A. Fery. 2010. Characterization of adhesion phenomena and contact of surfaces by soft colloidal probe AFM. *Soft Matter* 6:1432–37.

Ferri, J. K., P. Carl, N. Gorevski, et al. 2008. Separation membrane and surface tension contributions in Pickering droplet deformation. *Soft Matter* 4:2259–66.

Gingell, D., and I. Todd. 1979. Interference reflection microscopy: A quantitative theory for image interpretation and its application to cell–substratum separation measurement. *Biophys. J.* 26:507–26.

Graf, K., K. Büscher, M. Kappl, et al. 2010. The monolayer particle interaction apparatus: Interaction of colloidal probes and lipid-protein layers at the air-water interface. In *Biomimetics in biophysics: Model systems, experimental techniques and computation,* ed. J. L. Toca-Herrera. Kerala, India: Research Signpost.

Hecht, E. 2002. *Optics*, 4th ed. San Francisco: Pearson Education.

Hillner, P. E., M. Radmacher, and P. K. Hansma. 1995. Combined atomic force and scanning reflection interference contrast microscopy. *Scanning* 17:144–47.

Hlady, V., M. Pierce, and A. Pungor. 1996. Novel method of measuring cantilever deflection during an AFM force measurement. *Langmuir* 12:5244–46.

Kühner, M., and E. Sackmann. 1996. Ultrathin hydrated dextran films grafted on glass: Preparation and characterization of structural, viscous, and elastic properties by quantitative microinterferometry. *Langmuir* 12:4866–76.

McNamee, C. E., K. Graf, H.-J. Butt, K. Higashitani, and M. Kappl. 2010a. Interaction between a silica particle and the underside of a polymer monolayer at the air/water interface in the presence of an anionic surfactant. *Colloid. Surf. A Physicochem. Eng. Asp.* 383:32–40.

McNamee, C. E., M. Kappl, H.-J. Butt, H.-J., K. Higashitani, and K. Graf. 2010b. Interfacial forces between a silica particle and phosphatidylcholine monolayers at the air-water interface. *Langmuir* 26:14574–81.

Ploem, J. S. 1975. Reflection-contrast microscopy as a tool for investigation of the attachment of living cells to a glass surface. In *Monomolecular phagocytes in immunity, infection and pathology*, ed. R. V. Furth. Oxford, U.K.: Blackwell Scientific Publications.

Rädler, J., and E. Sackmann. 1992. On the measurement of weak repulsive and frictional colloidal forces by reflection interference contrast microscopy. *Langmuir* 8:848–53.

———. 1993. Imaging optical thicknesses and separation distances of phospholipid vesicles at solid surfaces. *J. Phys. II* 3:727–48.

Schilling, J., K. Sengupta, S. Goennenwein, A. R. Bausch, and E. Sackmann. 2004. Absolute interfacial distance measurements by dual-wavelength reflection interference contrast microscopy. *Phys. Rev. E* 69:021901-1/021901-9.

Stuart, J. K., and V. Hlady. 1999. Reflection interference contrast microscopy combined with scanning force microscopy verifies the nature of protein-ligand interaction force measurements. *Biophys. J.* 76:500–8.

Verschueren, H. 1985. Interference reflection microscopy in cell biology: Methodology and applications. *J. Cell Sci.* 75:279–301.

Wiegand, G., K. R. Neumaier, and E. Sackmann. 1998. Microinterferometry: Three-dimensional reconstruction of surface microtopography for thin-film and wetting studies by reflection interference contrast microscopy (RICM). *Appl. Opt.* 37:6892–6905.

Index

A

Abbe's theory, 69
Aberrations, 60
Absolute distance, 277–280
Acoustic impedance, 212
Acoustic oscillators, 230
Adhesion force, 271
Adhesion forces, 280
Airy disk, 71
Alkanethiols, 282
Amphiphilic molecules, 323–325
Amplitude reflection, 211
Ando, Toshio, 12
Anodes, 212
Anti-Stokes emission, 164
Anti-Stokes scattering, 162, 163
Atomic force microscope (AFM), 12
  @advantages of, 250
  @bright field and epifluorescence,
    combination with, 97
  @cantilevers, 30, 31, 99
  @feedback in intermittent-contact mode, 43
  @frequency modulation, 41
  @low-frequency noise, 49
  @multi-machined, 126
  @optical fluorescence microscopy,
    combination with, 98–101
  @position-sensitive detector (PSD), 26
  @probes, 26, 30, 31–32, 98, 132, 173, 245,
    318
  @pulsed force mode, 296
  @radius of curvature (ROC), 26
  @scan frequencies, 250
  @tip, position of, 36, 290–291
  @tips, 26, 30–31, 99, 245, 271
  @vibration amplitude, 43
Avalanche photodiodes (APDs), 90, 124, 173

B

Binnig, Gerd, 11, 12
Blodgett, Katherine, 321
Boltzmann's constant, 261, 285
Bond rupture, 283–286
Borosilicate, 86
Butterworth-van Dyke model (BvD), 231–233,
  236

C

Cantilevers, 30, 50
  @calibration, 273
  @deflection of, 269
  @mechanical equilibrium of, 271
  @micromachined, 173
  @spring constant, 271, 273–275
Carbon nanotubes, 31–32
  @single-wall, 186, 188–189
CARS. See Coherent anti-Stokes Raman
  scattering (CARS)
Cathodes, 212–213
Cells, living
  @basal membrane changes, 7
  @description, 6
  @processes within, 6–7
  @single cells, 7
Charge-coupled device (CCD), 65, 168, 313
Chemical force microscopy (CFM), 281–282
Chemical spectroscopy
  @infrared spectroscopy; see Infrared
    spectroscopy
  @overview, 153, 157–158
  @Raman spectroscopy; see Raman
    spectrometers
  @visible light transitions, 157
Circular polarization, 207
Coherent anti-Stokes Raman scattering (CARS),
  165–166
  @tip-enhanced, 173
Colloidal probe method, 291, 318
Compensators, 209–210, 225
  @rotating, 227
Complex modulus, 268
Conduction bands, 22
Confocal laser scanning microscopy (CLSM),
  81–83, 101–102, 104
Constant charge, 261
Constant potential, 261
Constant Stokes shift, 175
Constitutive equilibrium, 264
Coulomb's law, 43
Coulumbic interaction, 43
Counter electrodes, 213
Critical micellar concentration (CMC), 324–325
Curie temperature, 12, 36
Cyanine dye, 108

D

D'armate, Salvino, 56
Damped oscillators. *See* Oscillators
Delta functions, 310
Differential interference contrast (DIC)
    microscopy, 71
Dither piezoelectric devices, 90
DLVO theory, 262
Doped diamond coatings, 31
Drift, 272, 298
Dry mass, 240
Dynamometer, 259

E

Einstein, Albert, 119
Elastic bodies, Hooke's law for, 259
Electrochemistry, overview of, 212–213
Electrolytic cells, 213
Electromagnetic spectrum, 155–156
Electropolymerization, 243
Electrostatic double-layer forces, 260–261
Ellipsometry, 4, 203
    @basic equation of, 221–222
    @description, 221
    @development of, 204
    @element-rotating method, 226
    @films, use in tracking, 8
    @null, 226, 228–229
    @rotating analyzer method, 226, 227
    @rotating polarizer method, 226, 227
    @scanning near-field microscopy, combining
        with; *see* Scanning near-field ellipsometry
        microscopy (SNEM)
    @setup, basic, 225–226
    @single-layer model, 222–224
    @zones, 229
Elliptical polarization, 208
Emission filter, 78
Energy quantum, 56
Excitation filter, 78
Experimental science, 1

F

Far-field optics, 120
Faraday current, 43
Feedback gains, 49
Fermi level, 21
Film balance, 325–328, 329–330
Films
    @formation, process of, 8
Finite element analysis (FEA), 275
Finite-aperture theory, 307, 309

Fluorescence. *See also* Fluorescence microscopy
    @absorption, 72
    @definition, 72
    @emission, 72, 73, 74–75
    @lifetime, 72
    @processes of, 72, 75–76
    @quantum yield, 74–75
    @quenching, 74
    @spectra, 72–73, 76, 93
    @wavelengths, 73
Fluorescence correlation spectroscopy (FCS), 93,
    111–112
Fluorescence filters, 79
Fluorescence lifetime imaging microscopy
    (FLIM), 83–85
Fluorescence microscopy. *See also* Fluorescence
    @confocal laser scanning microscopy
        (CLSM); *see* Confocal laser scanning
        microscopy (CLSM)
    @dichroic mirror, 78–79
    @emission filter, 78
    @excitation filter, 78, 86
    @filter cube, 79
    @filters; *see* Fluorescence filters
    @fluorescence lifetime imaging microscopy
        (FLIM); *see* Fluorescence lifetime
        imaging microscopy (FLIM)
    @fluorescence scanning near-field optical
        microscopy (fluorescence SNOM); *see*
        Fluorescence scanning near-field optical
        microscopy (fluorescence SNOM)
    @light source, 78, 87
    @overview, 72
    @photobleaching, 80–81, 93–94
    @probes, 80
    @refractive index, 87
    @total internal reflection fluorescence
        (TIRF); *see* Total internal reflection
        fluorescence (TIRF)
Fluorescence scanning near-field optical
    microscopy (fluorescence SNOM)
    @(F)RET-SNOM, 94–95
    @aperture, 89, 90
    @apertureless, 90, 91–92
    @collection mode, 89–90
    @colocalization studies, 113
    @history of, 106
    @illumination mode, 89
    @life sciences, benefits in, 112
    @photobleaching, 93–94
    @potential of, 93
    @principles, 89
    @probe, 92–93
    @requirements, 89

@single fluorescent molecule detection, 107–108
@single-molecule detection, 106
@single-molecule imaging, 108
@single-polymer chains, use in conformation of, 109, 111
@time-resolved, 93
@tip, 91
@tip-enhanced two-photon excitation, 95
@tip-sample distance, 91
Fluorophores, 72, 75
Focal distance, 62
Focused ion beam (FIB) milling, 132
Force clamps, 290
Force curves, 280–281, 294–296
Force modulation microscopy (FMM), 45
Force spectroscopy, 257, 286
Fourier coefficients, 226
Fourier-transform infrared spectroscopy (FTIR), 160, 168
Free Gibbs' energy, 321
Free-electron laser (FEL), 179
Fresnel's equations, 211–212, 220
FRET-SNOM, 95. *See also* Fluorescence scanning near-field optical microscopy (fluorescence SNOM)
Friction force microscopy. *See* Lateral force microscopy

G

Galvanic cells, 213
Gated excitation, 100
Grating equation, 64

H

Hamaker constant, 262
Harmonic oscillators. *See* Oscillators
Heaviside functions, 310
High-aspect ratio probes, 31–32
Highest occupied molecular orbital (HOMO), 21
Hooke's law, 31, 259
Hooke, Robert, 303
Hydrophilic heads, 324
Hydrophobic forces, 263
Hydrophobic tails, 324
Hysteresis, loading-unloading, 291, 294
Hysteresis, piezo. *See* Piezo actuators

I

IBM, 11
Illumination side prism, 88
Image bow, 49

Image visualizing device, 65
Impedance, acoustic. *See* Acoustic impedance
Impedance, electrical, 212
Impedance, mechanical, 212
Infrared near-field spectroscopy
@aperture SNOM, with, 179–180
@apertureless SNOM, with, 180–182
@development, 178
@excitation range, 178
@inorganic materials, application to, 194
@photon excitation, 182–183
@polymers, application to, 194–195
@viruses and cells, application to, 195
@wave numbers, 182
Infrared spectroscopy
@disadvantages of, 168
@lateral resolution, 168
@near-field; *see* Infrared near-field spectroscopy
@wave number, use of, 156
Infrared vibrational contrast SNOM, 140, 142, 145–146
Interaction parameter Z, 261
Interfacial tension, water surface, 263, 321, 323, 326
Interference, 272
Interferograms, 306

K

Kelvin force microscopy, 45
Kelvin-Voigt model. *See* Voigt model
Koechler illumination, 70, 71
Kretschmann's configuration, 217, 218, 219

L

Laminar piezo actuators. *See* Piezo actuators
Langmuir isotherm, 326
Langmuir, Irving, 321
Langmuir-Blodgett technique, 328, 330
Langmuir-Schaeffer technique, 328–329, 330
Lateral force microscopy, 45
Lead (plumbum) zirconate titanate (PZT), 12, 33
@thermal stability, 35–36
Lenses
@aberrations caused by, 60
@definition, 60
@divergent, 60
@focal distance, 62
@objective lens, 88
@optical microscope, in; *see* Optical microscope
@optimum conditions, 60
@principal/optical axis, 60

@rays, assumptions regarding, 60
@surface, 60
Ligand-receptor interactions, 282–283
Light microscopy. *See* Optical microscopes
Light, visible
@circular polarization, 207
@elliptical polarization, 208
@external, 58, 305
@grid, traveling thru, 63–64
@interference from, 305
@internal, 58, 305
@linear polarization, 207
@narrow slit, traveling thru, 62–63
@overview, 56
@perpendicular, 58
@phase shifts, 305
@plane of incidence; *see* Plane of incidence
@propagation of; *see* Propagation, light
@reflection, 56, 58, 59
@refraction, 56, 59–60
@wavelengths, 56, 207, 305
Lightning rod effect, 130
Linear polarization, 207, 225

M

Macroscopic techniques, 4
Magnetic force microscopy, 32
Maxwell model, 266, 268
Maxwell's equations, 215–216
Mechanical equilibrium, 258, 323
Michelson interferometer, 161–162
Michelson, Albert Abraham, 161
Microinterferometry, 303–305
Microscopic techniques, 4
Microspectroscopy, 166
Modulus tensors, 265
Molecular orbitals, 21
Molecular recognition spectroscopy
@interference, circumventing, 297–298
@ligand–receptor unbinding events, 287–288
@molecular unfolding, 288–290
@negative controls, 286–287
Monolayer, 325, 328
Motion, Laws of, 258–259

N

Nanomechanics
@colloidal probe method, 291
@contact region, 291
@loading-unloading hysteresis, 291, 294
@overview, 290
@probe geometry of, 290–291
@stiffness, sample, 292, 295

Nature
@characteristic properties, 1
@complexity of, 1
@sequential observation, 2
@simultaneous observation, 2
Near-field optics, 123
Newton rings, 303, 306–307, 314, 318
Newton's Laws of Motion, 258–259
Newton, Isaac, 123, 258, 303
Noise
@high-frequency, 49
@interference, caused by, 272
@low-frequency, 49
@SPM images, appearance on, 49
@thermal, 272
Nonlocal theory, 307
Nonplanar interfaces, theory of, 311
Nonsaturated lipids, 325
Normal incidence theory, 307, 310, 311

O

Optical approach curves, 91
Optical beam detection, 37–38
Optical detectors, 120
Optical fibers, 124
Optical microscopes, 4
@apertures, 65, 67, 68
@condenser diaphragm, 65
@condenser lens, 64
@contrasts, optical, 69–70
@contrasts, phase, 70
@detection, limit of, 67
@eyepieces, 64, 66
@field stop diaphragm, 65, 66
@focal planes, 65–66
@image formation, 55
@immersion fluids, 67
@interference, 69
@lateral resolution, 68
@light diffraction, 66, 68
@light source, 65
@objective lens, 64, 88
@objectives, 66
@oculars, 64, 66
@overview, 55
@reflection mode, 55
@scanning probe techniques, implementation to, 95, 97
@transmission mode, 55
@vertical resolution, 68
Optical reflectometry, 4
Oscillators
@acoustic; *see* Acoustic oscillators
@cantilevered, 50

@damped, 22
@harmonic, 22–24
@large-amplitude, 272
@Q factors of, 24
@resonances of, 24

P

P-aminobenzoic acid (PABA), 174
Penetration depth, 87
Perturbation, 264, 265
PH, 325
Phase-contrast imaging, 45
Phonons, 172, 183
Photobleaching, 80–81, 93–94
Photoelastic modulators, 210–211, 227–228
Photoelasticity, 210
Photomultiplier tubes (PMTs), 90, 124
Photons, 56
Piezo actuators
@aging of, 19
@closed-loop, 19
@creep, 17, 19
@Curie temperatures, 36
@definition, 13
@deformation, 23
@displacement, 33, 273
@drift, 17, 19
@electrodes, 13–14
@hysteresis; *see* Piezo hysteresis
@laminar, 13, 15, 17
@low-noise amplifiers, use of, 36
@nonlinearity of, 17
@servo-control, use of, 19
@speed of, 36
@thermal stability, 36
Piezo hysteresis, 17, 19, 49
Piezo scanners, stack, 15
Piezo scanners, tube, 14–15
Piezo transducer, 210
Piezoelectricity. *See also* Piezo actuators
@overview, 12–14
@poling, 12–13
@transverse piezo effect, 15
Piezoresistive detection, 38
Plane of incidence, 58
Plasmons, 182
Plumbum zirconate titanate (PZT). *See* Lead
    (plumbum) zirconate titanate (PZT)
PMMA. *See* Poly(methylmethacrylate) (PMMA)
Poisson's ratio, 265, 292
Polarizability tensors, 162–163
Polarization effects, 186
Polarizers, 209, 225
Poling, 12–13

Poly(isobutyl-methacrylate) (PiBMA) chains, 111
Poly(methyl methacrylate) (PMMA) chains, 111
Poly(methylmethacrylate) (PMMA), 194, 195
Polyelectrolyte microcapsules, 319–321
Polyether ether ketone (PEEK), 330
Polystyrene (PS), 194
Polytetrafluoroethylene (PTFE), 330
Position-sensitive photodetector (PSPD), 37–38
Pressure isotherm, 326
Propogation, light, 56
Pull-off force, 271
PZT. *See* Lead (plumbum) zirconate titanate
    (PZT)

Q

Q factor, 23–24, 229
Q probes, 24
Quality factor. *See* Q factor
Quarter-wave plate, 210
Quartz crystal microbalance (QCM), 6, 204
@AT-cut, 230, 231, 233, 238
@Butterworth-van Dyke model (BvD); *see*
    Butterworth-van Dyke model (BvD)
@dissipation modeling, with, 236–237
@films, use in tracking, 8
@frequency shifts, 238, 241, 250
@impedance analysis, 235–236
@overtones, 231
@overview, 229–230
@rigid film in air, applied to, 238
@ring-down technique, 236–238
@shear-mode resonators, 241
@shear-wave resonator, 229–230
@small-load approximation, 233
@SPM, combined with, 248–251
@thickness-shear mode, 229
@viscoelastic film in air, applied to, 239–241
Quartz crystal microbalance (QCM-D), 236–237
Quartz, fused, 210

R

Raman effect, 154
Raman scanning near-field optical microscopy
    (SNOM)
@aperture near-field, 169
@description, 169
Raman scattering, 157, 162
@contrast C enhancement, 175
@cross sections, 173
@enhancing, 173–174
@excitation, 177
@near-field radiation, 177
@net enhancement, 177

@observed enhancement, 175
@signal-enhanced Raman scattering (SERS)
  factor, 165
@surface-enhanced, 165, 174, 177–178
@tip-enhanced; *see* Tip-enhanced near-field
  Raman spectroscopy (TERS)
Raman spectrometers, 130, 153, 166. *See also*
  Raman scanning near-field optical
  microscopy (SNOM)
@carbon nanotubes, application to, 186,
  188–189
@enhancement of Raman signal, 173–175
@laser spot size, 168
@silicon, application to, 187
@single-point, 166
@single-wall carbon nanotubes, application
  to, 185
@sub-monolayers of dyes, application to,
  with SPM, 185
@tip-enhanced near-field Raman
  spectroscopy (TERS); *see* Tip-enhanced
  near-field Raman spectroscopy (TERS)
@wave number, use of, 156
Raman, Chandrasekhara Venkata, 154
Raster scanning, 169
Rayleigh scattering, 162, 163, 164
Rayleigh's criterion, 71
Reference electrodes, 213
Reflection interference contrast microscopy
  (RICM), 7, 303
@applications, 304–305, 315–316
@bottom profile, 309
@dual wavelength, 314–315
@experimental setup, 312–313
@image processing, 313, 319
@lateral resolution, 311
@origins, 304
@principles of, 305, 307–311
@speed, application, 319
@SPM, combined technique with, 318–319
@vertical resolution, 311
Refractive index, 59–60, 67–68, 87, 308
Reproducibility, 2
Resonance, 22
Resonance energy transfer (RET), 75–77
Retarders. *See* Compensators
Ring-down technique, 236–238
Rohrer, Heinrich, 11, 12

S

Samples
@process of, 2
@properties, 2
Saturated lipids, 325
Sauerbery equations, 238–239

Sauerbery thickness, 239
Scalars, 258
Scanning capacitive microscopy, 43
Scanning electron microcopy (SEM), 291
Scanning near-field ellipsometry microscopy
  (SNEM), 244–245
@applications, 246
@optical contrasts, 246
Scanning near-field optical microscopy (SNOM),
  4. *See also* Fluorescence scanning near-
  field optical microscopy (fluorescence
  SNOM)
@aperture, 126, 128, 135
@aperture probes, 132, 134–135
@apertureless, 126, 128, 130
@apertureless probes, 135
@background suppression, 124, 126
@challenges, liquid environments, 137
@collection-mode, 173
@diffraction limit, 123, 124
@disadvantages, 169
@evanescent waves, 122
@far-field zone, 122
@feedback modes, 135, 137
@field enhancement, 130–131
@history of, 119–120, 131–132, 142
@illumination mode, 126
@imaging modes, 126
@infrared vibrational contrast, 140, 142,
  145–146
@lateral imaging, 119
@light transmission, 134
@lightning rod effect, 130
@living cells, use on, 144–145
@micromachined probes, 134–135
@near-field light, 119, 124
@near-field zone, 122
@opaque mask, 126
@plasmonic resonances in gold
  nanostructures, case example of SNOM
  use on, 146, 148
@polarization contrast, 137, 140–141
@probes, 131–132, 169, 171, 173
@Raman; *see* Raman scanning near-field
  optical microscopy (SNOM)
@reflection, 128
@refractive index, 137, 142
@resolution, 5, 124, 128
@scattering-type, 131
@sequential imaging, 111
@shear-force feedback, 137
@subsurface imaging, 145–146
@tip-enhanced, 131
@tip-on-aperture probes, 135
@tip-sample distance, 137

@tips, 175
@topological images, 119
Scanning probe microscope (SPM) scanners
@contact mode, 39, 41
@feedback mechanism, 39, 41
@intermittent contact mode, 39
@materials, 33, 35
@movement of, 33
@noncontact mode, 41–42
@object detection, 38–39
@responsiveness, 33
@set point, 39
@tip, position of, 36, 39
@tube piezo scanners; *see* Tube piezo
scanners
Scanning probe microscopes (SPM)
@applicability, 3
@artifacts in images, 46, 49
@choosing, 4
@commercial scan frequencies, 50
@confolcal laser scanning optical microscopy
(CLSM), combining with, 101–102, 104
@feedback mechanism, 26, 46
@fluorescence lifetime imaging, combining
with, 101
@fluorescence microscopy, combining with,
101
@history of use, 11–12
@image bow, 49
@images, three-dimensional, 11
@intermitten contact mode, 50
@intuitive nature of, 25
@Nobel Prize for, 12
@overview/description of, 25
@principles behind, 257
@probe-sample interaction, relative
displacement of, 268–269, 271–272
@probes, 25–26, 26, 27–28, 46
@recording of images on, 49–50
@resolution, 2
@sample preparation, 5–6
@scanners; *see* Scanning probe microscope
(SPM) scanners
@technique of, use of, 2
@time resolution in, 49–50
@tip, 25, 26, 27, 28, 46
@total internal reflection fluorescence
(TIRF), combining with, 106
@tunneling, 46
@X mapping Y, feedback on, 44–46
Scanning tunneling microscopy (STM), 12, 20–21
@current, tunneling, 37, 41
@probe, 26, 28, 30
@tip, position of, 36
Scanning tunneling optical microscopy (STOM),
130

Second-harmonic generation, 178
Self-assembled monolayers (SAMs), 282
Servo-control, 19
Shear modulus, 265
Signal-enhanced Raman scattering (SERS)
factor, 165
Silicon carbide, 196
Silver, 216
Snell-Descartes law, 60
Solvation (hydration) forces, 263–264
Spectrometers, 159–161. *See also* Raman
spectrometers
@Fourier-transform infrared spectroscopy
(FTIR); *see* Fourier-transform infrared
spectroscopy (FTIR)
Spectroscopy. *See also* Chemical; Spectroscopy
@defining, 153
@overview, 153
Specular reflection, 59
Spring constants, cantilever. *See under*
Cantilevers
Stack piezo scanners, 15. *See also* Piezo actuators
Stern-Helmholtz layer, 260
Stokes emission, 164
Stokes scattering, 162, 163
Surface phonon polaritons, 182, 183
Surface plasmon polaritons, 182, 214–215
@electromagnetic field of, 214, 215
@evanescence, 214, 215
@Kretschmann's configuration; *see*
Kretschmann's configuration
@*p*-polarized light, 217
@prism, light, 217, 218
@sensitivity, 219
@skin depths, 214
@wave vectors, 215
Surface plasmon resonance (SPR), 4, 6
@experimental setup, 219–220, 243
@films, use in tracking, 8
@history of, 203–204
@prisms, 243
@sensors, 243
@usage, 203
Surface pressure, 326
Surface tension, 321, 323, 326, 363
Synge, E. H., 119–120

T

Tapered fiber probes, 132
Terrylene diimide (TDI) molecules, 108
TERS. *See* Tip-enhanced near-field Raman
spectroscopy (TERS)
Time-correlated single-photon counting
(TCSPC) detector., 93

Tip-enhanced coherent anti-Stokes Raman
      scattering (CARS), 173, 178
Tip-enhanced near-field Raman spectroscopy
      (TERS), 169, 171, 174–175
   @bacteria, application to, 188
   @biomolecules, application to, 187–188
   @combined enhanced (with surface
      enhancement), 177–178
   @human cells, application to, 188
   @intensity ratios, 177
   @nucleic acid sequencing, application to,
      188–191
   @proof-of-concept, 185
   @viruses, application to, 188, 191
Tobacco mosaic virus (TMV), 195
Total internal reflection fluorescence (TIRF), 7,
      86–87, 88
   @scanning probe microscopy (SPM)
      combining with, 106
Transmission axis, 209
Transmission coefficients, 211
Transversal wave, 207
Transverse piezo effect, 15
Tube piezo scanners, 14–15. *See also* Piezo
      actuators
   @SPM scanners, 35
Tuning forks, 90
Tunneling effect, 19–21, 36–37

U

Undulators, 179

V

Valence bands, 22
Van der Waals forces, 261–262
Vectors, 258
Vibration (IR) spectroscopy, 130
Viscoelastic film, 239
Viscoelasticity
   @creep, 266–267
   @dynamic experiments, 266, 268
   @overview, 266
   @stress relaxation, 266–267
   @transient experiments, 266–267
Voigt model, 266, 268
Voltage signals, 273
Voltammograms, 213, 250

W

Wave numbers, 156, 182
Wilhelmy plate, 323, 325
Working electrodes, 213

Y

Young's modulus, 275, 292, 296

Z

Z-scanners, 51
Zero force, extrapolated, 277

T - #0363 - 101024 - C16 - 234/156/20 - PB - 9781138374584 - Gloss Lamination